LABORATORY MANUAL

HOLE'S ESSENTIALS OF HUMAN ANATOMY & PHYSIOLOGY

Fourteenth Edition

Charles J. Welsh

DUQUESNE UNIVERSITY

CONTRIBUTOR

Cynthia Prentice-Craver

CHEMEKETA COMMUNITY COLLEGE

PREVIOUS EDITION AUTHORS

| **David Shier** | **Jackie Butler** | **Ricki Lewis** |
| WASHTENAW COMMUNITY COLLEGE | GRAYSON COLLEGE | ALBANY MEDICAL COLLEGE |

Phillip Snider

GADSDEN STATE COMMUNITY COLLEGE

Terry Martin

KISHWAUKEE COLLEGE

Mc Graw Hill

LABORATORY MANUAL TO ACCOMPANY HOLE'S ESSENTIALS OF HUMAN ANATOMY & PHYSIOLOGY, FOURTEENTH EDITION

Published by McGraw-Hill Education, 2 Penn Plaza, New York, NY 10121. Copyright ©2021 by McGraw-Hill Education. All rights reserved. Printed in the United States of America. Previous editions ©2018, 2015, and 2012. No part of this publication may be reproduced or distributed in any form or by any means, or stored in a database or retrieval system, without the prior written consent of McGraw-Hill Education, including, but not limited to, in any network or other electronic storage or transmission, or broadcast for distance learning.

Some ancillaries, including electronic and print components, may not be available to customers outside the United States.

This book is printed on acid-free paper.

4 5 6 7 8 LKV 26 25 24 23 22

ISBN 978-1-260-42584-0
MHID 1-260-42584-3

Portfolio Manager: *Matt Garcia*
Product Developer: *Krystal Faust*
Marketing Manager: *Valerie Kramer*
Content Project Managers: *Ann Courtney/Brent dela Cruz*
Buyer: *Sandra Ludovissy*
Design: *David W. Hash*
Content Licensing Specialist: *Beth Cray*
Cover Image: *©LightField Studios/Shutterstock*
Compositor: *MPS Limited*

All credits appearing on page are considered to be an extension of the copyright page.

Some of the laboratory experiments included in this text may be hazardous if materials are handled improperly or if procedures are conducted incorrectly. Safety precautions are necessary when you are working with chemicals, glass test tubes, hot water baths, sharp instruments, and the like, or for any procedures that generally require caution. Your school may have set regulations regarding safety procedures that your instructor will explain to you. Should you have any problems with materials or procedures, please ask your instructor for help.

The Internet addresses listed in the text were accurate at the time of publication. The inclusion of a website does not indicate an endorsement by the authors or McGraw-Hill Education, and McGraw-Hill Education does not guarantee the accuracy of the information presented at these sites.

mheducation.com/highered

Contents

Preface

This laboratory manual was prepared to supplement the textbook *Hole's Essentials of Human Anatomy and Physiology*, Fourteenth Edition, by Dr. Charles Welsh. As in the textbook, the laboratory manual is designed for students with minimal backgrounds in the physical and biological sciences pursuing careers in professional health fields.

The laboratory manual contains forty-nine laboratory exercises and reports closely integrated with the chapters of the textbook. The exercises are planned to illustrate and review the anatomical and physiological facts and principles presented in the textbook and to help students investigate some of these ideas in greater detail.

Many of the laboratory exercises are short or are divided into several separate procedures. This allows an instructor to select those exercises or parts of exercises that will best meet the needs of a particular program. Also, exercises requiring a minimal amount of laboratory equipment have been included.

The laboratory exercises have a variety of special features designed to stimulate interest in the subject matter, to involve students in the learning process, and to guide them through the planned activities. These special features include the following:

Materials Needed

This section lists the laboratory materials required to complete the exercise and to perform the demonstrations and learning extensions.

⚠ Safety

A list of safety guidelines is included as Appendix 1. Each laboratory session that requires special safety guidelines has a safety section following "Materials Needed." The instructor might require some modifications of these guidelines.

Purpose of the Exercise

The purpose provides a statement about the intent of the exercise—that is, what will be accomplished.

Learning Outcomes ①

The learning outcomes list what a student should be able to do after completing the exercise. Each learning outcome has a matching assessment indicated by a corresponding icon ⚠ in the laboratory exercise or the laboratory report.

Introduction

The introduction briefly describes the subject of the exercise or the ideas that will be investigated.

Procedure

The procedure provides a set of detailed instructions for accomplishing the planned laboratory activities. Each procedure heading includes an explore icon indicating ways to accomplish the learning outcomes. Usually, these instructions are presented in outline form, so that a student can proceed through the exercise in stepwise fashion. Often, the student is referred to particular sections of the textbook for necessary background information or for review of subject matter previously presented.

The procedures include a wide variety of laboratory activities and, from time to time, direct the student to complete various tasks in the laboratory reports.

Laboratory Reports

A laboratory report to be completed by the student immediately follows each exercise. These reports include various types of review activities, spaces for sketches of microscopic objects, tables for recording observations and experimental results, and questions dealing with the analysis of such data. There are also several labeling exercises included in many of the lab reports that utilize images of commonly used anatomical models and cadaver images from the Practice Atlas for Anatomy and Physiology and the Anatomy & Physiology Revealed resources.

As a result of these activities, students will develop a better understanding of the structural and functional characteristics of their bodies and will increase their skills in gathering information by observation and experimentation. Some of the exercises also include demonstrations, learning extensions, and useful illustrations. By completing all of the assessments in the laboratory report, students are able to determine if they have accomplished all of the learning outcomes.

Demonstrations

Demonstrations appear in separate boxes. They describe specimens, specialized laboratory equipment, or other materials of interest that an instructor may want to display to enrich the student's laboratory experience.

Learning Extensions

Learning extensions also appear in separate boxes. They encourage students to extend their laboratory experiences. Some of these activities are open-ended in that they suggest the student plan an investigation or experiment and carry it out after receiving approval from the laboratory instructor. Some of the figures are illustrated as line art or in grayscale. This allows colored pencils to be used as a visual learning activity to distinguish various structures.

The Use of Animals in Biology Education*

The National Association of Biology Teachers (NABT) believes that the study of organisms, including nonhuman animals, is essential to the understanding of life on Earth. NABT

---

*Adopted by the Board of Directors in October 1995. This policy supersedes and replaces all previous NABT statements regarding animals in biology education.

recommends the prudent and responsible use of animals in the life science classroom. NABT believes that biology teachers should foster a respect for life. Biology teachers also should teach about the interrelationship and interdependency of all things.

Classroom experiences that involve nonhuman animals range from observation to dissection. NABT supports these experiences as long as they are conducted within the long-established guidelines of proper care and use of animals, as developed by the scientific and educational community.

As with any instructional activity, the use of nonhuman animals in the biology classroom must have sound educational objectives. Any use of animals, whether for observation or for dissection, must convey substantive knowledge of biology. NABT believes that biology teachers are in the best position to make this determination for their students.

NABT acknowledges that no alternative can substitute for the actual experience of dissection or other use of animals and urges teachers to be aware of the limitations of alternatives. When the teacher determines that the most effective means to meet the objectives of the class do not require dissection, NABT accepts the use of alternatives to dissection, including models and the various forms of multimedia. The Association encourages teachers to be sensitive to substantive student objections to dissection and to consider providing appropriate lessons for those students where necessary.

To implement this policy, NABT endorses and adopts the "Principles and Guidelines for the Use of Animals in Precollege Education" of the Institute of Laboratory Animals Resources (National Research Council). Copies of the "Principles and Guidelines" may be obtained from the ILAR (2101 Constitution Avenue, NW, Washington, DC 20418; 202–334-2590).

Illustrations

Diagrams from the textbook and diagrams similar to those in the textbook often are used as aids for reviewing subject matter. Other illustrations provide visual instructions for performing steps in procedures or are used to identify parts of instruments or specimens. Micrographs are included to help students identify microscopic structures or to evaluate student understanding of tissues.

In some exercises, the figures include line drawings suitable for students to color with colored pencils. This activity may motivate students to observe the illustrations more carefully and help them to locate the special features represented in the figures. Students can check their work by referring to the corresponding full-color illustrations in the textbook.

Frequent variations exist in anatomical structures among humans. The illustrations in the textbook and the laboratory manual represent normal (normal means the most common variation) anatomy. Variations from normal anatomy do not represent abnormal anatomy unless some function is impaired.

Features of This Edition

Many of the changes in this new edition of the laboratory manual are a result of evaluations and suggestions from anatomy and physiology students. Many suggestions from users and reviewers of previous editions have been incorporated.

Some unique features of this laboratory manual include the following:

1. Learning outcomes with icons ① have matching assessments with icons Ⓐ so students can be sure they have accomplished all of the content of the laboratory exercise.
2. "Blood Testing" is a supplemental lab available at the text's Connect site www.mcgrawhillconnect.com instead of within the printed laboratory manual.
3. "Critical Thinking Applications" are incorporated within most of the laboratory exercises to enhance valuable critical thinking skills that students will need throughout their lives.
4. The *Instructor's Manual for Laboratory Manual to Accompany Hole's Essentials of Human Anatomy and Physiology* has been revised and updated. It contains a "Student Safety Contract" and a "Student Informed Consent Form" for possible inclusion in the class. The forms could be adapted for a specific institution based upon school policy. The instructor's manual is located online at on the text's Connect site at www.mcgrawhillconnect.com.
5. Two assessment tools (rubrics) for laboratory reports are included in Appendix 4.

Specific Changes to Individual Laboratory Exercises

Lab 3: New figure 3.2, added revised labeling of test tubes in procedure C

Lab 5: New figure 5.2 added

Lab 20: New figure 20.4 added, Part C Assessments

Lab 21: New figures 21.5 a, b, and c, added Part C Assessments

Lab 23: New figures 23.8 and 23.9 added, Part C Assessments

Lab 26: New figure 26.4 added, Part C Assessments

Lab 28: New figure 28.6 added, Part C Assessments

Lab 37: Revised the Learning Outcomes, deleted procedure on Paths of Circulation

Lab 40: New figure 40.13 added, Part F Assessments

Lab 44: Revised Procedure C instructions, new figure 44.8 added, Part E Assessments

Lab 45: Revised Procedure A instructions, revised table for Part A Assessments

Lab 46: Revised Procedures A and B, new figure 46.7 added

Lab 47: New figure 47.7 added, Part C Assessments

Acknowledgments

I would like to thank the wonderful people at McGraw-Hill Education, including Matthew Garcia, Krystal Faust, Jim Connely, Valerie Kramer, Fran Simon, Dr. Charles Welsh, Cynthia Prentice-Craver, Mary Powers, Ann Courtney, and Beth Cray, for your help and support and for allowing me the opportunity to help revise this lab manual.

I would like to say a special thank you to Terry Martin. Terry and I met for the first time in 2008 at a Human Anatomy and Physiology Society (HAPS) conference. It was an instant friendship. Not only has Terry been a friend to me but I also consider him my mentor. I am eternally grateful for our friendship.

I would also like to thank all the other instructors I work with for their dedication to their profession. A special thank you to my department chair person, Ms. Shirley Colvin for her support and encouragement.

Gratitude is extended to John W. Hole, Jr., for his years of dedicated effort in early editions of this classic work.

About the Authors

Phillip Snider

Courtesy of Shanae Snider

This fourteenth edition is the first revision by Phillip Snider. Phillip has taught Human Anatomy and Physiology at the college level since 2001. He has also taught Human Gross Anatomy/Pathophysiology since 2008. His student-centered classes are designed to encourage the student's fascination and knowledge of the human body to further their professional careers.

Phillip taught the very first human-cadaver-based gross anatomy course at the community college level in the state of Alabama. He was instrumental in constructing the curriculum for the course and continues his own education in gross anatomy whenever the opportunity arises. Among Phillip's awards are Who's Who Among America's Teachers, the Brenda Crowe Exceptional Achievement in Teaching Award, and the state of Alabama Chancellor's Award. Phillip's professional memberships include the Human Anatomy and Physiology Society (HAPS), the Alabama Conference of Educators (ACOE), and the National Institute for Staff and Organizational Development (NISOD). He was also a contributor for the Terry Martin *Laboratory Manual for Human Anatomy & Physiology,* cat, main, and fetal pig versions, third editions. Phillip was also elected to be a faculty fellow at his college to help provide professional development and training to other faculty members.

Phillip has been married to his high school sweetheart, Shae, since 1991. He has two successful children, Collin and Camden, whom he is extremely proud of. Phillip enjoys spending time with his family, camping, and fishing.

Terry R. Martin

J and J Photography

This fourteenth edition is the ninth revision by Terry R. Martin of Kishwaukee College. Terry's teaching experience of over forty years, his interest in students and love for college instruction, and his innovative attitude and use of technology-based learning enhance the solid tradition of John Hole's laboratory manual. Among Terry's awards are the Kishwaukee College Outstanding Educator, Phi Theta Kappa Outstanding Instructor Award, Kishwaukee College ICCTA Outstanding Educator Award, Kishwaukee College Faculty Board of Trustees Award of Excellence, and John C. Roberts Community Service Award. Terry's professional memberships include the National Association of Biology Teachers, Human Anatomy and Physiology Society, and Nature Conservancy. In addition to writing many publications, he co-produced with Hassan Rastcgar a videotape entitled *Introduction to the Human Cadaver and Prosection,* published by Wm. C. Brown Publishers. Terry revised the *Laboratory Manual to Accompany Hole's Human Anatomy and Physiology,* Fifteenth Edition, and authored *Laboratory Manual for Human Anatomy & Physiology,* cat, main, and fetal pig versions, third editions. Cadaver dissection experiences have been provided for his students for over thirty-five years. Terry also was a faculty exchange member in Ireland.

To the Student

The exercises in this laboratory manual will provide you with opportunities to observe various anatomical parts and to investigate certain physiological phenomena. Such experiences should help you relate specimens, models, microscope slides, and your body to what you have learned in the lecture and read about in the textbook.

The following list of suggestions may help make your laboratory activities more effective and profitable.

1. Prepare yourself before attending the laboratory session by reading the assigned exercise and reviewing the related sections of the textbook. It is important to have some understanding of what will be done in the laboratory before you come to class.

2. Bring your laboratory manual and textbook to each laboratory session. These books are closely integrated and will help you complete most of the exercises.

3. Be on time. During the first few minutes of the laboratory meeting, the instructor often will provide verbal instructions. Also listen carefully for information about special techniques to be used and precautions to be taken.

4. Keep your work area clean and your materials neatly arranged so that you can quickly locate needed items. This will enable you to proceed efficiently and will reduce the chances of making mistakes.

5. Pay particular attention to the purpose of the exercise, which states what you are to accomplish in general terms, and to the learning outcomes, which list what you should be able to do as a result of the laboratory experience. Then, before you leave the class, review the outcomes and make sure that you can meet all of the assessments.

6. Precisely follow the directions in the procedure and proceed only when you understand them clearly. Do not improvise procedures unless you have the approval of the laboratory instructor. Ask questions if you do not understand exactly what you are supposed to do and why you are doing it.

7. Handle all laboratory materials with care. Some of the materials are fragile and expensive to replace.

8. Treat all living specimens humanely and try to minimize any discomfort they might experience.

9. Although at times you might work with a laboratory partner or a small group, try to remain independent when you are making observations, drawing conclusions, and completing the activities in the laboratory reports.

10. Read the instructions for each section of the laboratory report before you begin to complete it. Think about the questions before you answer them. Your responses should be based on logical reasoning and phrased in clear and concise language.

11. Use the time allowed in the lab wisely. In some of your lab periods you will have extra time to study and observe models, review concepts, finish lab reports, etc. Take advantage of this time.

12. At the end of each laboratory period, clean your work area and the instruments you have used. Return all materials to their proper places and dispose of wastes, including glassware or microscope slides that have become contaminated with human blood or body fluids, as directed by the laboratory instructor. Wash your hands thoroughly before leaving the laboratory.

Study Skills for Anatomy and Physiology

Everyone has his or her own learning styles. There are techniques that work well for most students enrolled in Human Anatomy and Physiology. Using some of the skills listed here can make your course more enjoyable and rewarding.

1. **Note taking:** Look for the main ideas and briefly express them in your own words. Organize, edit, and review your notes soon after the lecture. Add textbook information to your notes as you reorganize them. Underline or highlight with different colors the important points, major headings, and key terms. Study your notes daily, as they provide sequential building blocks of the course content.

2. **Note cards/flash cards:** Make your own. Add labels and colors to enhance the material. Keep them with you; study them often and for short periods. Concentrate on a small number of cards at one time. Shuffle your cards and have someone quiz you on their content. As you become familiar with the material, you can set aside cards that don't require additional mastery.

3. **Chunking:** Organize information into logical groups or categories. Study and master one chunk of information at a time. For example, study the bones of the upper limb, lower limb, trunk, and head as separate study tasks.

4. **Mnemonic devices:** An *acrostic* is a combination of association and imagery to aid your memory. It is often in the form of a poem, rhyme, or jingle in which the first letter of each word corresponds to the first letters of the words you need to remember. **S**o **L**ong **T**op **P**art, **H**ere **C**omes **T**he **T**humb is an example of such a mnemonic device for remembering the eight carpals in the correct sequence. *Acronyms* are words formed by the first letters of the items to remember. *IPMAT* is an example of this type of mnemonic device to help you remember the phases of the cell cycle in the correct sequence. Try to create some of your own.

5. **Time management:** Prepare monthly, weekly, and daily schedules. Include dates of quizzes, exams, and projects on the calendar. On your daily schedule, budget several short study periods. Daily repetition alleviates cramming for exams. Prioritize your tasks so that you still have time for work and leisure activities. Find an appropriate study atmosphere with minimum distractions.

6. **Recording and recitation:** An auditory learner can benefit by recording lectures and review sessions with a digital recorder. Many students listen to the recorded sessions as they drive or just before going to bed. Reading your notes aloud can help also. Explain the material to anyone (even if there are no listeners). Talk about anatomy and physiology in everyday conversations.

7. **Study groups:** Small study groups that meet periodically to review course material and compare notes have helped and encouraged many students. However, keep the group on the task at hand. It's okay to socialize but don't forget to study! This group often becomes a support group.

8. **Use all available resources:** Utilize the resources associated with your lab manual and textbook (i.e., Connect, Learnsmart, Anatomy & Physiology Revealed, Practice Atlas for Anatomy and Physiology) to help master the concepts covered in your class.

Correlation of Textbook Chapters and Laboratory Exercises

Textbook Chapters	Related Laboratory Exercises		
Chapter 1 Introduction to Human Anatomy and Physiology	Exercise 1 Scientific Method and Measurements Exercise 2 Body Organization and Terminology		
Chapter 2 Chemical Basis of Life	Exercise 3 Chemistry of Life		
Chapter 3 Cells	Exercise 4 Care and Use of the Microscope Exercise 5 Cell Structure and Function	Exercise 6 Movements Through Membranes Exercise 7 Cell Cycle	
Chapter 4 Cellular Metabolism			
Chapter 5 Tissues	Exercise 8 Epithelial Tissues Exercise 9 Connective Tissues	Exercise 10 Muscle and Nervous Tissues	
Chapter 6 Integumentary System	Exercise 11 Integumentary System		
Chapter 7 Skeletal System	Exercise 12 Bone Structure Exercise 13 Organization of the Skeleton Exercise 14 Skull Exercise 15 Vertebral Column and Thoracic Cage	Exercise 16 Pectoral Girdle and Upper Limb Exercise 17 Pelvic Girdle and Lower Limb Exercise 18 Joint Structure and Movements	
Chapter 8 Muscular System	Exercise 19 Skeletal Muscle Structure and Function Exercise 20 Muscles of the Head and Neck Exercise 21 Muscles of the Chest, Shoulder, and Upper Limb Exercise 22 Muscles of the Abdominal Wall and Pelvic Floor Exercise 23 Muscles of the Hip and Lower Limb Exercise 24 Surface Anatomy		
Chapter 9 Nervous System	Exercise 25 Nervous Tissue and Nerves Exercise 26 Spinal Cord and Meninges Exercise 27 Reflex Arc and Reflexes	Exercise 28 Brain and Cranial Nerves Exercise 29 Dissection of the Sheep Brain	
Chapter 10 The Senses	Exercise 30 Ear and Hearing Exercise 31 Eye Structure	Exercise 32 Visual Tests and Demonstrations	
Chapter 11 Endocrine System	Exercise 33 Endocrine Histology and Diabetic Physiology		
Chapter 12 Blood	Exercise 34 Blood Cells and Blood Typing	Exercise 49 Blood Testing (available online)	
Chapter 13 Cardiovascular System	Exercise 35 Heart Structure Exercise 36 Cardiac Cycle Exercise 37 Blood Vessel, Structure, Arteries, and Veins Exercise 38 Pulse Rate and Blood Pressure		
Chapter 14 Lymphatic System and Immunity	Exercise 39 Lymphatic System		
Chapter 15 Digestive System and Nutrition	Exercise 40 Digestive Organs	Exercise 41 Action of a Digestive Enzyme	
Chapter 16 Respiratory System	Exercise 42 Respiratory Organs Exercise 43 Breathing and Respiratory Volumes		
Chapter 17 Urinary System	Exercise 44 Urinary Organs	Exercise 45 Urinalysis	
Chapter 18 Water, Electrolyte, and Acid-Base Balance			
Chapter 19 Reproductive Systems	Exercise 46 Male Reproductive System	Exercise 47 Female Reproductive System	
Chapter 20 Pregnancy, Growth, Development, and Genetics	Exercise 48 Genetics		

1

Scientific Method and Measurements

MATERIALS NEEDED

Meterstick
Calculator
Human skeleton

PURPOSE OF THE EXERCISE

To become familiar with the scientific method of investigation, to learn how to formulate sound conclusions, and to provide opportunities to use the metric system of measurements.

LEARNING OUTCOMES

After completing this exercise, you should be able to

1. Convert English measurements to the metric system, and vice versa.
2. Calculate expected upper limb length and actual percentage of height from recorded upper limb lengths and heights.
3. Apply the scientific method to test the validity of a hypothesis concerning the direct, linear relationship between human upper limb length and height.
4. Design an experiment, formulate a hypothesis, and test it using the scientific method.

Scientific investigation involves a series of logical steps to arrive at explanations for various biological phenomena. It reflects a long history of asking questions and searching for knowledge. This technique, called the *scientific method,* is used in all disciplines of science. It allows scientists to draw logical and reliable conclusions about phenomena.

The scientific method begins with making *observations* related to the topic under investigation. This step commonly involves the accumulation of previously acquired information and/or your observations of the phenomenon. These observations are used to formulate a tentative explanation known as the *hypothesis.* An important attribute of a hypothesis is that it must be testable. The testing of the proposed hypothesis involves designing and performing a carefully controlled *experiment* to obtain data that can be used to support, reject, or modify the hypothesis.

During the experiment to test the proposed hypothesis, it is important to be able to examine only a single changeable factor, known as a *variable.* An *independent variable* is one that can be changed but is determined before the experiment occurs; a *dependent variable* is determined from the results of the experiment.

An *analysis of data* is conducted using sufficient information collected during the experiment. Data analysis may include organization and presentation of data as tables, graphs, and drawings. From the interpretation of the data analysis, conclusions are drawn. (If the data do not support the hypothesis, you must reexamine the experimental design and the data and, if needed, develop a new hypothesis.) The final presentation of the information is made from the conclusions. Results and conclusions are presented to the scientific community for evaluation through peer reviews, presentations at professional meetings, and published articles. If many investigators working independently can validate the hypothesis by arriving at the same conclusions, the explanation can become a *theory.* A theory serves as the explanation from a summary of known experiments and supporting evidence, unless it is disproved by new information. The five components of the scientific method are summarized as

Observations
↓
Hypothesis
↓
Experiment
↓
Analysis of data
↓
Conclusions

Metric measurements are characteristic tools of scientific investigations. The English system of measurements is often used in the United States, so the investigator must make conversions from the English system to the metric system. Use table 1.1 for the conversion of English units of measure to metric units for length, mass, volume, time, and temperature.

EXPLORE

PROCEDURE A—Using the Steps of the Scientific Method

1. Refer to Appendix B of the textbook for additional information on the scientific method.
2. Many people have observed a correlation between the length of the upper and lower limbs and the height (stature) of an individual. For example, a person who has long upper limbs (the arm, forearm, and hand combined) tends to be tall. Make some visual observations of other people in your class to observe a possible correlation.
3. From such observations, the following hypothesis is formulated: the length of a person's upper limb is equal to 0.4 (40%) of the height of the person. To test this hypothesis, perform the following experiment.
4. In this experiment, use a meterstick (fig. 1.1) to measure an upper limb length of ten subjects. For each measurement, place the meterstick in the axilla (armpit) and record the length in centimeters to the end of the longest finger (fig. 1.2). Obtain the height of each person in centimeters by measuring them without shoes against a wall (fig. 1.3). The height of each person can also be calculated by multiplying each individual's height in inches by 2.54 to obtain his or her height in centimeters. Record all your measurements in Part A of Laboratory Report 1.

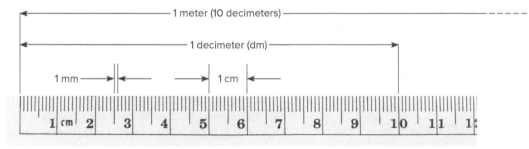

Metric ruler

FIGURE 1.1 Metric ruler with metric lengths indicated. A meterstick length would be 100 centimeters (10 decimeters). (The image size is approximately to scale.)

FIGURE 1.2 Measurement of upper limb length. ©J and J Photography

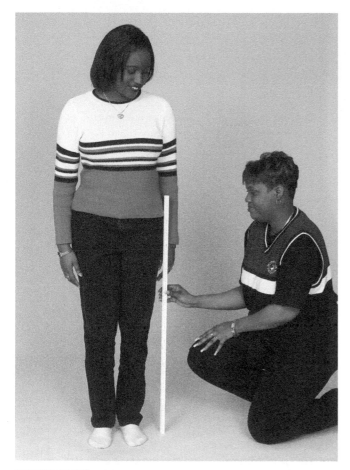

FIGURE 1.3 Measurement of height.
©J and J Photography

5. The data collected from all of the measurements can now be analyzed. The expected (predicted) correlation between upper limb length and height is determined using the following equation:

$$\text{Height} \times 0.4 = \text{expected upper limb length}$$

The observed (actual) correlation to be used to test the hypothesis is determined by

$$\text{Length of upper limb/height} = \text{actual \% of height}$$

6. A graph is an excellent way to display a visual representation of the data. Plot the subjects' data in Part A of the laboratory report. Plot the upper limb length of each subject on the x-axis (independent variable) and the height of each person on the y-axis (dependent variable). A line is already located on the graph that represents a hypothetical relationship of 0.4 (40%) upper limb length compared to height. This is a graphic representation of the original hypothesis.

7. Compare the distribution of all of the points (actual height and upper limb length) that you placed on the graph with the distribution of the expected correlation represented by the hypothesis.

8. Complete Part A of the laboratory report.

EXPLORE

PROCEDURE B—Design an Experiment

Critical Thinking Application

You have probably concluded that there is some correlation of the length of body parts to height. Often when a skeleton is found by a paleontologist, it is not complete. It is occasionally feasible to use the length of a single bone to estimate the height of an individual. Observe human skeletons and locate the humerus bone in an upper limb or the femur bone in a lower limb. Use your observations to identify a mathematical relationship between the length of the humerus or the femur and height. Formulate a hypothesis that can be tested. Make measurements, analyze data, and develop a conclusion from your experiment. Complete Part B of the laboratory report.

TABLE 1.1 Metric Measurement System and Conversions

Measurement	Unit & Abbreviation	Metric Equivalent	Conversion Factor Metric to English (approximate)	Conversion Factor English to Metric (approximate)
LENGTH	1 kilometer (km)	1,000 (10^3) m	1 km = 0.62 mile	1 mile = 1.61 km
	1 meter (m)	100 (10^2) cm 1,000 (10^3) mm	1 m = 1.1 yards = 3.3 feet = 39.4 inches	1 yard = 0.9 m 1 foot = 0.3 m
	1 decimeter (dm)	0.1 (10^{-1}) m	1 dm = 3.94 inches	1 inch = 0.25 dm
	1 centimeter (cm)	0.01 (10^{-2}) m	1 cm = 0.4 inches	1 foot = 30.5 cm 1 inch = 2.54 cm
	1 millimeter (mm)	0.001 (10^{-3}) m 0.1 (10^{-1}) cm	1 mm = 0.04 inches	
	1 micrometer (μm)	0.000001 (10^{-6}) m 0.001 (10^{-3}) mm		
MASS	1 metric ton (t)	1,000 (10^3) kg	1 t = 1.1 ton	1 ton = 0.91 t
	1 kilogram (kg)	1,000 (10^3) g	1 kg = 2.2 pounds	1 pound = 0.45 kg
	1 gram (g)	1,000 (10^3) mg	1 g = 0.04 ounce	1 pound = 454 g 1 ounce = 28.35 g
	1 milligram (mg)	0.001 (10^{-3}) g		
VOLUME (LIQUIDS AND GASES)	1 liter (L)	1,000 (10^3) mL	1 L = 1.06 quarts	1 gallon = 3.78 L 1 quart = 0.95 L
	1 milliliter (mL)	0.001 (10^{-3}) L 1 cubic centimeter (cc or cm^3)	1 mL = 0.03 fluid ounce 1 mL = 1/5 teaspoon 1 mL = 15–16 drops	1 quart = 946 mL 1 fluid ounce = 29.6 mL 1 teaspoon = 5 mL
TIME	1 second (s)	1/60 minute	same	same
	1 millisecond (ms)	0.001 (10^{-3}) s	same	same
TEMPERATURE	Degrees Celsius (°C)		°F = 9/5 °C + 32	°C = 5/9 (°F − 32)

Name _____

Date _____

Section _____

The Ⓐ corresponds to the indicated Learning Outcome(s) found at the beginning of the
laboratory exercise.

Scientific Method
and Measurements

PART A ASSESSMENTS

1. Record measurements for the upper limb length and height of ten subjects. Use a calculator to determine the expected
   upper limb length and the actual percentage (as a decimal or a percentage) of the height for the ten subjects. Record your
   results in the following table. Ⓐ2

Subject	Measured Upper Limb Length (cm)	Height* (cm)	Height × 0.4 = Expected Upper Limb Length (cm)	Actual % of Height = Measured Upper Limb Length (cm)/Height (cm)
1.				
2.				
3.				
4.				
5.				
6.				
7.				
8.				
9.				
10.				

*The height of each person can be calculated by multiplying each individual's height in inches by 2.54 to obtain his or her height in centimeters. Ⓐ1

2. Plot the distribution of data (upper limb length and height) collected for the ten subjects on the following graph. The line located on the graph represents the expected 0.4 (40%) ratio of upper limb length to measured height (the original hypothesis). (The x-axis represents upper limb length, and the y-axis represents height.) Draw a line of *best fit* through the distribution of points of the plotted data of the ten subjects. Compare the two distributions (expected line and the distribution line drawn for the ten subjects). ⚠3

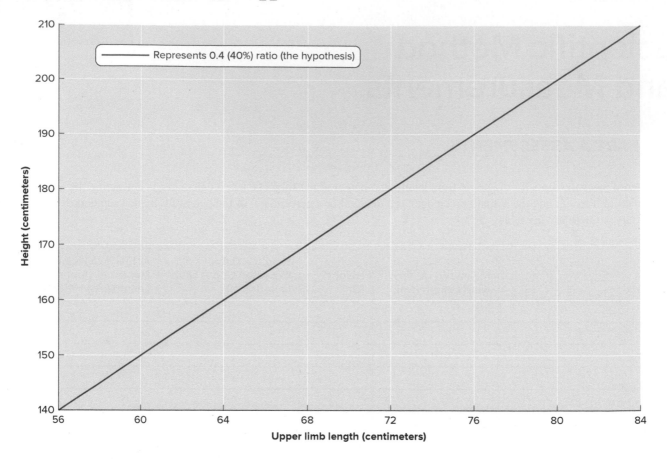

3. Does the distribution of the ten subjects' measured upper limb lengths support or reject the original hypothesis? _____ Explain your answer. ⚠3

PART B ASSESSMENTS

1. Describe your observations of a possible correlation between the humerus or femur length and height. ⚠4

2. Write a hypothesis based on your observations. ⚠4

3. Describe the design of the experiment that you devised to test your hypothesis. ⚠

4. Place your analysis of the data in this space in the form of a table and a graph. ⚠
   a. Table:

   b. Graph:

5. Based on an analysis of your data, what can you conclude? Did these conclusions confirm or refute your original hypothesis? Ⓐ

6. Discuss your results and conclusions with classmates. What common conclusion can the class formulate about the correlation between the humerus or femur length and height? Ⓐ

2

Body Organization and Terminology

MATERIALS NEEDED

Textbook
Dissectible human torso model (manikin)
Variety of specimens or models sectioned along
    various planes

For Learning Extension:
Colored pencils

PURPOSE OF THE EXERCISE

To review the organizational pattern of the human body; to review its organ systems and the organs included in each system; and to become acquainted with the terms used to describe the relative position of body parts, body sections, and body regions.

LEARNING OUTCOMES APR

After completing this exercise, you should be able to

1. Locate and name the body cavities and identify the organs and membranes associated with each cavity.
2. Differentiate the general functions of the organ systems of the human body.
3. Associate the organs included within each system and locate the organs in a dissectible human torso model.
4. Select the terms used to describe the relative positions of body parts.
5. Match the terms used to identify body sections and identify the plane along which a particular specimen is cut.
6. Label body regions and associate the terms used to identify body regions.
7. Recognize the anatomical terms pertaining to body organization.

The major features of the human body include certain cavities, a set of membranes associated with these cavities, and a group of organ systems composed of related organs. To communicate effectively with each other about the body, scientists have devised names to describe these body features. They also have developed terms to represent the relative positions of body parts, imaginary planes passing through these parts, and body regions.

EXPLORE

PROCEDURE A—Body Cavities and Membranes

1. Review the concept headings of "Body Cavities" and "Thoracic and Abdominopelvic Membranes" of section 1.6 in chapter 1 of the textbook.
2. As a review activity, label figures 2.1 and 2.2.
3. Locate the following features on the reference plates near the end of chapter 1 of the textbook and on the dissectible human torso model (fig. 2.3):

body cavities
    cranial cavity
    vertebral canal (spinal cavity)
    thoracic cavity
        mediastinum (region between the lungs; includes
            pericardial cavity)
        pleural cavities
    abdominopelvic cavity
        abdominal cavity
        pelvic cavity

diaphragm
smaller cavities within the head
    oral cavity
    nasal cavity with connected sinuses
    orbital cavity
    middle ear cavity

membranes and cavities
    pleural cavity
        parietal pleura
        visceral pleura
    pericardial cavity
        parietal pericardium (covered by fibrous
            pericardium)
        visceral pericardium (epicardium)

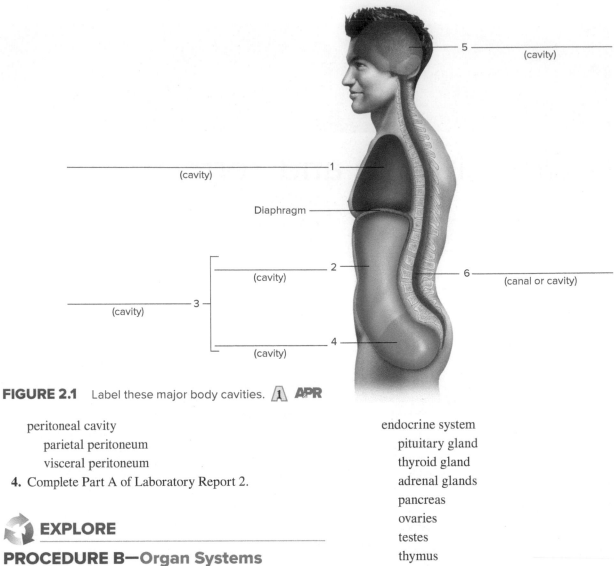

_____ 5 _____
                    (cavity)

_____ 1 _____
(cavity)

Diaphragm _____

        ┌─ _____ 2 _____   _____ 6 _____
        │   (cavity)   (canal or cavity)
_____ 3 ─┤
(cavity) │
        └─ _____ 4 _____
               (cavity)

FIGURE 2.1 Label these major body cavities.

peritoneal cavity
    parietal peritoneum
    visceral peritoneum
4. Complete Part A of Laboratory Report 2.

EXPLORE

PROCEDURE B—Organ Systems

1. Review the concept heading entitled "Organ Systems" of section 1.6 in chapter 1 of the textbook.
2. Use the reference plates near the end of chapter 1 of the textbook and the dissectible human torso model (fig. 2.3) to locate the following organs:

integumentary system
    skin
    accessory organs such as hair and nails
skeletal system
    bones
    ligaments
muscular system
    skeletal muscles
    tendons
nervous system
    brain
    spinal cord
    nerves

endocrine system
    pituitary gland
    thyroid gland
    adrenal glands
    pancreas
    ovaries
    testes
    thymus
cardiovascular system
    heart
    arteries
    veins
lymphatic system
    lymphatic vessels
    lymph nodes
    thymus
    spleen
digestive system
    mouth
    tongue
    teeth
    salivary glands
    pharynx
    esophagus
    stomach
    liver
    gallbladder

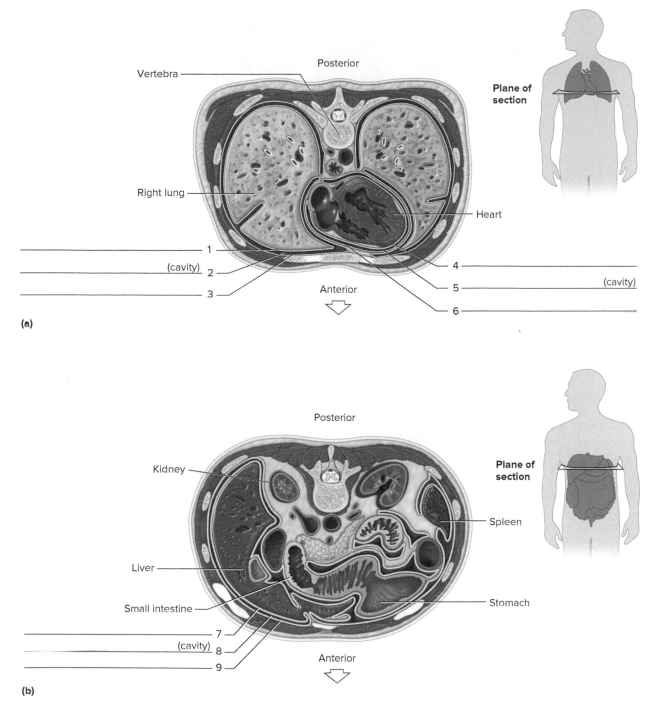

Vertebra

Posterior

Right lung

Heart

Plane of
section

1

(cavity) 2

3

Anterior

4

5 (cavity)

6

(a)

Posterior

Kidney

Spleen

Liver

Small intestine

Stomach

7

(cavity) 8

9

Anterior

(b)

Plane of
section

FIGURE 2.2 Label the thoracic membranes and cavities in (*a*) and the abdominopelvic membranes and cavity in (*b*), as shown in these superior views of transverse sections. Ⓐ **APR**

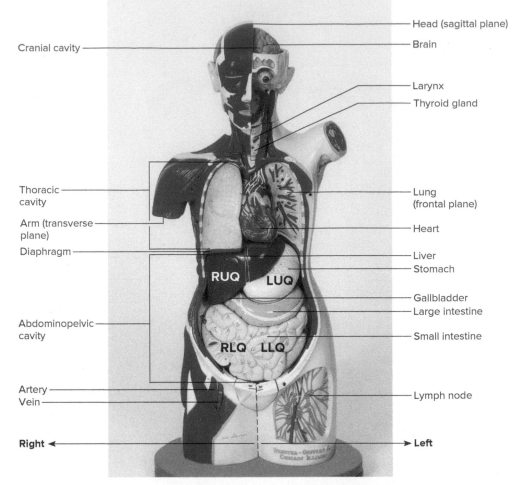

Cranial cavity

Head (sagittal plane)

Brain

Larynx

Thyroid gland

Thoracic cavity

Arm (transverse plane)

Diaphragm

Lung (frontal plane)

Heart

Liver

Stomach

RUQ   LUQ

Gallbladder

Large intestine

Abdominopelvic cavity

Small intestine

RLQ   LLQ

Artery

Vein

Lymph node

Right ⟵   ⟶ Left

FIGURE 2.3 Dissectible human torso model with body cavities, abdominopelvic quadrants, body planes, and major organs indicated. The abdominopelvic quadrants include right upper quadrant (RUQ), left upper quadrant (LUQ), right lower quadrant (RLQ), and left lower quadrant (LLQ). **APR** ©J and J Photography

pancreas
small intestine
large intestine
respiratory system
    nasal cavity
    pharynx
    larynx
    trachea
    bronchi
    lungs
urinary system
    kidneys
    ureters
    urinary bladder
    urethra
male reproductive system
    scrotum
    testes
    penis
    urethra

female reproductive system
    ovaries
    uterine tubes (oviducts; fallopian tubes)
    uterus
    vagina

3. Complete Part B of the laboratory report.

EXPLORE

PROCEDURE C—Relative Positions, Planes, Sections, and Regions

1. Observe the person standing in anatomical position (fig. 2.4). Anatomical terminology assumes the body is in anatomical position even though a person is often observed differently.
2. Review section 1.7 entitled "Anatomical Terminology" in chapter 1 of the textbook.
3. As a review activity, label figures 2.5, 2.6, and 2.7.
4. Examine the sectioned specimens on the demonstration table, and identify the plane along which each is cut. Cylindrical structures, such as a long bone or a blood

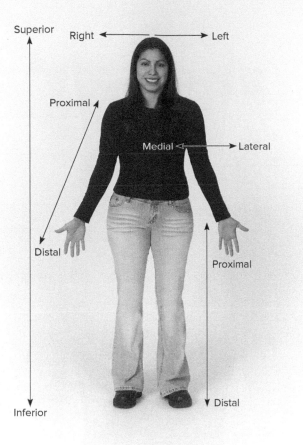

FIGURE 2.4 Anatomical position with directional terms indicated. The body is standing erect, face forward, with upper limbs at the sides and palms forward. When the palms are forward (supinated), the radius and ulna in the forearm are nearly parallel. This results in an anterior view of the body. Terms of relative position are used to describe the location of one body part with respect to another. **APR** ©J and J Photography

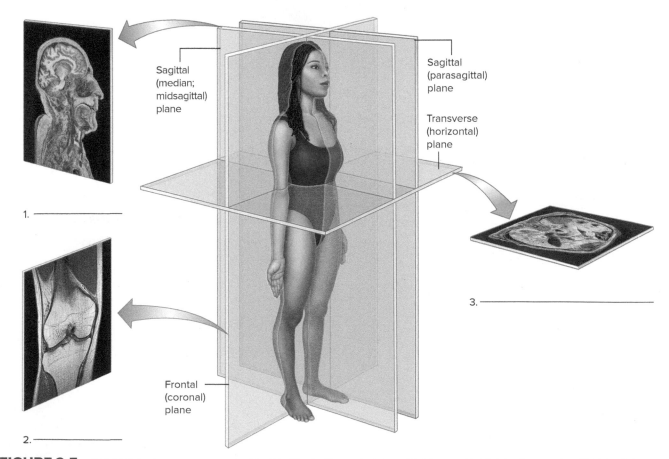

FIGURE 2.5 Label the planes represented in this illustration. **(1):** McGraw-Hill Education/Karl Rubin, photographer; **(2):** Living Art Enterprises/Science Source; **(3):** Karl Rubin/McGraw-Hill Education; **(center):** Joe DeGrandis/McGraw-Hill Education

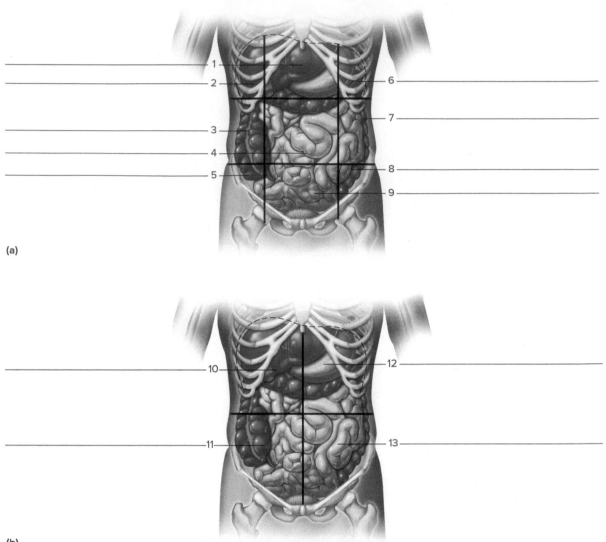

(a)

(b)

FIGURE 2.6 The abdominopelvic cavity is subdivided into either nine regions or four quadrants for purposes of locating organs, injuries, pain, or performed medical procedures. Label (*a*) the 9 regions and (*b*) the 4 quadrants of the abdominopelvic cavity. **A APR**

vessel, may be cut in cross section, oblique section, or longitudinal section. The same three sections can be demonstrated by three cuts of a banana (fig. 2.8).

5. Complete Parts C, D, E, F, and G of the laboratory report.

LEARNING EXTENSION

Use different colored pencils to distinguish the body regions in figure 2.7.

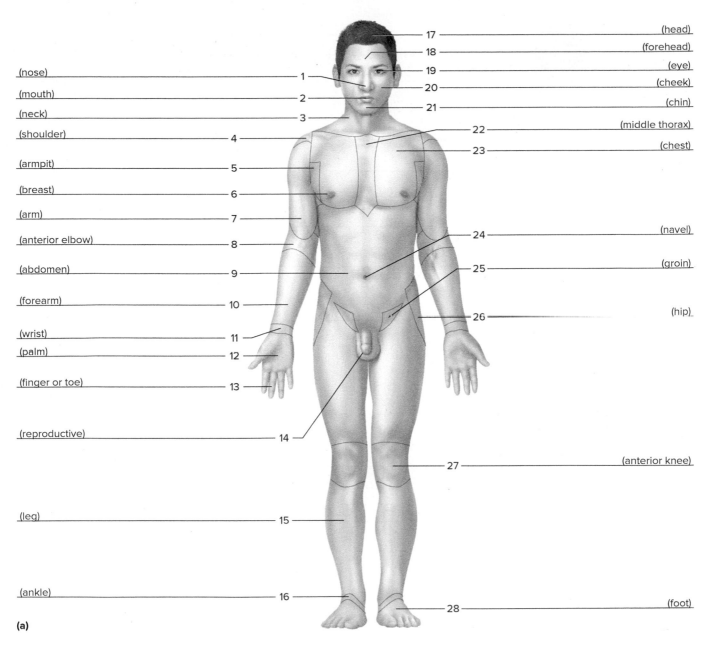

(nose) ——————————— 1
(mouth) ——————————— 2
(neck) ——————————— 3
(shoulder) ——————————— 4
(armpit) ——————————— 5
(breast) ——————————— 6
(arm) ——————————— 7
(anterior elbow) ——————————— 8
(abdomen) ——————————— 9
(forearm) ——————————— 10
(wrist) ——————————— 11
(palm) ——————————— 12
(finger or toe) ——————————— 13
(reproductive) ——————————— 14
(leg) ——————————— 15
(ankle) ——————————— 16

17 ——————————— (head)
18 ——————————— (forehead)
19 ——————————— (eye)
20 ——————————— (cheek)
21 ——————————— (chin)
22 ——————————— (middle thorax)
23 ——————————— (chest)
24 ——————————— (navel)
25 ——————————— (groin)
26 ——————————— (hip)
27 ——————————— (anterior knee)
28 ——————————— (foot)

(a)

FIGURE 2.7 Label these diagrams with terms used to describe body regions: (*a*) anterior regions; (*b*) posterior regions. 🄰 **APR**

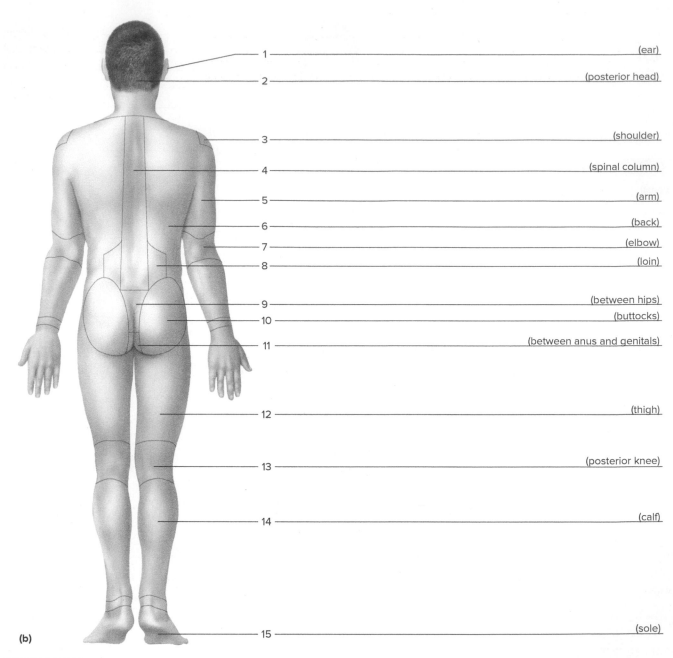

1 ————————————— (ear)

2 ————————————— (posterior head)

3 ————————————— (shoulder)

4 ————————————— (spinal column)

5 ————————————— (arm)

6 ————————————— (back)

7 ————————————— (elbow)

8 ————————————— (loin)

9 ————————————— (between hips)

10 ————————————— (buttocks)

11 ————————————— (between anus and genitals)

12 ————————————— (thigh)

13 ————————————— (posterior knee)

14 ————————————— (calf)

15 ————————————— (sole)

(b)

FIGURE 2.7 *Continued.*

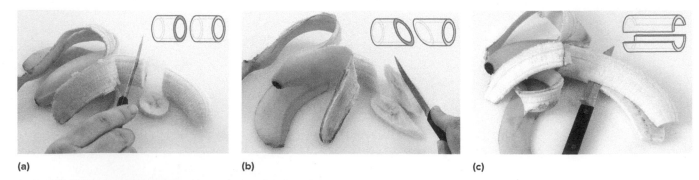

(a) (b) (c)

FIGURE 2.8 Three possible cuts of a banana: (*a*) cross section; (*b*) oblique section; (*c*) longitudinal section. Sections through an organ, as a body tube, frequently produce views similar to the cut banana. **(a–c):** ©J and J Photography

The A corresponds to the indicated Learning Outcome(s) found at the beginning of the laboratory exercise.

Body Organization and Terminology

PART A ASSESSMENTS

Match the cavities in column A with the organs contained in the cavities in column B. Place the letter of your choice in the space provided. A

Column A	Column B
a. Abdominal cavity	_____ **1.** Liver
b. Cranial cavity	_____ **2.** Lungs
c. Pelvic cavity	_____ **3.** Spleen
d. Thoracic cavity	_____ **4.** Stomach
e. Vertebral canal (spinal cavity)	_____ **5.** Brain
	_____ **6.** Trachea and esophagus
	_____ **7.** Urinary bladder
	_____ **8.** Small intestine
	_____ **9.** Spinal cord
	_____ **10.** Internal reproductive organs
	_____ **11.** Heart
	_____ **12.** Mediastinum

PART B ASSESSMENTS

Match the organ systems in column A with the general functions in column B. Place the letter of your choice in the space provided.

Column A

a. Cardiovascular system
b. Digestive system
c. Endocrine system
d. Integumentary system
e. Lymphatic system
f. Muscular system
g. Nervous system
h. Reproductive system
i. Respiratory system
j. Skeletal system
k. Urinary system

Column B

_____ 1. Main system that secretes hormones that function in the integration and coordination of the body

_____ 2. Provides an outer covering of the body for protection

_____ 3. Produces gametes (eggs and sperm) and new organisms

_____ 4. Stimulates muscles to contract, interprets information from sensory units, and functions in the integration and coordination of the body

_____ 5. Provides a framework and support for soft tissues and produces blood cells in red marrow

_____ 6. Exchanges gases between air and blood

_____ 7. Transports excess fluid from tissues to blood and helps defend the body against infections

_____ 8. Involves contractions and movement of the joints of the body, and creates most body heat

_____ 9. Removes liquid wastes from blood and transports them to the outside of the body

_____ 10. Breaks down and converts food molecules into absorbable forms

_____ 11. Transports nutrients, wastes, and gases throughout the body

PART C ASSESSMENTS

Indicate whether each of the following sentences makes correct or incorrect usage of the word in boldface type (assume that the body is in the anatomical position as observed in fig. 2.4). If the sentence is incorrect, in the space provided, supply a term that will make it correct. 4

1. The mouth is **superior** to the nose. _____

2. The stomach is **inferior** to the diaphragm. _____

3. The trachea is **anterior** to the spinal cord. _____

4. The larynx is **posterior** to the esophagus. _____

5. The heart is **medial** to the lungs. _____

6. The kidneys are **inferior** to the lungs. _____

7. The hand is **proximal** to the elbow. _____

8. The knee is **proximal** to the ankle. _____

9. The thumb is the **lateral** digit of the hand. _____

10. The popliteal surface region is **anterior** to the patellar region. _____

PART D ASSESSMENTS

Match the body regions in column A with the body parts in column B. Place the letter of your choice in the space provided. 6

Column A	Column B
a. Antecubital	_____ **1.** Wrist
b. Axillary	_____ **2.** Reproductive organs
c. Brachial	_____ **3.** Armpit
d. Buccal	_____ **4.** Posterior region of elbow
e. Carpal	_____ **5.** Buttocks
f. Cephalic	_____ **6.** Finger or toe
g. Cervical	_____ **7.** Neck
h. Crural	_____ **8.** Arm
i. Cubital	_____ **9.** Cheek
j. Digital	_____ **10.** Leg
k. Genital	_____ **11.** Head
l. Gluteal	_____ **12.** Anterior region of elbow

PART E ASSESSMENTS

Match the body regions in column A with the locations in column B. Place the letter of your choice in the space provided. 6

Column A	Column B
a. Inguinal	_____ **1.** Ankle
b. Lumbar	_____ **2.** Breasts
c. Mammary	_____ **3.** Between anus and reproductive organs
d. Occipital	_____ **4.** Sole of foot
e. Palmar	_____ **5.** Middle of thorax
f. Pectoral	_____ **6.** Chest
g. Pedal	_____ **7.** Posterior region of knee
h. Perineal	_____ **8.** Foot
i. Plantar	_____ **9.** Inferior posterior region of head
j. Popliteal	_____ **10.** Abdominal wall near thigh (groin)
k. Sternal	_____ **11.** Lower back
l. Tarsal	_____ **12.** Palm

PART F ASSESSMENTS

 ## Critical Thinking Application

State the quadrant of the abdominopelvic cavity in which the pain or sound would be located for each of the six conditions listed. In some cases, there may be more than one correct answer, and pain is sometimes referred to another region. This phenomenon, called *referred pain,* occurs when pain is interpreted as originating from some area other than the parts being stimulated. When referred pain is involved in the patient's interpretation of the pain location, the proper diagnosis of the ailment is more challenging. For the purpose of this exercise, assume the pain is interpreted as originating from the organ involved. ⚠3

1. Stomach ulcer _____

2. Appendicitis _____

3. Bowel sounds _____

4. Gallbladder attack _____

5. Kidney stone in left ureter _____

6. Ruptured spleen _____

PART G ASSESSMENTS

Locate the 16 anatomical terms pertaining to "Body Organization and Terminology." ⚠7

E	N	D	O	C	R	I	N	E	B	C	R
Q	L	A	N	I	M	O	D	B	A	X	E
C	R	A	N	I	A	L	U	P	E	L	P
S	L	A	T	T	I	G	A	S	Y	A	R
C	R	U	L	M	V	E	R	Q	O	R	O
I	L	G	L	U	T	E	A	L	K	O	D
C	A	C	F	H	V	W	L	L	T	G	U
A	T	O	C	S	B	H	U	I	W	T	C
R	N	X	N	B	U	C	C	A	L	K	T
O	O	A	D	H	L	A	S	R	A	T	I
H	R	L	F	L	A	R	U	S	K	O	V
T	F	W	G	Z	R	A	M	L	A	P	E

Find and circle words pertaining to Laboratory Exercise 2. The words may be horizontal, vertical, diagonal, or backward.

8 body surface regions (buccal, coxal, gluteal, oral, otic, palmar, sural, tarsal)

2 organ system names; 3 body cavities; 3 planes of the body

3

Chemistry of Life

MATERIALS NEEDED

For pH Tests:
Chopped fresh red cabbage
Beaker (250 mL)
Distilled water
Tap water
Vinegar
Baking soda
Laboratory scoop for measuring
Pipets for measuring
Full-range pH test papers
7 assorted common liquids clearly labeled in closed
    bottles on a tray or in a tub
Droppers labeled for each liquid

For Organic Tests:
Test tubes
Test-tube rack
Test-tube clamps
China marker
Hot plate
Beaker for hot water bath (500 mL)
Pipets for measuring
Benedict's solution
Biuret reagent (or 10% NaOH and 1% $CuSO_4$)
Iodine-potassium-iodide (IKI) solution
Sudan IV dye
Egg albumin
10% glucose solution
Clear carbonated soft drink
10% starch solution
Potatoes for potato water
Distilled water
Vegetable oil
Brown paper
Numbered unknown organic samples

⚠ SAFETY

- Review all Laboratory Safety Guidelines in Appendix 1 of your laboratory manual.
- Clean laboratory surfaces before and after laboratory procedures using soap and water.
- Use extreme caution when working with chemicals.
- Wear safety goggles.
- Wear disposable gloves while working with the chemicals.
- Take precautions to prevent chemicals from contacting your skin.
- Do not mix any of the chemicals together unless instructed to do so.
- Clean up any spills immediately and notify the instructor at once.
- Wash your hands before leaving the laboratory.

↻ LEARNING OUTCOMES

After completing this exercise, you should be able to

(1) Associate and illustrate the basic organization of atoms and molecules.

(2) Measure pH values of various substances through testing methods.

(3) Determine categories of organic compounds with basic colorimetric tests.

(4) Discover the organic composition of an unknown solution.

PURPOSE OF THE EXERCISE

To review the organization of atoms and molecules, types of chemical interactions, and basic categories of organic compounds and to differentiate between types of organic compounds.

The complexities of the human body arise from the organization and interactions of chemicals. Organisms are made of matter, and the most basic unit of matter is the chemical element. The smallest unit of an element is an atom, and two or more of those can unite to form a molecule. All processes that occur within the body involve chemical reactions—interactions between atoms and molecules. We breathe to supply oxygen to our cells for energy. We eat and drink to bring chemicals into our bodies that our cells need. Water fills and bathes all of our cells and allows an amazing

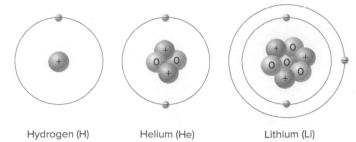

Hydrogen (H)   Helium (He)   Lithium (Li)

FIGURE 3.1 The single electron of a hydrogen atom is located in its first shell. The two electrons of a helium atom fill its first shell. Two of the three electrons of a lithium atom are in the first shell, and one is in the second shell. **APR**

array of reactions to occur, all of which are designed to keep us alive. Chemistry forms the basis of life and thus forms the foundation of anatomy and physiology.

EXPLORE

PROCEDURE A—Matter, Molecules, Bonding, and pH

1. Review section 2.1 entitled "Fundamentals of Chemistry" in chapter 2 of the textbook. Use the Periodic Table of Elements, located in Appendix 2 of the laboratory manual, as a reference.
2. Review figures 3.1, 3.2, and 3.3, which show molecular diagrams, formation of an ionic bond, and formation of a covalent bond, respectively.
3. Complete Part A of Laboratory Report 3.

EXPLORE

PROCEDURE B—The pH Scale

1. Review section 2.5 entitled "Acids and Bases" in chapter 2 of the textbook.
2. Review figure 3.4, which shows the pH scale. Note the range of the scale and the pH value considered neutral.
3. **Cabbage water tests.** Many tests can determine a pH value, but among the more interesting are colorimetric tests in which an indicator chemical changes color when it reacts. Many plant pigments, especially anthocyanins (which give plants color, from blue to red), can be used as colorimetric pH indicators. One that works well is the red pigment in red cabbage.
   a. Prepare cabbage water to be used as a general pH indicator. Fill a 250 mL beaker to the 100 mL level with chopped red cabbage. Add water to make 150 mL. Place the beaker on a hot plate and simmer the mixture until the pigments come out of the cabbage and the water turns deep purple. Allow the water to cool. (You may proceed to step 4 while you wait for this to finish.)
   b. Label three clean test tubes, one for water, one for vinegar, and one for baking soda.

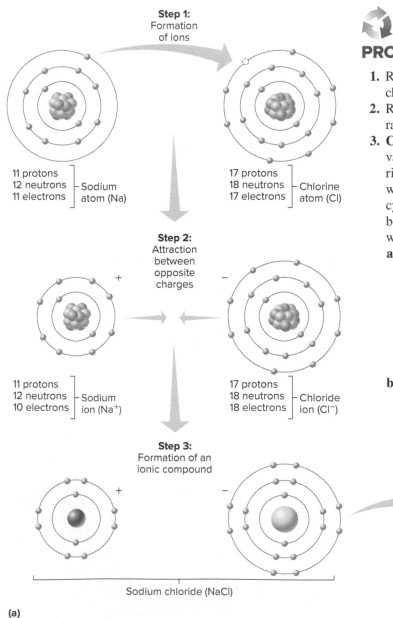

Step 1:
Formation of ions

11 protons
12 neutrons — Sodium
11 electrons   atom (Na)

17 protons
18 neutrons — Chlorine
17 electrons   atom (Cl)

Step 2:
Attraction between opposite charges

11 protons
12 neutrons — Sodium
10 electrons   ion (Na⁺)

17 protons
18 neutrons — Chloride
18 electrons   ion (Cl⁻)

Step 3:
Formation of an ionic compound

Sodium chloride (NaCl)

(a)

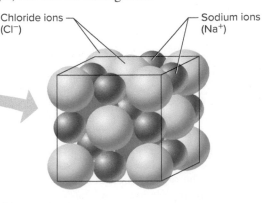

Chloride ions (Cl⁻) — — Sodium ions (Na⁺)

(b)

FIGURE 3.2 Formation of an ionic bond. **APR**

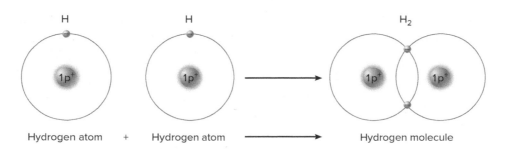

FIGURE 3.3 A hydrogen molecule forms when two hydrogen atoms share a pair of electrons and join by a covalent bond. **APR**

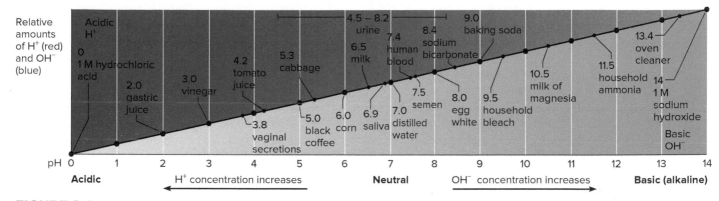

FIGURE 3.4 As the concentration of hydrogen ions (H⁺) increases, a solution becomes more acidic and the pH value decreases. As the concentration of ions that combine with hydrogen ions (such as hydroxide ions) increases, a solution becomes more basic (alkaline) and the pH value increases. The pH values of some common substances are shown. **APR**

c. Place 2 mL of cabbage water into each test tube.

d. To the first test tube, add 2 mL of distilled water and swirl the mixture. Record the color in Part B of the laboratory report.

e. Repeat this procedure for test tube 2, adding 2mL of vinegar and swirling the mixture. Record the results in Part B of the laboratory report.

f. Repeat this procedure for test tube 3, adding one laboratory scoop of baking soda. Swirl the mixture. Record the results in Part B of the laboratory report.

4. **Testing with pH paper.** Many commercial pH indicators are available. A simple one to use is pH paper, which comes in small strips. Don gloves for this procedure so your skin secretions do not contaminate the paper and to protect you from the chemicals you are testing.

a. Test the pH of distilled water by dropping one or two drops of distilled water onto a strip of pH paper and note the color. Compare the color to the color guide on the container. Note the results of the test as soon as color appears, as drying of the paper will change the results. Record the pH value in Part B of the laboratory report.

b. Repeat this procedure for tap water. Record the results.

c. Individually test the various common substances found on your lab table. Record your results.

d. Complete Part B of the laboratory report.

EXPLORE

PROCEDURE C—Organic Molecules

In this section, you will perform some simple tests to check for the presence of the following categories of organic molecules (biomolecules): protein, carbohydrate (sugar and starch), and lipid. Specific color changes will occur if the target compound is present. Please note the original color of the indicator being added so you can tell if the color really changed. For example, Biuret reagent and Benedict's solution are both initially blue, so an end color of blue would indicate no change. Use caution when working with these chemicals, and follow all directions. Carefully label the test tubes and avoid contaminating any of your samples. Only the Benedict's test requires heating and a time delay for accurate results. Do not heat any other tubes. To prepare for the Benedict's test, fill a 500 mL beaker half full with water and place it on the hot plate. Turn the hot plate on and bring the water to a boil. To save time, start the water bath before doing the Biuret test.

1. **Biuret test for protein.** In the presence of protein, Biuret reagent reacts with peptide bonds and changes to violet or purple. A pinkish color indicates that shorter polypeptides are present. The color intensity is proportional to the number of peptide bonds; thus, the intensity reflects the length of polypeptides (amount of protein).

a. Label six test tubes with numbers 1 to 6.

b. To each test tube, add 2 mL of one of the samples as follows:
   1—2 mL distilled water
   2—2 mL egg albumin
   3—2 mL 10% glucose solution
   4—2 mL carbonated soft drink
   5—2 mL 10% starch solution
   6—2 mL potato juice

c. To each, add 2 mL of Biuret reagent, swirl the tube to mix it, and note the final color. [*Note:* If Biuret reagent is not available, add 2 mL of 10% NaOH (sodium hydroxide) and about 10 drops of 1% $CuSO_4$ (copper sulfate).] **Be careful—NaOH is very caustic.**

d. Record your results in Part C of the laboratory report.

e. Mark a "+" on any tubes with positive results, and retain them for comparison while testing your unknown sample. Put remaining test tubes aside so they will not get confused with future trials.

2. **Benedict's test for sugar (monosaccharides).** In the presence of sugar, Benedict's solution changes from its initial blue to a green, yellow, orange, or reddish color, depending on the amount of sugar present. Orange and red indicate greater amounts of sugar.

a. Label six test tubes with numbers 1 to 6.

b. To each test tube, add 2 mL of one of the samples as follows:
   1—2 mL distilled water
   2—2 mL egg albumin
   3—2 mL 10% glucose solution
   4—2 mL carbonated soft drink
   5—2 mL 10% starch solution
   6—2 mL potato juice

c. To each, add 2 mL of Benedict's solution, swirl the tube to mix it, and place all tubes into the boiling water bath for minutes.

d. Note the final color of each tube after heating and record the results in Part C of the laboratory report.

e. Mark a "+" on any tubes with positive results and retain them for comparison while testing your unknown sample. Put remaining test tubes aside so they will not get confused with future trials.

3. **Iodine test for starch.** In the presence of starch, iodine turns a dark purple or blue-black color. Starch is a long chain formed by many glucose units linked together side by side. This regular organization traps the iodine molecules and produces the dark color.

a. Label six test tubes with numbers 1 to 6.

b. To each test tube, add 2 mL of one of the samples as follows:
   1—2 mL distilled water
   2—2 mL egg albumin
   3—2 mL 10% glucose solution
   4—2 mL carbonated soft drink
   5—2 mL 10% starch solution
   6—2 mL potato juice

c. To each, add 0.5 mL of IKI (iodine solution) and swirl the tube to mix it.

d. Record the final color of each tube in Part C of the laboratory report.

e. Mark a "+" on any tubes with positive results and retain them for comparison while testing your unknown sample. Put remaining test tubes aside so they will not get confused with future trials.

4. **Tests for lipids.**

a. Label two separate areas of a piece of brown paper as "water" or "oil."

b. Place a drop of water on the area marked "water" and a drop of vegetable oil on the area marked "oil."

c. Let the spots dry several minutes; then record your observations in Part C of the laboratory report. Upon drying, oil leaves a translucent stain (grease spot) on brown paper; water does not leave such a spot.

d. In a test tube, add 2 mL of water and 2 mL of vegetable oil and observe. Shake the tube vigorously; then let it sit for minutes and observe again. Record your observations in Part C of the laboratory report.

e. Sudan IV is a dye that is lipid-soluble but not water-soluble. If lipids are present, the Sudan IV will stain them pink or red. Add a small amount of Sudan IV to the test tube that contains the oil and water. Swirl it; then let it sit for a few minutes. Record your observations in Part C of the laboratory report.

EXPLORE

PROCEDURE D—Identifying Unknown Compounds

Now you will apply the information you gained with the tests in the previous section. You will retrieve an unknown sample that contains none, one, or any combination of the following types of organic compounds: protein, sugar, starch, or lipid. You will test your sample using each test from the previous section and record your results in Part D of the laboratory report.

1. Label four test tubes (1, 2, 3, 4).

2. Add 2 mL of your unknown sample to each test tube.

3. **Test for protein.** Add 2 mL of Biuret reagent to tube 1. Swirl the tube and observe the color. Record your observations in Part D of the laboratory report.

4. **Test for sugar.** Add 2 mL of Benedict's solution to tube 2. Swirl the tube and place it in a boiling water bath for 3 to 5 minutes. Record your observations in Part D of the laboratory report.

5. **Test for starch.** Add several drops of iodine solution to tube 3. Swirl the tube and observe the color. Record your observations in Part D of the laboratory report.

6. **Test for lipid.** Add 2 mL of water to tube 4. Swirl the tube and note if there is any separation.

7. Add a small amount of Sudan IV to test tube 4, swirl the tube, and record your observations in Part D of the laboratory report.

8. Based on the results of these tests, determine if your unknown sample contains any organic compounds and, if so, what they are. Record and explain your identification in Part D of the laboratory report.

Name _____

Date _____

Section _____

The Ⓐ corresponds to the indicated Learning Outcome(s) found at the beginning of the laboratory exercise.

Chemistry of Life

PART A ASSESSMENTS

Match the terms in column A with the descriptions in column B. Place the letter of your choice in the space provided. Ⓐ

Column A	Column B
a. Atomic number	_____ **1.** The number of protons plus the number of neutrons
b. Atomic weight	
c. Base	_____ **2.** A combination of two or more atoms of different elements
d. Catalyst (enzyme)	_____ **3.** A small, negatively charged particle that orbits the nucleus
e. Compound	
f. Covalent bond	_____ **4.** An atom that has gained or lost electrons and thus carries an electrical charge
g. Electrolyte	
h. Electron	_____ **5.** Atoms are held together by sharing electrons
i. Elements	_____ **6.** The most fundamental substances of matter
j. Ion	
k. Isotopes	_____ **7.** A substance that combines with hydrogen ions
l. Nucleus	_____ **8.** A molecule that influences the rate of chemical reactions but is not consumed in the reaction
	_____ **9.** Two atoms with the same atomic number but different atomic weights
	_____ **10.** A substance that releases ions in water
	_____ **11.** The number of protons in an atom
	_____ **12.** The area of an atom where protons and neutrons are located

MOLECULES AND BONDING

After reviewing figures 3.1, 3.2, and 3.3, complete the following:

1. The atomic number of hydrogen is _____, and it has _____ electron(s) in its outer shell. Ⓐ

2. The atomic number of chlorine is _____, and it has _____ electron(s) in its outer shell. Ⓐ

3. Using the style shown in figure 3.2, draw an atom of hydrogen and one of chlorine, clearly indicating the numbers and positions of the protons, neutrons, and electrons. Ⓐ

Hydrogen   Chlorine

4. Is the hydrogen atom stable?  _____

Is the chlorine atom stable?  _____

5. What type of bond is likely to form the compound HCl (hydrochloric acid)?  _____

6. Repeat your drawings here, but also show and explain how this bond of HCl would form. Clearly indicate the positions of protons, neutrons, and electrons.  _____

PART B ASSESSMENTS

1. Results from cabbage water tests:

Substance	Cabbage Water	Distilled Water	Vinegar	Baking Soda
Color				
Acid, base, or neutral?				

2. Results from pH paper tests:

Substance Tested	Distilled Water	Tap Water	Sample 1:_____	Sample 2:_____	Sample 3:_____	Sample 4:_____	Sample 5:_____	Sample 6:_____	Sample 7:_____
pH value									

3. Are the pH values the same for distilled water and tap water?  _____

4. If not, what might explain this difference? _____

_____

5. Draw the pH scale here, and indicate the following values: 0, 7 (neutral), and 14. Now label the scale with the names, and indicate the location of the pH values for each substance you tested.

PART C ASSESSMENTS

1. **Biuret test results for protein.** Enter your results from the Biuret test for protein. A color change to purple indicates that protein is present; pink indicates that short polypeptides are present. **3**

Tube	Contents	Color	Protein Present (+) or Absent (–)
1	Distilled water		
2	Egg albumin		
3	Glucose solution		
4	Soft drink		
5	Starch solution		
6	Potato juice		

2. **Benedict's test results for most sugars (monosaccharides).** Enter your results from the Benedict's test for sugars. A color change to green, yellow, orange, or red indicates that sugar is present. Note the color after the mixture has been heated for 3 to 5 minutes. **3**

Tube	Contents	Color	Sugar Present (+) or Absent (–)
1	Distilled water		
2	Egg albumin		
3	Glucose solution		
4	Soft drink		
5	Starch solution		
6	Potato juice		

3. **Iodine test results for starch.** Enter your results from the iodine test for starch. A color change to dark blue or black indicates that starch is present. **3**

Tube	Contents	Color	Starch Present (+) or Absent (–)
1	Distilled water		
2	Egg albumin		
3	Glucose solution		
4	Soft drink		
5	Starch solution		
6	Potato juice		

4. **Lipid test results.** What did you observe when you allowed the drops of oil and water to dry on the brown paper?_____

What did you observe when you mixed the oil and water together?_____

_____

What did you observe when you added the Sudan IV dye? **3** _____

Critical Thinking Application

If a person were on a low-carbohydrate, high-protein diet, which of the six substances tested would the person want to increase? _____

Explain your answer. _____

_____

_____

_____

PART D ASSESSMENTS

1. What is the number of your sample? _____

2. Record the results of your tests on the unknown sample here: 4️⃣

Test Performed	Results
Biuret test for protein	
Benedict's test for sugars	
Iodine test for starch	
Water test for lipid	
Sudan IV test for lipid	

3. Based on these results, what organic compound(s) does your unknown sample contain? 4️⃣ _____

4. Explain your answer. 4️⃣ _____

_____

_____

4

Care and Use of the Microscope

MATERIALS NEEDED

Compound light microscope
Lens paper
Microscope slides
Coverslips
Transparent plastic millimeter ruler
Prepared slide of letter *e*
Slide of three colored threads
Medicine dropper
Dissecting needle (needle probe)
Specimen examples for wet mounts
Methylene blue (dilute) or iodine-potassium-iodide
    stain

For Demonstrations:
Micrometer scale
Stereomicroscope (dissecting microscope)

PURPOSE OF THE EXERCISE

To become familiar with the major parts of a compound light microscope and their functions and to make use of the microscope to observe small objects.

LEARNING OUTCOMES

After completing this exercise, you should be able to

1. Locate and identify the major parts of a compound light microscope and differentiate the functions of these parts.

2. Calculate the total magnification produced by various combinations of eyepiece and objective lenses.

3. Demonstrate proper use of the microscope to observe and measure small objects.

4. Prepare a simple microscope slide and sketch the objects you observed.

The human eye cannot perceive objects less than 0.1 mm in diameter, so a microscope is an essential tool for the study of small structures such as cells. The microscope used for this purpose is the *compound light microscope*. It is called compound because it uses two sets of lenses: an eyepiece, or ocular, lens system and an objective lens system. The eyepiece lens system magnifies, or compounds, the image reaching it after the image is magnified by the objective lens system. Such an instrument can magnify images of small objects up to about 1,000 times.

EXPLORE

PROCEDURE A—Microscope Basics

1. Familiarize yourself with the following list of rules for care of the microscope:
   a. Keep the microscope under its *dustcover* and in a cabinet when it is not being used.
   b. Handle the microscope with great care. It is an expensive and delicate instrument. To move it or carry it, hold it by its *arm* with one hand and support its *base* with the other hand (fig. 4.1).
   c. Always store the microscope with the scanning or lowest power objective in place. Always start with this objective when using the microscope.
   d. To clean the lenses, rub them gently with *lens paper* or a high-quality cotton swab. If the lenses need additional cleaning, follow the directions in the "Lens-Cleaning Technique" section that follows.

LENS-CLEANING TECHNIQUE

1. Moisten one end of a high-quality cotton swab with one drop of lens cleaner. Keep the other end dry.
2. Clean the optical surface with the wet end. Dry it with the other end, using a circular motion.
3. Use a hand aspirator to remove lingering dust particles.
4. Start with the scanning objective and work upward in magnification, using a new cotton swab for each objective.
5. When cleaning the eyepiece, do not open the lens unless it is absolutely necessary.
6. Use alcohol for difficult cleaning.

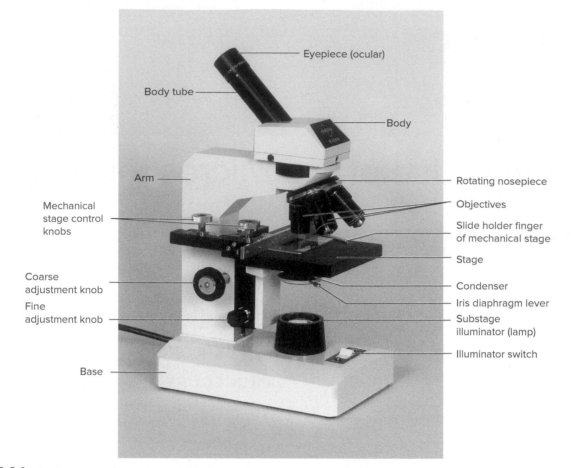

FIGURE 4.1 Major parts of a compound light microscope with a monocular (one eyepiece) body and a mechanical stage. Some compound microscopes are equipped with a binocular (two eyepieces) body. ©J and J Photography

e. If the microscope has a substage lamp, be sure the electric cord does not hang off the laboratory table where someone might trip over it. The bulb life can be extended if the lamp is cool before the microscope is moved.

f. Never drag the microscope across the laboratory table after you have placed it down for use.

g. Never remove parts of the microscope or try to disassemble the eyepiece or objective lenses.

h. If your microscope is not functioning properly, report the problem to your laboratory instructor immediately.

2. Observe a compound light microscope and study figure 4.1 to learn the names of its major parts. Your microscope might be equipped with a binocular body and two eyepieces (fig. 4.2). The lens system of a compound microscope includes three parts—the condenser, objective lens, and eyepiece (ocular).

   Light enters this system from a *substage illuminator* (*lamp*) or *mirror* and is concentrated and focused by a *condenser* onto a microscope slide (fig. 4.3). The condenser, which contains a set of lenses, usually is kept in its highest position possible.

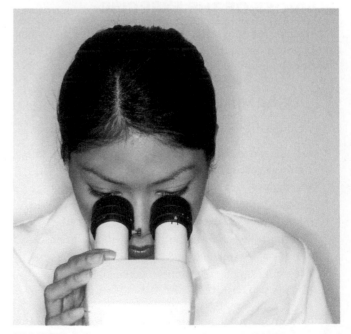

FIGURE 4.2 Scientist using a compound light microscope equipped with a binocular body and two eyepieces. George Doyle/Getty Images

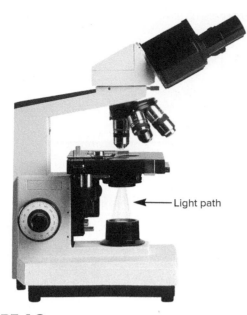

FIGURE 4.3 Microscope, showing the path of light through it. Comstock/Alamy

(a)

The *iris diaphragm,* located between the light source and the condenser, can be used to increase or decrease the intensity of the light entering the condenser and to control the contrast of the image. Locate the lever that operates the iris diaphragm beneath the *stage* and move it back and forth. Note how this movement causes the size of the opening in the diaphragm to change. (Some microscopes have a revolving plate called a *disc diaphragm* beneath the stage instead of an iris diaphragm. Disc diaphragms have different-sized holes to admit varying amounts of light.) Which way do you move the diaphragm to increase the light intensity? _____ Which way to decrease it?_____

After light passes through a specimen mounted on a microscope slide, it enters an *objective lens system.* This lens projects the light upward into the *body tube,* where it produces a magnified image of the object being viewed.

The *eyepiece (ocular) lens* system then magnifies this image to produce another image, seen by the eye. Typically, the eyepiece lens magnifies the image ten times (10×). Look for the number in the metal of the eyepiece that indicates its power (fig. 4.4). What is the eyepiece power of your microscope? _____

The objective lenses are mounted in a *rotating nosepiece* so that different magnifications can be achieved by rotating any one of several objective lenses into position above the specimen. Commonly, this set of lenses includes a scanning objective (4×), a low-power objective (10×), and a high-power objective,

(b)

FIGURE 4.4 The powers of this 10× eyepiece (*a*) and this 40× objective (*b*) are marked in the metal. DIN is an international optical standard on quality optics. The 0.65 on the 40× objective is the numerical aperture, a measure of the light-gathering capabilities. **(a, b):** ©J and J Photography

also called a high-dry-power objective (about 40×). Sometimes an oil immersion objective (about 100×) is present. Look for the number printed on each objective that indicates its power. What are the objective lens powers of your microscope? _____

To calculate the *total magnification* achieved when using a particular objective, multiply the power of the eyepiece by the power of the objective used. Thus, the 10× eyepiece and the 40× objective produce a total magnification of 10 × 40, or 400×. See a summary of microscope lenses in table 4.1.

TABLE 4.1   Microscope Lenses

Objective Lens Name	Common Objective Lens Magnification	Common Eyepiece Lens Magnification	Total Magnification
Scan	4×	10×	40×
Low-power (LP)	10×	10×	100×
High-power (HP)	40×	10×	400×
Oil immersion	100×	10×	1,000×

Note: If you wish to observe an object under LP, HP, or oil immersion, locate and then center and focus the object first under scan magnification.

3. Complete Part A of Laboratory Report 4.
4. Turn on the substage illuminator and look through the eyepiece. You will see a lighted, circular area called the *field of view*.

   You can measure the diameter of this field of view by focusing the lenses on the millimeter scale of a transparent plastic ruler. To do this, follow these steps:

   a. Place the ruler on the microscope stage in the spring clamp of a slide holder finger on a *mechanical stage* or under the *stage (slide) clips.* (*Note:* If your microscope is equipped with a mechanical stage, it may be necessary to use a short section cut from a transparent plastic ruler. The section should be several millimeters long and can be mounted on a microscope slide for viewing.)

   b. Center the millimeter scale in the beam of light coming up through the condenser, and rotate the scanning objective into position.

   c. While you watch from the side to prevent the lens from touching anything, raise the stage until the objective is as close to the ruler as possible, using the *coarse adjustment knob* (fig. 4.5). (*Note:* The adjustment knobs on some microscopes move the body and objectives downward and upward for focusing.)

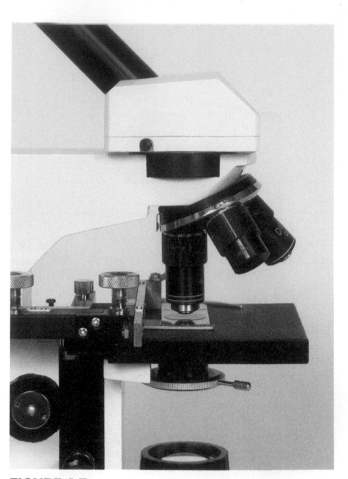

FIGURE 4.5   When you focus using a particular objective, you can prevent it from touching the specimen by watching from the side. ©J and J Photography

   d. Look into the eyepiece and use the *fine adjustment knob* to raise the stage until the lines of the millimeter scale come into sharp focus.

   e. Adjust the light intensity by moving the *iris diaphragm lever* so that the field of view is brightly illuminated but comfortable to your eye. At the same time, take care not to overilluminate the field because transparent objects tend to disappear in bright light.

   f. Position the millimeter ruler so that its scale crosses the greatest diameter of the field of view. Also, move the ruler so that one of the millimeter marks is against the edge of the field of view.

   g. In millimeters, measure the distance across the field of view.

5. Complete Part B of the laboratory report.
6. Most microscopes are designed to be *parfocal*. This means that when a specimen is in focus with a lower-power objective, it will be in focus (or nearly so) when a higher-power objective is rotated into position. Always center the specimen in the field of view before changing to higher objectives.

Rotate the low-power objective into position, and then look at the millimeter scale of the transparent plastic ruler. If you need to move the low-power objective to sharpen the focus, use the fine adjustment knob.

Adjust the iris diaphragm so that the field of view is properly illuminated. Once again, adjust the millimeter ruler so that the scale crosses the field of view through its greatest diameter, and position the ruler so that a millimeter mark is against one edge of the field. Measure the distance across the field of view in millimeters.

7. Rotate the high-power objective into position while you watch from the side, and then observe the millimeter scale on the plastic ruler. *All focusing using high-power magnification should be done only with the fine adjustment knob. Never use the coarse adjustment knob while observing with high-power magnification.* If you use the coarse adjustment knob with the high-power objective, you can accidentally force the objective into the coverslip and break the slide. This is because the *working distance* (the distance from the objective lens to the slide on the stage) is much shorter when using higher magnifications.

Adjust the iris diaphragm for proper illumination. When using higher magnifications, more illumination usually will help you view the objects more clearly. Try to measure the distance across the field of view in millimeters.

8. Locate the numeral 4 (or 9) on the plastic ruler and focus on it using the scanning objective. Note how the number appears in the field of view. Move the plastic ruler to the right and note which way the image moves. Slide the ruler away from you and again note how the image moves.

9. Observe a letter *e* slide in addition to the numerals on the plastic ruler. Note the orientation of the letter *e* using the scan, LP, and HP objectives. As you increase magnifications, note that the amount of the letter *e* shown is decreased. Move the slide to the left, then away from you, and note the direction in which the observed image moves. If a *pointer* is visible in your field of view, you can manipulate the pointer to a location (structure) within the field of view by moving the slide or rotating the eyepiece.

10. Examine the slide of the three colored threads using the low-power objective and then the high-power objective. Focus on the location where the three threads cross. By using the fine adjustment knob, determine the order from top to bottom by noting which color is in focus at different depths. The other colored threads will still be visible, but they will be blurred. Be sure to notice whether the stage or the body tube moves up and down with the adjustment knobs of the microscope being used for this depth determination. The vertical depth of a specimen clearly in focus is called the *depth of field (focus).*

Whenever specimens are examined, continue to use the fine adjustment focusing knob to determine relative depths of structures clearly in focus within cells, giving a three-dimensional perspective. The depth of field is less at higher magnifications.

Critical Thinking Application

What was the sequence of the three colored threads from top to bottom? Explain how you came to that conclusion. 3

_____

_____

_____

11. Complete Parts C and D of the laboratory report.

DEMONSTRATION

A compound light microscope is sometimes equipped with a micrometer scale mounted in the eyepiece. Such a scale is subdivided into 50 to 100 equal divisions (fig. 4.6). These arbitrary divisions can be calibrated against the known divisions of a micrometer slide placed on the microscope stage. Once the values of the divisions are known, the length and width of a microscopic object can be measured by superimposing the scale over the magnified image of the object.

Observe the micrometer scale in the eyepiece of the demonstration microscope. Focus the low-power objective on the millimeter scale of a micrometer slide (or a plastic ruler) and measure the distance between the divisions on the micrometer scale in the eyepiece. What is the distance between the finest divisions of the scale in micrometers?

EXPLORE

PROCEDURE B—Slide Preparation

1. Prepare several temporary *wet mounts,* using any small, transparent objects of interest, and examine the specimens using the low-power objective and then a high-power objective to observe their details. To prepare a wet mount, follow these steps (fig. 4.7):
   a. Obtain a precleaned microscope slide.
   b. Place a tiny, thin piece of the specimen you want to observe in the center of the slide, and use a medicine dropper to put a drop of water over it. Consult with your instructor if a drop of stain might enhance the

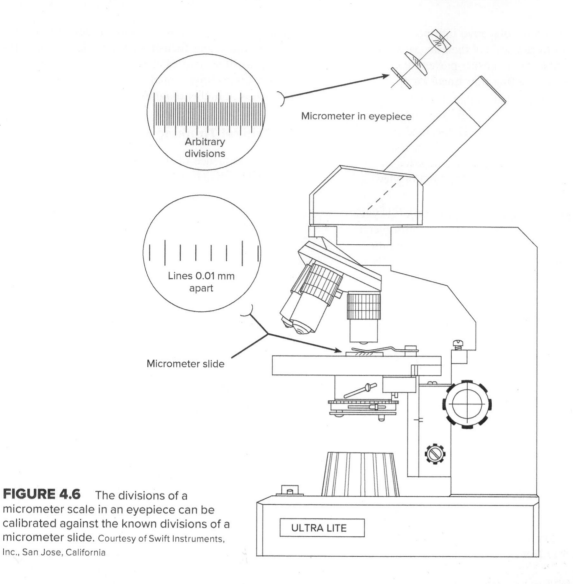

FIGURE 4.6 The divisions of a micrometer scale in an eyepiece can be calibrated against the known divisions of a micrometer slide. Courtesy of Swift Instruments, Inc., San Jose, California

image of any cellular structures of your specimen. If the specimen is solid, you might want to tease some of it apart with dissecting needles. In any case, the specimen must be thin enough that light can pass through it. Why is it necessary for the specimen to be so thin?

c. Cover the specimen with a coverslip. Try to avoid trapping bubbles of air beneath the coverslip by slowly lowering it at an angle into the drop of water.

d. Remove any excess water from the edge of the coverslip with absorbent paper. If your microscope has an inclination joint, do not tilt the microscope while observing wet mounts because the fluid will flow.

e. Place the slide under the stage (slide) clips or in the slide holder on a mechanical stage and position the slide so that the specimen is centered in the light beam passing up through the condenser.

f. Focus on the specimen using the scanning objective first. Next focus using the low-power objective, and then examine it with the high-power objective.

2. If an oil immersion objective is available, use it to examine the specimen. To use the oil immersion objective, follow these steps:

a. Center the object you want to study under the high-power field of view.

b. Rotate the high-power objective away from the microscope slide, place a small drop of immersion oil on the coverslip, and swing the oil immersion objective into position. To achieve sharp focus, use the fine adjustment knob only.

c. You will need to open the iris diaphragm more fully for proper illumination. More light is needed

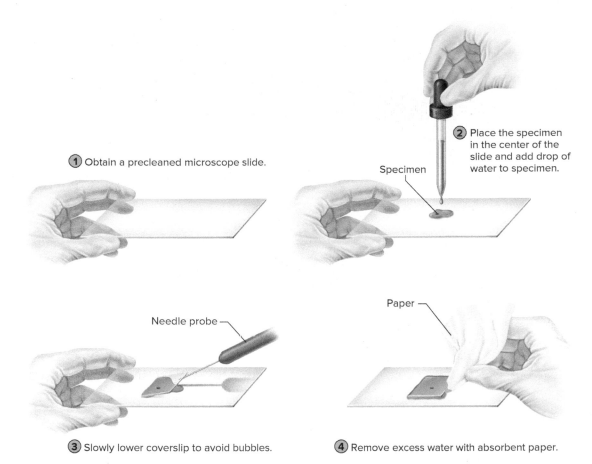

① Obtain a precleaned microscope slide.

② Place the specimen in the center of the slide and add drop of water to specimen.

Specimen

Needle probe

Paper

③ Slowly lower coverslip to avoid bubbles.

④ Remove excess water with absorbent paper.

FIGURE 4.7   Steps in the preparation of a wet mount.

because the oil immersion objective covers a very small lighted area of the microscope slide.

d. The oil immersion objective must be very close to the coverslip to achieve sharp focus, so care must be taken to avoid breaking the coverslip or damaging the objective lens. For this reason, never lower the objective when you are looking into the eyepiece. Instead, always raise the objective to achieve focus, or prevent the objective from touching the coverslip by watching the microscope slide and coverslip from the side if the objective needs to be lowered. Usually when using the oil immersion objective, only the fine adjustment knob needs to be used for focusing. Never switch back to one of the other three objectives while using oil immersion. This could cause damage to the microscope. The other objectives were not designed for oil immersion.

3. When you have finished working with the microscope, remove the microscope slide from the stage and wipe any oil from the objective lens with lens paper or a high-quality cotton swab. Swing the scanning objective or the low-power objective into position. Wrap the electric cord around the base of the microscope and replace the dustcover.

4. Complete Part E of the laboratory report.

DEMONSTRATION

A stereomicroscope (dissecting microscope) (fig. 4.8) is useful for observing the details of relatively large, opaque specimens. Although this type of microscope achieves less magnification than a compound light microscope, it has the advantage of producing a three-dimensional image rather than the flat, two-dimensional image of the compound microscope. In addition, the image produced by the stereomicroscope is positioned in the same manner as the specimen, rather than being reversed and inverted, as it is by the compound light microscope.

Observe the stereomicroscope. The eyepieces can be pushed apart or together to fit the distance between your eyes. Focus the microscope on the end of your finger. Which way does the image move when you move your finger to the right? _____ When you move it away from you? _____

If the instrument has more than one objective, change the magnification to higher power. Use the instrument to examine various small, opaque objects available in the laboratory.

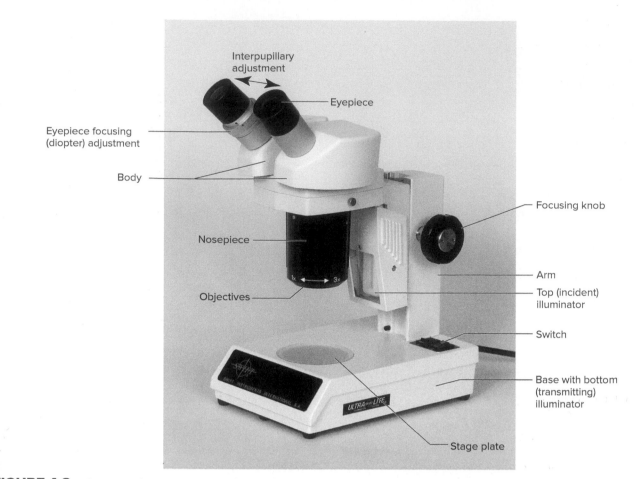

Interpupillary
adjustment

Eyepiece

Eyepiece focusing
(diopter) adjustment

Body

Focusing knob

Nosepiece

Arm

Objectives

Top (incident)
illuminator

Switch

Base with bottom
(transmitting)
illuminator

Stage plate

FIGURE 4.8 A stereomicroscope; also called a dissecting microscope. ©J and J Photography

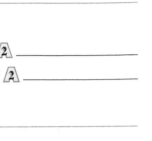
Name _____

Date _____

Section _____

The $\triangle$ corresponds to the indicated Learning Outcome(s) found at the beginning of the laboratory exercise.

Care and Use of the Microscope

🔄 PART A ASSESSMENTS

Revisit Procedure A, number 2; then complete the following:

1. What total magnification will be achieved if the 10× eyepiece and the 10× objective are used? $\triangle$2 _____

2. What total magnification will be achieved if the 10× eyepiece and the 100× objective are used? $\triangle$2 _____

🔄 PART B ASSESSMENTS

Revisit Procedure A, number 2; then complete the following:

1. Sketch the millimeter scale as it appears under the scanning objective magnification. (The circle represents the field of view through the microscope.)

2. In millimeters, what is the diameter of the scanning field of view? $\triangle$3 _____

3. Microscopic objects often are measured in *micrometers*. A micrometer equals 1/1,000 of a millimeter and is symbolized by μm. In micrometers, what is the diameter of the scanning power field of view? $\triangle$3 _____

4. If a circular object or specimen extends halfway across the scanning field, what is its diameter in millimeters? $\triangle$3 _____

5. In micrometers, what is its diameter? $\triangle$3 _____

🔄 PART C ASSESSMENTS

Complete the following:

1. Sketch the millimeter scale as it appears using the low-power objective.

2. What do you estimate the diameter of this field of view to be in millimeters? $\triangle$3 _____

3. How does the diameter of the scanning power field of view compare with that of the low-power field? _____

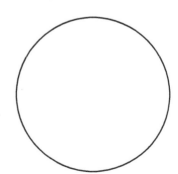

4. Why is it more difficult to measure the diameter of the high-power field of view than that of the low-power field?

_____

_____

5. What change occurred in the intensity of the light in the field of view when you exchanged the low-power objective for the high-power objective? _____

6. Sketch the numeral 4 (or 9) as it appears through the scanning objective of the compound microscope.

7. What has the lens system done to the image of the numeral? (Is it right side up, upside down, or what?) _____

8. When you moved the ruler to the right, which way did the image move? _____

9. When you moved the ruler away from you, which way did the image move? _____

♻ PART D ASSESSMENTS

Match the names of the microscope parts in column A with the descriptions in column B. Place the letter of your choice in the space provided.

Column A		Column B	
a. Adjustment knob (coarse)	_f_	1. Increases or decreases the light intensity	IRIS DIAPRAM
b. Arm	_i_	2. Platform that supports a microscope slide	STAGE
c. Condenser	_C_	3. Concentrates light onto the specimen	CONDENSER
d. Eyepiece (ocular)	_A_	4. Causes stage (or objective lens) to move upward or downward	KNOB
e. Field of view	_h_	5. After light passes through specimen, it next enters this lens system	OBJECT
f. Iris diaphragm	_J_	6. Holds a microscope slide in position	Clip
g. Nosepiece	_d_	7. Contains a lens at the top of the body tube	evereice
h. Objective lens system	_b_	8. Serves as a handle for carrying the microscope	ARM
i. Stage	_g_	9. Part to which the objective lenses are attached	nose piece
j. Stage (slide) clip	_E_	10. Circular area seen through the eyepiece	field of view

♻ PART E ASSESSMENTS

Prepare sketches of the objects you observed using the microscope. For each sketch, include the name of the object, the magnification you used to observe it, and its estimated dimensions in millimeters and micrometers.

5

Cell Structure and Function

MATERIALS NEEDED

Textbook
Animal cell model
Clean microscope slides
Coverslips
Flat toothpicks
Medicine dropper
Methylene blue (dilute) or iodine-potassium-iodide stain
Prepared microscope slides of human tissues
Compound light microscope

For Learning Extensions:
Single-edged razor blade
Plant materials such as leaves, soft stems, fruits,
    onion peel, and vegetables
Cultures of *Amoeba* and *Paramecium*

⚠ SAFETY

- Review all the Laboratory Safety Guidelines in Appendix 1.
- Clean laboratory surfaces before and after laboratory procedures.
- Wear disposable gloves for the wet-mount procedures of the cells lining the inside of the cheek.
- Work only with your own materials when preparing the slide of cheek cells. Observe the same precautions as with all body fluids.
- Dispose of laboratory gloves, slides, coverslips, and toothpicks as instructed.
- Take precautions to prevent cellular stains from contacting your clothes and skin.
- Use the biohazard container to dispose of items used during the cheek cells procedure.
- Wash your hands before leaving the laboratory.

PURPOSE OF THE EXERCISE

To review the structure and functions of major cellular components and to observe examples of human cells.

LEARNING OUTCOMES APR

After completing this exercise, you should be able to

1. Name and locate the components of a cell.
2. Differentiate functions of cellular components.
3. Prepare a wet mount of cells lining the inside of the cheek; stain the cells; and identify the cell membrane, nucleus, and cytoplasm.
4. Examine cells on prepared slides of human tissues and identify their major components.

Cells are the smallest, most basic units of life. Their arrangement and interactions result in the shape, organization, and construction of the body, and cells are responsible for carrying on its life processes. Under the light microscope, with a properly applied stain to make structures visible, the *cell (plasma) membrane,* the *cytoplasm,* and a *nucleus* are easily seen. The cytoplasm is composed of a clear fluid, the *cytosol,* and numerous *cytoplasmic organelles* suspended in the cytosol.

The cell membrane, composed of lipids and proteins, composes the cell boundary and functions in various methods of membrane transport. Chromatin within the nucleus contains fine strands of DNA and protein. Various cytoplasmic organelles, including mitochondria, endoplasmic reticulum, and Golgi apparatus, provide specialized metabolic functions.

EXPLORE

PROCEDURE—Cell Structure and Function

1. Review section 3.2 entitled "Composite Cell" in chapter 3 of the textbook.
2. Observe an animal cell model and identify its major structures.
3. As a review activity, label figure 5.1 and study figure 5.2.
4. Complete Parts A and B of Laboratory Report 5.

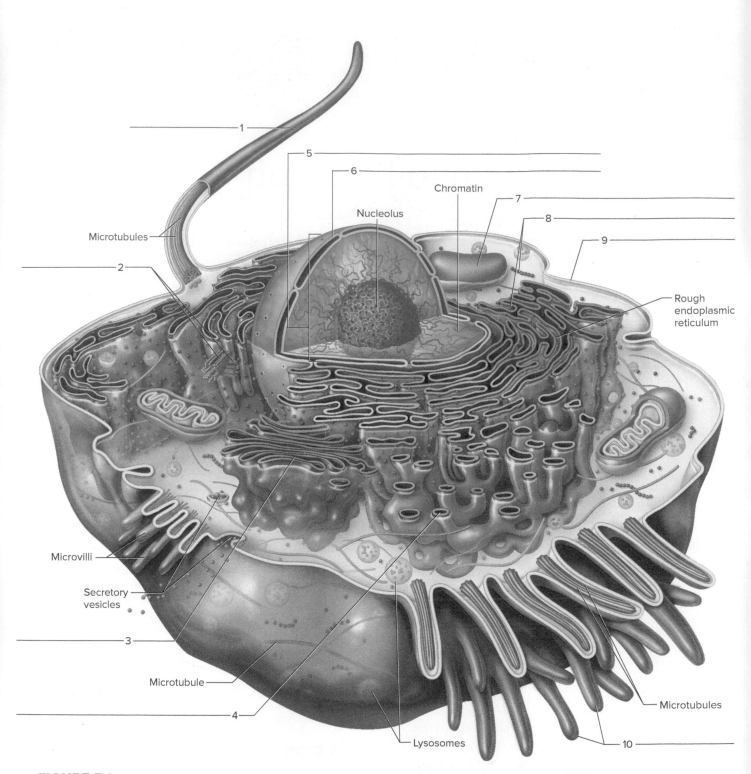

Microtubules

Nucleolus

Chromatin

Rough
endoplasmic
reticulum

Microvilli

Secretory
vesicles

Microtubule

Lysosomes

Microtubules

FIGURE 5.1 Label the structures of this composite cell. The structures are not drawn to scale. **APR**

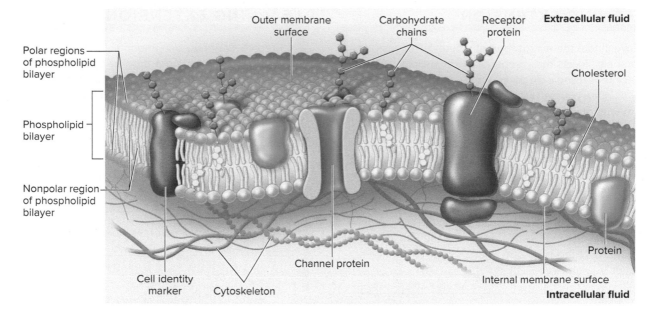

FIGURE 5.2 Structures of the cell membrane. **APR**

5. Prepare a wet mount of cells lining the inside of the cheek. To do this, follow these steps:
   a. Gently scrape (force is not necessary and should be avoided) the inner lining of your cheek with the broad end of a flat toothpick.
   b. Stir the toothpick in a drop of water on a clean microscope slide and dispose of the toothpick as directed by your instructor.
   c. Cover the drop with a coverslip.
   d. Observe the cheek cells by using the microscope. Compare your image with figure 5.3. To report what you observe, sketch a single cell in the space provided in Part C of the laboratory report.
6. Prepare a second wet mount of cheek cells, but this time, add a drop of dilute methylene blue or iodine-potassium-iodide stain to the cells. Cover the liquid with a coverslip and observe the cells with the microscope. Add to your sketch any additional structures you observe in the stained cells.
7. Complete Part C of the laboratory report.
8. Observe all the safety guidelines for disposal of items used during this procedure.
9. Using the microscope, observe each of the prepared slides of human tissues. To report what you observe, sketch a single cell of each type in the space provided in Part D of the laboratory report.
10. Complete Part D of the laboratory report.

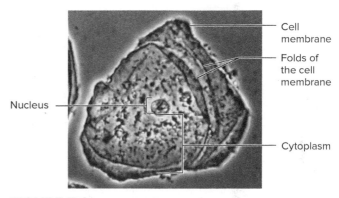

FIGURE 5.3 Iodine-stained cell from inner cheek, as viewed through the compound light microscope using the high-power objective (400×). ©Ed Reschke

 Critical Thinking Application

The cells lining the inside of the cheek are frequently removed for making observations of basic cell structure. The cells are from stratified squamous epithelium. Explain why these cells are used instead of outer body surface tissue.

_____

_____

_____

LEARNING EXTENSION

Investigate the microscopic structure of various plant materials. To do this, prepare tiny, thin slices of plant specimens using a single-edged razor blade. (*Take care not to injure yourself with the blade.*) Keep the slices in a container of water until you are ready to observe them. To observe a specimen, place it into a drop of water on a clean microscope slide and cover it with a coverslip. Use the microscope and view the specimen using low- and high-power magnifications. Observe near the edges where your section of tissue is most likely to be one cell thick. Add a drop of dilute methylene blue or iodine-potassium-iodide stain, and note if any additional structures become visible. How are the microscopic structures of the plant specimens similar to the human tissues you observed? _____

_____

How are they different? _____

_____

LEARNING EXTENSION

Prepare separate wet mounts of the *Amoeba* and *Paramecium* by putting a drop of each of the cultures on a clean glass slide. Gently cover each sample with a clean coverslip. Observe the movements of the *Amoeba* with pseudopodia and the *Paramecium* with cilia. Try to locate cellular components such as the cell membrane, nuclear envelope, nucleus, mitochondria, and contractile vacuoles.

Describe the movement of the *Amoeba*.

_____

_____

Describe the movement of the *Paramecium*.

_____

_____

Name _____

Date _____

Section _____

The Ⓐ corresponds to the indicated Learning Outcome(s) found at the beginning of the laboratory exercise.

Cell Structure and Function

PART A ASSESSMENTS

Match the cellular components in column A with the descriptions in column B. Place the letter of your choice in the space provided.

Column A	Column B
a. Chromatin	_____ 1. Loosely coiled fiber containing DNA and protein within a nucleus
b. Cytoplasm	
c. Endoplasmic reticulum	_____ 2. Location of ATP production from digested food molecules
d. Golgi apparatus	_____ 3. Small RNA-containing particle for the synthesis of proteins
e. Lysosome	
f. Microtubule	_____ 4. Membranous sac that stores or transports substances
g. Mitochondrion	_____ 5. Dense body of RNA within a nucleus
h. Nuclear envelope	_____ 6. Slender tubes that provide movement in cilia and flagella
i. Nucleolus	
j. Nucleus	_____ 7. Composed of membrane-bound canals and sacs for tubular transport throughout the cytoplasm
k. Ribosome	
l. Vesicle	_____ 8. Occupies space between the cell membrane and the nucleus
	_____ 9. Flattened, membranous sacs that package a secretion
	_____ 10. Membranous sac that contains digestive enzymes
	_____ 11. Separates nuclear contents from the cytoplasm
	_____ 12. Spherical organelle that contains chromatin and the nucleolus

PART B ASSESSMENTS

Complete the following:

1. The cell membrane is composed mainly of _____
   _____. Ⓐ

2. The basic framework of a cell membrane can be described as _____
   _____. Ⓐ

3. A cell membrane is relatively impermeable to substances that are _____. Ⓐ

4. Molecules of _____ are responsible for special functions of cell membranes as receptor sites and transport of ions or molecules across the membrane. Ⓐ

PART C ASSESSMENTS

Complete the following:

1. Sketch a single cell from inside the cheek that has been stained. Label the cellular components you recognize. (The circle represents the field of view through the microscope.)

Magnification ____ ×

2. After comparing the wet mount and the stained cheek cells, describe the advantage gained by staining cells.

_____

3. Are the stained cheek cells nearly the same size and shape? _____

Explain your answer. _____

PART D ASSESSMENTS

Complete the following:

1. Sketch a single cell of each type you observed in the prepared slides of human tissues. Name the tissue, indicate the magnification used, and label the cellular components you recognize. 4

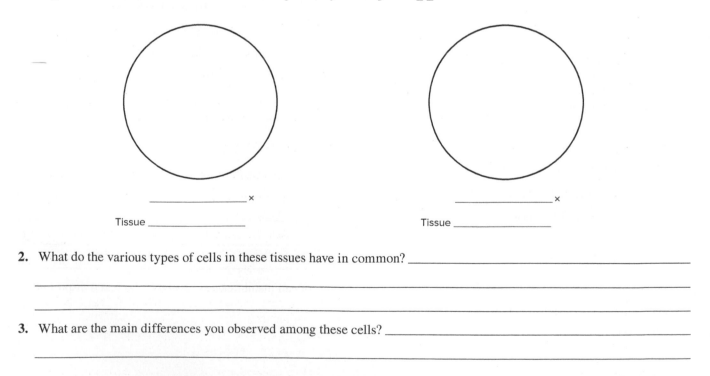

_____ ×   _____ ×

Tissue _____   Tissue _____

2. What do the various types of cells in these tissues have in common? _____

_____

_____

3. What are the main differences you observed among these cells? _____

_____

_____

6

Movements Through Membranes

MATERIALS NEEDED

For Procedure A—Diffusion:
Textbook
Petri dish
White paper
Forceps
Potassium permanganate crystals
Millimeter ruler (thin and transparent)

For Procedure B—Osmosis:
Thistle tube
Molasses (or Karo dark corn syrup)
Selectively permeable (semipermeable) membrane
    (presoaked dialysis tubing of 1⁵⁄₁₆" or greater
    diameter)
Support stand and clamp
Beaker
Rubber band
Millimeter ruler

*For Procedure C—Hypertonic, Hypotonic, and
    Isotonic Solutions:*
Test tubes
Marking pen
Test-tube rack
10 mL graduated cylinder
Medicine dropper
Uncoagulated animal blood
Distilled water
0.9% NaCl (aqueous solution)
3.0% NaCl (aqueous solution)
Clean microscope slides
Coverslips
Compound light microscope

For Procedure D—Filtration:
Glass funnel
Filter paper
Support stand and ring
Beaker
Powdered charcoal or ground black pepper
1% glucose (aqueous solution)
1% starch (aqueous solution)
Test tubes
10 mL graduated cylinder
Water bath (boiling water)
Benedict's solution

Iodine-potassium-iodide solution
Medicine dropper

For Alternative Osmosis Activity:
Fresh chicken egg
Beaker
Laboratory balance
Spoon
Vinegar
Corn syrup (Karo)

⚠ SAFETY

- Clean laboratory surfaces before and after laboratory procedures.
- Wear disposable gloves when handling chemicals and animal blood.
- Wear safety glasses when using chemicals.
- Dispose of laboratory gloves and blood-contaminated items as instructed.
- Wash your hands before leaving the laboratory.

PURPOSE OF THE EXERCISE

To demonstrate some of the physical processes by which substances move through cell membranes.

LEARNING OUTCOMES APR

After completing this exercise, you should be able to

1. Demonstrate the process of diffusion and identify examples of diffusion.
2. Explain diffusion by preparing and interpreting a graph.
3. Demonstrate the process of osmosis and identify examples of osmosis.
4. Distinguish among hypertonic, hypotonic, and isotonic solutions and observe the effects of these solutions on animal cells.
5. Demonstrate the process of filtration and identify examples of filtration.
6. Recognize the anatomical terms pertaining to movements through membranes.

A cell membrane functions as a gateway through which chemical substances and small particles may enter or leave a cell. These substances move through the membrane by physical processes such as diffusion, osmosis, and filtration, or by physiological processes such as active transport, phagocytosis, or pinocytosis.

EXPLORE

PROCEDURE A—Diffusion A&PR

1. Review the the concept of diffusion under the heading "Passive Mechanisms" in section 3.3 of chapter 3 in the textbook.
2. To demonstrate *diffusion,* refer to figure 6.1 as you follow these steps:
   a. Place a petri dish, half filled with water, on a piece of white paper that has a millimeter ruler positioned on the paper. Wait until the water surface is still. Allow approximately 3 minutes. *Note:* The petri dish should remain level. A second millimeter ruler may be needed under the petri dish as a shim to obtain a level amount of the water inside the petri dish.
   b. Using forceps, place one crystal of potassium permanganate near the center of the petri dish and near the millimeter ruler (fig. 6.1).
   c. Measure the radius of the purple circle at 1-minute intervals for 10 minutes.
3. Complete Part A of Laboratory Report 6.

LEARNING EXTENSION

Repeat the demonstration of diffusion using a petri dish filled with ice-cold water and a second dish filled with very hot water. At the same moment, add a crystal of potassium permanganate to each dish and observe the circle as before. What difference do you note in the rate of diffusion in the two dishes? How do you explain this difference?

_____

_____

_____

EXPLORE

PROCEDURE B—Osmosis A&PR

1. Review the the concept of osmosis under the heading "Passive Mechanisms" in section 3.3 of chapter 3 in the textbook.
2. To demonstrate *osmosis,* refer to figure 6.2 as you follow these steps:
   a. One person plugs the tube end of a thistle tube with a finger.
   b. Another person then fills the bulb with molasses until it is about to overflow at the top of the bulb. Allow the molasses to enter the first centimeter of the stem, leaving the rest filled with trapped air.
   c. Cover the bulb opening with a single thickness piece of moist selectively permeable (semipermeable) membrane. Dialysis tubing that has been soaked

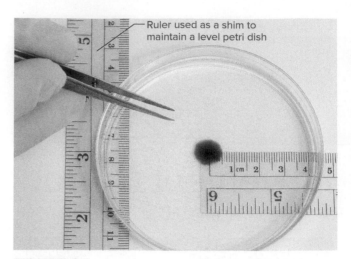
Ruler used as a shim to maintain a level petri dish

FIGURE 6.1 To demonstrate diffusion, place one crystal of potassium permanganate in the center of a petri dish containing water. Place the crystal near the millimeter ruler (positioned under the petri dish). ©J and J Photography

for 30 minutes can easily be cut open because it becomes pliable.
   d. Tightly secure the membrane in place with several wrappings of a rubber band.
   e. Immerse the bulb end of the tube in a beaker of water. If leaks are noted, repeat the procedures.
   f. Support the upright portion of the tube with a clamp on a support stand. Folded paper between the stem and clamp will protect the thistle tube stem from breakage.
   g. Mark the meniscus level of the molasses in the tube. *Note:* The best results will occur if the mark of the molasses is a short distance up the stem of the thistle tube when the experiment starts.
   h. Measure the level changes after 10 minutes and 30 minutes.
3. Complete Part B of the laboratory report.

ALTERNATIVE PROCEDURE

Eggshell membranes possess selectively permeable properties. To demonstrate osmosis using a natural membrane, soak a fresh chicken egg in vinegar for about 24 hours to remove the shell. Use a spoon to carefully handle the delicate egg. Place the egg in a hypertonic solution (corn syrup) for about 24 hours. Remove the egg, rinse it, and using a laboratory balance, weigh the egg to establish a baseline weight. Place the egg in a hypotonic solution (distilled water). Remove the egg and weigh it every 15 minutes for an elapsed time of 75 minutes. Explain any weight changes that were noted during this experiment.

_____

_____

_____

_____

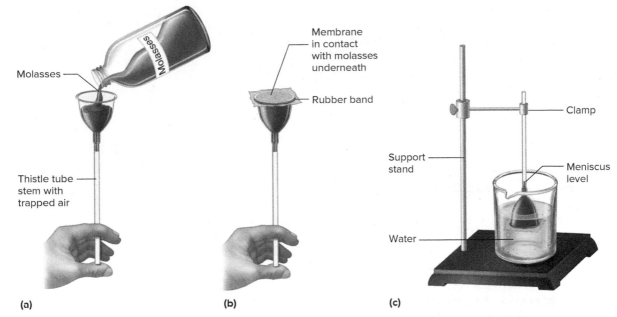

FIGURE 6.2 (*a*) Fill the bulb of the thistle tube with molasses; (*b*) tightly secure a piece of selectively permeable (semipermeable) membrane over the bulb opening; and (*c*) immerse the bulb in a beaker of water. *Note:* These procedures require the participation of two people.

 **EXPLORE**

PROCEDURE C—Hypertonic, Hypotonic, and Isotonic Solutions **A&PR**

1. Review the the concept of osmosis under the heading "Passive Mechanisms" in section 3.3 of chapter 3 in the textbook.
2. To demonstrate the effect of *hypertonic, hypotonic,* and *isotonic* solutions on animal cells, follow these steps:
   a. Place three test tubes in a rack and mark them as *tube 1, tube 2,* and *tube 3.* (*Note:* One set of tubes can be used to supply test samples for the entire class.)
   b. Using 10 mL graduated cylinders, add 3 mL of distilled water to tube 1; add 3 mL of 0.9% NaCl to tube 2; and add 3 mL of 3.0% NaCl to tube 3.
   c. Place 3 drops of fresh, uncoagulated animal blood into each of the tubes, and gently mix the blood with the solutions. Wait 5 minutes.
   d. Using three separate medicine droppers, remove a drop from each tube and place the drops on three separate microscope slides marked *1, 2,* and *3.*
   e. Cover the drops with coverslips and observe the blood cells, using the high power of the microscope.
3. Complete Part C of the laboratory report.

ALTERNATIVE PROCEDURE

Various substitutes for blood can be used for Procedure C. Onion cells, cucumber cells, and cells lining the inside of the cheek represent three possible options.

EXPLORE

PROCEDURE D—Filtration

1. Review the concept of filtration under the heading "Passive Mechanisms" in section 3.3 of chapter 3 in the textbook.
2. To demonstrate *filtration,* follow these steps:
   a. Place a glass funnel in the ring of a support stand over an empty beaker. Fold a piece of filter paper in half and then in half again. Open one thickness of the filter paper to form a cone. Wet the cone, and place it in the funnel. The filter paper is used to demonstrate how movement across membranes is limited by size of the molecules, but it does not represent a working model of biological membranes.
   b. Prepare a mixture of 5 cc (approximately 1 teaspoon) powdered charcoal (or ground black pepper) and equal amounts of 1% glucose solution and 1% starch solution in a beaker. Pour some of the mixture into the funnel until it nearly reaches the top of the filter-paper cone. Care should be taken to prevent the mixture from spilling over the top of the filter paper. Collect the filtrate in the beaker below the funnel (fig. 6.3).
   c. Test some of the filtrate in the beaker for the presence of glucose. To do this, place 1 mL of filtrate in a clean test tube and add 1 mL of Benedict's solution. Place the test tube in a water bath of boiling water for 2 minutes and then allow the liquid to cool slowly. If the color of the solution changes to green, yellow, or red, glucose is present (fig. 6.4).

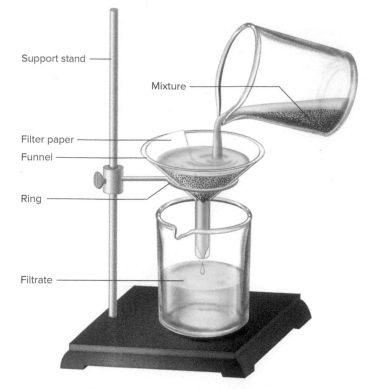

FIGURE 6.3 Apparatus used to illustrate filtration.

Support stand

Mixture

Filter paper

Funnel

Ring

Filtrate

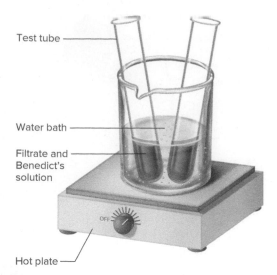

Test tube

Water bath

Filtrate and
Benedict's
solution

Hot plate

FIGURE 6.4 Heat the filtrate and Benedict's solution in a boiling water bath for 2 minutes.

d. Test some of the filtrate in the beaker for the presence of starch. To do this, place a few drops of filtrate in a test tube and add 1 drop of iodine-potassium-iodide solution. If the color of the solution changes to blue-black, starch is present.

e. Observe any charcoal in the filtrate.

3. Complete Parts D and E of the laboratory report.

Name _____

Date _____

Section _____

The Ⓐ corresponds to the indicated Learning Outcome(s) found at the beginning of the laboratory exercise.

Movements Through Membranes

![recycle icon] **PART A ASSESSMENTS**

Complete the following:

1. Enter data for changes in the movement of the potassium permanganate. Ⓐ1

Elapsed Time	Radius of Purple Circle in Millimeters
Initial	_____
1 minute	_____
2 minutes	_____
3 minutes	_____
4 minutes	_____
5 minutes	_____
6 minutes	_____
7 minutes	_____
8 minutes	_____
9 minutes	_____
10 minutes	_____

2. Prepare a graph that illustrates the diffusion distance of potassium permanganate in 10 minutes. Ⓐ2

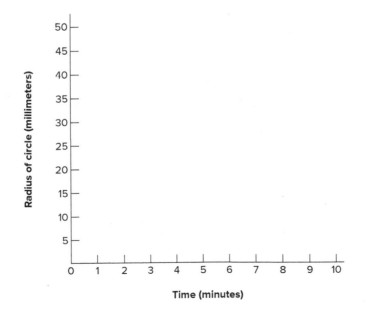

3. Interpret your graph. [2] _____

_____

4. Define *diffusion.* _____

_____

Critical Thinking Application

By answering yes or no, indicate which of the following provides an example of diffusion. [1]

1. A perfume bottle is opened, and soon the odor can be sensed in all parts of the room. _____

2. A tea bag is dropped into a cup of hot water, and, without being stirred, all of the liquid becomes the color of the tea leaves. _____

3. Water molecules move from a faucet through a garden hose when the faucet is turned on. _____

4. A person blows air molecules into a balloon by forcefully exhaling. _____

5. A crystal of blue copper sulfate is placed in a test tube of water. The next day, the solid is gone, but the water is evenly colored. _____

PART B ASSESSMENTS

Complete the following:

1. What was the change in the level of molasses in 10 minutes? [3] _____

2. What was the change in the level of molasses in 30 minutes? [3] _____

3. How do you explain this change? [3] _____

_____

4. Define *osmosis.* _____

_____

Critical Thinking Application

By answering yes or no, indicate which of the following involves osmosis. [3]

1. A fresh potato is peeled, weighed, and soaked in a strong salt solution. The next day, it is discovered that the potato has lost weight. _____

2. Garden grass wilts after being exposed to dry chemical fertilizer. _____

3. Air molecules escape from a punctured tire as a result of high pressure inside. _____

4. Plant seeds soaked in water swell and become several times as large as before soaking. _____

5. When the bulb of a thistle tube filled with water is sealed by a selectively permeable membrane and submerged in a beaker of molasses, the water level in the tube falls. _____

![recycle icon] **PART C ASSESSMENTS**

Complete the following:

1. In the spaces, sketch a few blood cells from each of the test tubes, and indicate the magnification. /4\

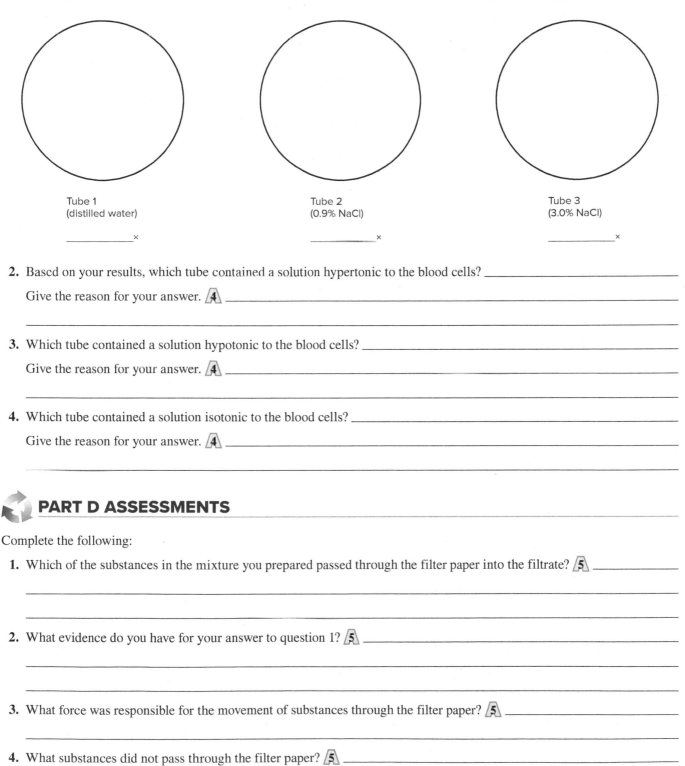

Tube 1
(distilled water)

_____×

Tube 2
(0.9% NaCl)

_____×

Tube 3
(3.0% NaCl)

_____×

2. Based on your results, which tube contained a solution hypertonic to the blood cells? _____

   Give the reason for your answer. /4\ _____

   _____

3. Which tube contained a solution hypotonic to the blood cells? _____

   Give the reason for your answer. /4\ _____

   _____

4. Which tube contained a solution isotonic to the blood cells? _____

   Give the reason for your answer. /4\ _____

   _____

![recycle icon] **PART D ASSESSMENTS**

Complete the following:

1. Which of the substances in the mixture you prepared passed through the filter paper into the filtrate? /5\ _____

   _____

   _____

2. What evidence do you have for your answer to question 1? /5\ _____

   _____

   _____

3. What force was responsible for the movement of substances through the filter paper? /5\ _____

   _____

4. What substances did not pass through the filter paper? /5\ _____

   _____

5. What factor prevented these substances from passing through? /5\ _____

_____

6. Define *filtration*. _____

_____

Critical Thinking Application

By answering yes or no, indicate which of the following involves filtration. /5\

1. Oxygen molecules move into a cell and carbon dioxide molecules leave a cell because of differences in the concentrations of these substances on either side of the cell membrane. _____

2. Blood pressure forces water molecules and small, dissolved solutes from the blood outward through the thin wall of a blood capillary. _____

3. Urine is forced from the urinary bladder through the tubular urethra by muscular contractions. _____

4. Air molecules enter the lungs through the airways when air pressure is greater outside these organs than inside. _____

5. Noninstant coffee is made using a coffeemaker. _____

PART E ASSESSMENTS

Locate the 16 anatomical terms pertaining to "Movements Through Membranes." /6\

E	N	I	D	O	I	B	E	A	K	E	R
R	N	H	E	R	U	T	X	I	M	C	M
D	F	G	Y	O	S	M	O	S	I	S	E
I	B	L	K	P	W	Y	P	N	D	V	N
F	L	U	B	H	O	S	O	T	W	L	I
F	A	C	I	S	O	T	O	N	I	C	S
U	O	O	F	B	R	A	O	X	G	Z	C
S	C	S	G	E	Y	R	K	N	N	Q	U
I	R	E	P	G	E	C	P	B	I	S	S
O	A	Y	H	J	B	H	U	T	R	C	V
N	H	D	B	R	U	L	E	R	K	M	Y
J	C	N	O	I	T	A	R	T	L	I	F

Find and circle words pertaining to Laboratory Exercise 6. The words may be horizontal, vertical, diagonal, or backward.

10 terms (beaker, charcoal, glucose, iodine, meniscus, mixture, ring, ruler, starch, tube)

3 physical processes of membrane passage; 3 tonicity terms

7

Cell Cycle

MATERIALS NEEDED

Textbook
Models of animal mitosis
Microscope slides of whitefish mitosis (blastula)
Compound light microscope

For Demonstration:

Microscope slide of human chromosomes from leukocytes in mitosis
Oil immersion objective

PURPOSE OF THE EXERCISE

To review the phases in the cell cycle and to observe cells in various phases of their life cycles.

LEARNING OUTCOMES APR

After completing this exercise, you should be able to

1. Describe the cell cycle and locate structures involved with the process.
2. Identify and sketch the phases (stages) in the life cycle of a particular cell.
3. Arrange into a correct sequence a set of models or drawings of cells in various phases of their life cycles.
4. Recognize the anatomical terms pertaining to the cell cycle.

The cell cycle consists of the series of changes a cell undergoes from the time it is formed until it divides. Typically, a newly formed diploid cell with 46 chromosomes grows to a certain size and then divides to form two new cells *(daughter cells),* each with 46 chromosomes. This cell division process involves two major steps: (1) division of the cell's nuclear parts, *mitosis,* and (2) division of the cell's cytoplasm, *cytokinesis.* Before the cell divides, it must synthesize biochemicals and other contents. This period of preparation is called *interphase.* The extensive period of interphase is divided into three phases. The S phase, when DNA synthesis occurs, is between two gap phases (G_1 and G_2), when cell growth occurs and cytoplasmic organelles duplicate. Eventually, some specialized cells, such as skeletal muscle cells and most nerve cells, cease further cell division, but remain alive.

A special type of cell division, called *meiosis,* occurs in the reproductive system to produce haploid gametes with 23 chromosomes. Meiosis is not included in this laboratory exercise.

EXPLORE

PROCEDURE—Cell Cycle

1. Review section 3.4 entitled "The Cell Cycle" in chapter 3 of the textbook.
2. As a review activity, study the various phases of the cell's life cycle represented in figures 7.1, 7.2, and 7.3.
3. Label the structures indicated in figure 7.4.
4. Using figure 7.1 as a guide, observe the animal mitosis models, and review the major events in a cell's life cycle represented by each of them. Be sure you can arrange these models in correct sequence if their positions are changed. The acronym IPMAT can help you arrange the correct order of phases in the cell cycle. This includes interphase followed by the four phases of mitosis. Cytokinesis overlaps anaphase and telophase.
5. Complete Part A of Laboratory Report 7.
6. Obtain a slide of the whitefish mitosis (blastula).
   a. Examine the slide using the high-power objective of a microscope. The tissue on this slide was obtained from a developing embryo (blastula) of a fish, and many of the embryonic cells are undergoing mitosis. The chromosomes of these dividing cells are darkly stained (fig. 7.3).
   b. Search the tissue for cells in various phases of cell division. There are several sections on the slide. If you cannot locate different phases in one section, examine the cells of another section because the phases occur randomly on the slide.
   c. Each time you locate a cell in a different phase, sketch it in an appropriate circle in Part B of the laboratory report.

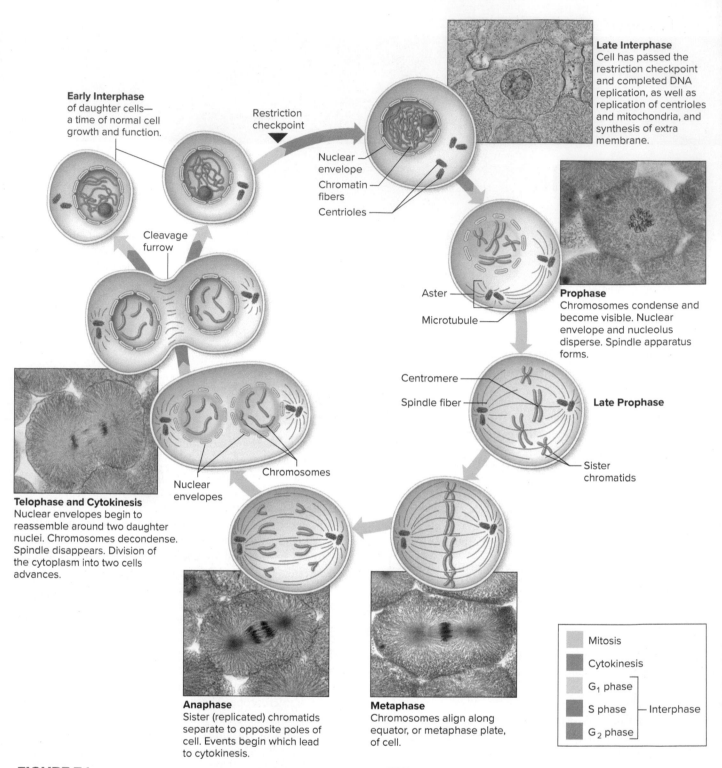

Early Interphase
of daughter cells—
a time of normal cell
growth and function.

Restriction
checkpoint

Cleavage
furrow

Nuclear
envelope

Chromatin
fibers

Centrioles

Late Interphase
Cell has passed the
restriction checkpoint
and completed DNA
replication, as well as
replication of centrioles
and mitochondria, and
synthesis of extra
membrane.

Aster

Microtubule

Prophase
Chromosomes condense and
become visible. Nuclear
envelope and nucleolus
disperse. Spindle apparatus
forms.

Centromere

Spindle fiber

Late Prophase

Sister
chromatids

Chromosomes

Nuclear
envelopes

Telophase and Cytokinesis
Nuclear envelopes begin to
reassemble around two daughter
nuclei. Chromosomes decondense.
Spindle disappears. Division of
the cytoplasm into two cells
advances.

Anaphase
Sister (replicated) chromatids
separate to opposite poles of
cell. Events begin which lead
to cytokinesis.

Metaphase
Chromosomes align along
equator, or metaphase plate,
of cell.

Mitosis

Cytokinesis

G_1 phase

S phase ⎤ Interphase

G_2 phase ⎦

FIGURE 7.1 The cell cycle: interphase, mitosis, and cytokinesis. **APR** ©Ed Reschke

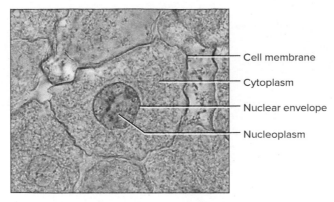

FIGURE 7.2 Cell in interphase (400×). The nucleoplasm contains a fine network of chromatin. ©Ed Reschke

Cell membrane
Cytoplasm
Nuclear envelope
Nucleoplasm

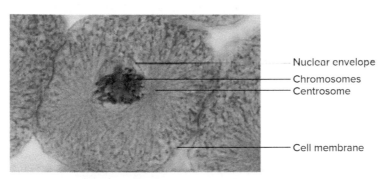

FIGURE 7.3 Cell in prophase (400×) ©Ed Reschke

Nuclear envelope
Chromosomes
Centrosome
Cell membrane

 **Critical Thinking Application**

Which phase (stage) of the cell cycle was the most numerous in the blastula?_____

Explain your answer._____

_____

_____

   7. Complete Parts C and D of the laboratory report.

DEMONSTRATION

Using the oil immersion objective of a microscope, see if you can locate some human chromosomes by examining a prepared slide of human chromosomes from leukocytes. The cells on this slide were cultured in a special medium and were stimulated to undergo mitosis. The mitotic process was arrested in metaphase by exposing the cells to a chemical called colchicine, and the cells were caused to swell osmotically. As a result of this treatment, the chromosomes were spread apart. A complement of human chromosomes should be visible when they are magnified about 1,000×. Each chromosome is double-stranded and consists of two chromatids joined by a common centromere (fig. 7.5).

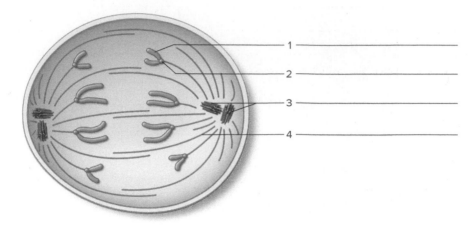

FIGURE 7.4 Label the structures indicated in the dividing cell during anaphase. 🅐

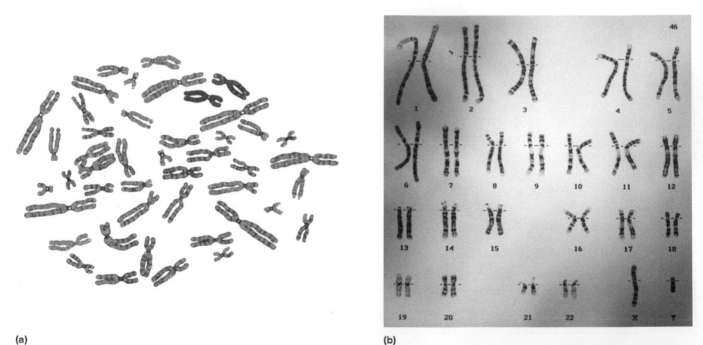

(a)

(b)

FIGURE 7.5 (*a*) A complement of human chromosomes of a female (1,000×). (*b*) A *karyotype* can be constructed by arranging the homologous chromosome pairs together in a chart. A karyotype can aid in diagnosis of genetic conditions and abnormalities. The completed karyotype indicates a normal male. **APR** (a): ©James Cavallini/Science Source (b): ©Randy Allbritton/Getty Images RF

Name _____

Date _____

Section _____

The corresponds to the indicated Learning Outcome(s) found at the beginning of the laboratory exercise.

Cell Cycle

🔄 PART A ASSESSMENTS

Complete the table by listing the major events of each phase in the second column. ⚠️1

Phase	Major Events Occurring
Interphase (G₁, S, and G₂)	
Mitosis Prophase	
Metaphase	
Anaphase	
Telophase	
Cytokinesis	

PART B ASSESSMENTS

Sketch an interphase cell and cells in different phases (stages) of mitosis to illustrate the whitefish cell's life cycle. Label the major cellular structures represented in the sketches and indicate cytokinesis locations. (The circles represent fields of view through the microscope.) 2

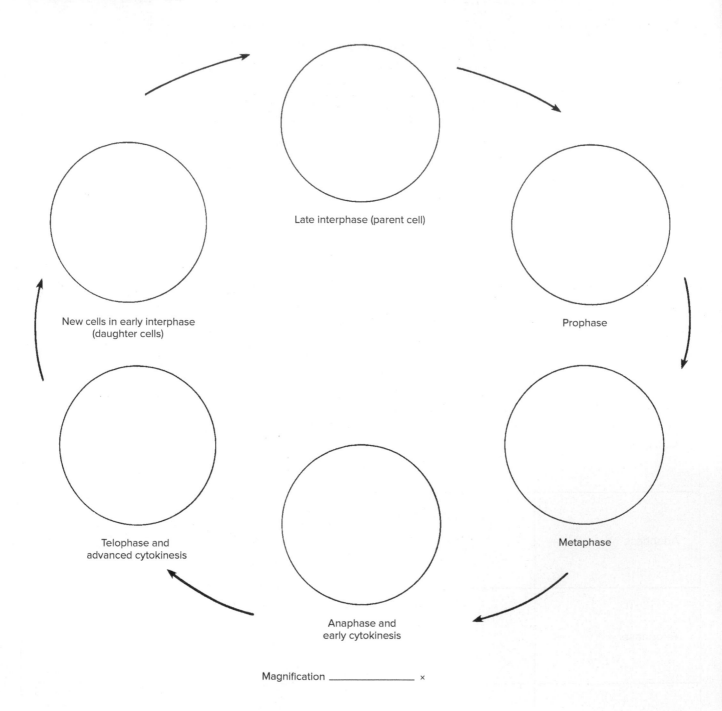

Late interphase (parent cell)

New cells in early interphase
(daughter cells)

Prophase

Telophase and
advanced cytokinesis

Metaphase

Anaphase and
early cytokinesis

Magnification _____ ×

PART C ASSESSMENTS

1. Identify the mitotic phase represented by each of the micrographs in figure 7.6(a–d). 3

   a. _____   c. _____

   b. _____   d. _____

2. Identify the structures indicated by numbers in figure 7.6 by placing the correct numbers in the spaces provided. 1

   _____ Cell membrane   _____ Metaphase plate

   _____ Centrosome   _____ Nuclear envelope

   _____ Cleavage furrow   _____ Spindle fiber

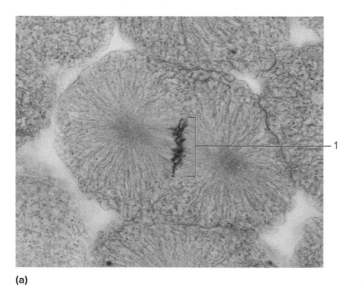

(a)

(b)

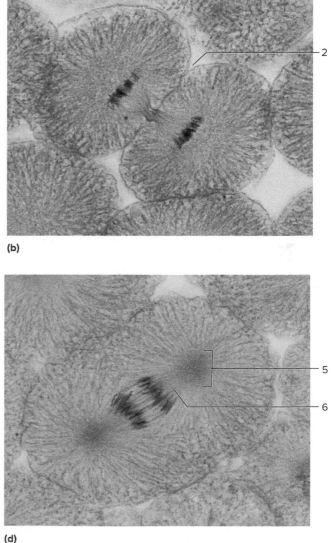

(c)

(d)

FIGURE 7.6   Identify the mitotic phase and structures of the cell in each of these micrographs of the whitefish blastula (250×). **(a–d):** ©Ed Reschke

Locate the 18 anatomical terms pertaining to the "Cell Cycle." ◢4◣

E	S	A	H	P	A	T	E	M	W	P	R
C	E	N	T	R	O	M	E	R	E	L	E
E	S	A	H	P	R	E	T	N	I	A	B
N	U	C	L	E	O	L	U	S	B	T	I
T	K	P	A	N	A	P	H	A	S	E	F
R	E	L	U	B	U	T	O	R	C	I	M
I	D	K	S	Y	N	T	H	E	S	I	S
O	E	S	A	H	P	O	L	E	T	X	M
L	S	I	S	E	N	I	K	O	T	Y	C
E	P	R	O	P	H	A	S	E	Z	V	K
R	E	T	S	A	N	I	E	L	C	U	N
U	B	F	G	D	S	P	I	N	D	L	E

Find and circle words pertaining to Laboratory Exercise 7. The words may be horizontal, vertical, diagonal, or backward.

7 terms (aster, cytokinesis, fiber, mitosis, nuclei, plate, synthesis)

6 cellular structures involved during mitosis; 5 phases of the cell cycle

8

Epithelial Tissues

PURPOSE OF THE EXERCISE

To review the characteristics of epithelial tissues and to observe examples.

LEARNING OUTCOMES APR

After completing this exercise, you should be able to

(1) Differentiate the special characteristics of each type of epithelial tissue.

(2) Sketch and label the characteristics of epithelial tissues that you were able to observe.

(3) Indicate a location and function of each type of epithelial tissue.

(4) Identify the major types of epithelial tissues on microscope slides.

A tissue is composed of a layer or group of cells similar in size, shape, and function. A study of tissues is called *histology.* Within the human body, there are four major types of tissues: (1) *epithelial*, which cover the body's external and internal surfaces and compose most glands; (2) *connective*, which bind and support parts; (3) *muscle*, which make movement possible; and (4) *nervous*, which conduct impulses from one part of the body to another and help to control and coordinate body activities.

Epithelial tissues are tightly packed single (simple) to multiple (stratified) layers of cells that provide protective barriers. The underside of this tissue contains an acellular basement membrane layer composed of adhesive cellular secretions and collagen through which the epithelial cells anchor to an underlying connective tissue. The cells readily divide and lack blood vessels. Epithelial cells always have a free (apical) surface exposed to the outside or to an open space internally and a basal surface that attaches to the basement membrane. The functions of epithelial cells include protection, filtration, secretion, and absorption. Many epithelial cell shapes are used to name and identify the variations. Many of the prepared slides contain more than the tissue to be studied, so be certain that your view matches the correct tissue. Also be aware that stained colors of all tissues might vary.

EXPLORE

PROCEDURE—Epithelial Tissues APR

1. Review section 5.2 entitled "Epithelial Tissues" in chapter 5 of the textbook.
2. Complete Part A of Laboratory Report 8.
3. Use the microscope to observe the prepared slides of types of epithelial tissues. As you observe each tissue, look for its special distinguishing features as described in the textbook, such as cell size, shape, and arrangement. Compare your prepared slides of epithelial tissues to the micrographs in figure 8.1. As you observe each type of epithelial tissue, prepare a labeled sketch of a representative portion of the tissue in Part B of the laboratory report.
4. Complete Part B of the laboratory report.
5. Test your ability to recognize each type of epithelial tissue. To do this, have a laboratory partner select one of the prepared slides, cover its label, and focus the microscope on the tissue. Then see if you can correctly identify the tissue. (4)

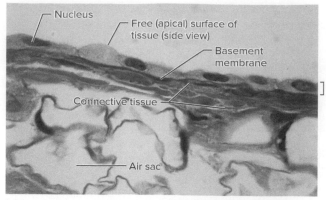

(a) Simple squamous epithelium (side view) (from lung) (250×)

Nucleus

Free (apical) surface of tissue (side view)

Basement membrane

Connective tissue

Air sac

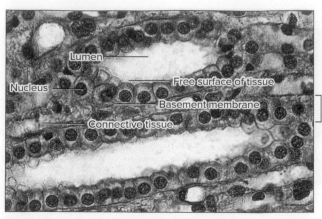

(b) Simple cuboidal epithelium (from kidney) (165×)

Lumen

Nucleus

Free surface of tissue

Basement membrane

Connective tissue

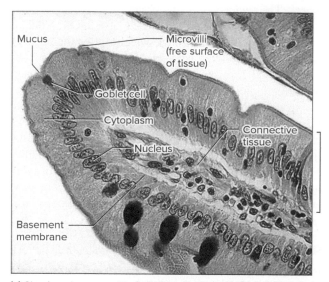

(c) Simple columnar epithelium (nonciliated) (from intestine) (40×)

Mucus

Microvilli (free surface of tissue)

Goblet cell

Cytoplasm

Nucleus

Connective tissue

Basement membrane

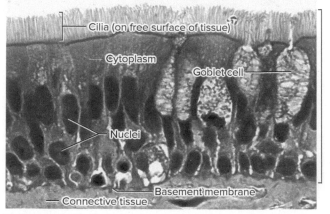

(d) Pseudostratified columnar epithelium with cilia (from trachea) (1,000×)

Cilia (on free surface of tissue)

Cytoplasm

Goblet cell

Nuclei

Basement membrane

Connective tissue

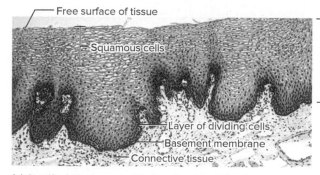

(e) Stratified squamous epithelium (nonkeratinized) (from esophagus) (100×)

Free surface of tissue

Squamous cells

Layer of dividing cells

Basement membrane

Connective tissue

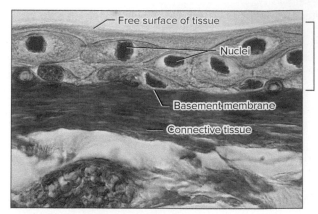

(f) Transitional epithelium (stretched) (from urinary bladder) (675×)

Free surface of tissue

Nuclei

Basement membrane

Connective tissue

FIGURE 8.1 Micrographs of epithelial tissues *(a–f)*. *Note:* A bracket to the right of a micrograph indicates the tissue layer. **APR**

(a), (d), (f): ©Ed Reschke; (b–c): Victor P. Eroschenko; (e): Al Telser/McGraw-Hill Education

The Ⓐ corresponds to the indicated Learning Outcome(s) found at the beginning of the laboratory exercise.

Epithelial Tissues

PART A ASSESSMENTS

Match the tissues in column A with the characteristics in column B. Place the letter of your choice in the space provided. (Some answers may be used more than once.) Ⓐ₁ Ⓐ₃

Column A	Column B
a. Simple columnar epithelium	_____ **1.** Consists of several layers of cells, allowing an expandable lining
b. Simple cuboidal epithelium	_____ **2.** Commonly possesses cilia that move dust and mucus out of the respiratory airways
c. Simple squamous epithelium	
d. Pseudostratified columnar epithelium	_____ **3.** Single layer of flattened cells
e. Stratified squamous epithelium	_____ **4.** Nuclei located at different levels within a single row of aligned cells
f. Transitional epithelium	
	_____ **5.** Forms walls of capillaries and air sacs of lungs
	_____ **6.** Appears layered (statified) but is a single layer of cells (simple)
	_____ **7.** Deeper cells cuboidal, or columnar; older cells flattened nearest the free surface
	_____ **8.** Forms inner lining of urinary bladder
	_____ **9.** Lines kidney tubules and ducts of salivary glands
	_____ **10.** Forms lining of stomach and intestines
	_____ **11.** Elongated cells with elongated nuclei located near basement membrane
	_____ **12.** Forms lining of oral cavity, esophagus, anal canal, and vagina

In the space that follows, sketch a few cells of each type of epithelium you observed. For each sketch, label the major characteristics, indicate the magnification used, write an example of a location in the body, and provide a function. **1** **2** **3**

Simple squamous epithelium (_____ ×)
Location example: _____
Function: _____

Simple cuboidal epithelium (_____ ×)
Location example: _____
Function: _____

Simple columnar epithelium (_____ ×)
Location example: _____
Function: _____

Pseudostratified columnar epithelium with cilia (_____ ×)
Location example: _____
Function: _____

Stratified squamous epithelium (_____ ×)
Location example: _____
Function: _____

Transitional epithelium (_____ ×)
Location example: _____
Function: _____

LEARNING EXTENSION

Use colored pencils to differentiate various cellular structures in Part B. Select a different color for a nucleus, cytoplasm, cell membrane, basement membrane, goblet cell, and cilia whenever visible.

 **Critical Thinking Application**

As a result of your observations of epithelial tissues, which one(s) provide(s) the best protection? Explain your answer. **3** _____

_____

9

Connective Tissues

Textbook
Compound light microscope
Prepared slides of the following:
    Areolar tissue
    Adipose tissue
    Dense connective tissue (regular type)
    Hyaline cartilage
    Elastic cartilage
    Fibrocartilage
    Bone (compact, ground, cross section)
    Blood (human smear)

For Learning Extension:
Colored pencils

PURPOSE OF THE EXERCISE

To review the characteristics of connective tissues and to observe examples of the major types.

LEARNING OUTCOMES **APR**

After completing this exercise, you should be able to

(1) Differentiate the special characteristics of each of the major types of connective tissue.

(2) Sketch and label the characteristics of connective tissues that you were able to observe.

(3) Indicate a location and function of each type of connective tissue.

(4) Identify the major types of connective tissues on microscope slides.

Connective tissues contain a variety of cell types and occur in all regions of the body. They bind structures together, provide support and protection, fill spaces, store fat, and produce blood cells.

Connective tissue cells are often widely scattered in an abundance of extracellular matrix. The matrix consists of fibers and a ground substance of various densities and consistencies. Many of the prepared slides contain more than the tissue to be studied, so be certain that your view matches the correct tissue. Additional study of bone and blood will be found in Laboratory Exercises 12 and 34.

EXPLORE

PROCEDURE —Connective Tissues **APR**

1. Review section 5.3 entitled "Connective Tissues" in chapter 5 of the textbook.
2. Complete Part A of Laboratory Report 9.
3. Use a microscope to observe the prepared slides of various connective tissues. As you observe each tissue, look for its special distinguishing features as described in the textbook. Compare your prepared slides of connective tissues to the micrographs in figure 9.1. As you observe each type of connective tissue, prepare a labeled sketch of a representative portion of the tissue in Part B of the laboratory report.
4. Complete Part B of the laboratory report.
5. Test your ability to recognize each of these connective tissues by having a laboratory partner select a slide, cover its label, and focus the microscope on this tissue. Then see if you correctly identify the tissue. (4)

FIGURE 9.1

Micrographs of connective tissues (a–h). **APR**

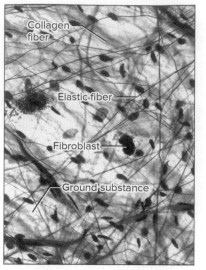

(a) Areolar tissue (from beneath the skin) (800×)

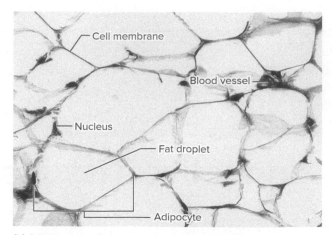

(b) Adipose tissue (from subcutaneous layer) (400×)

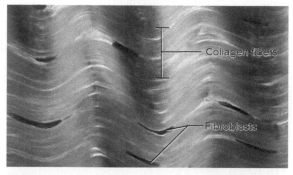

(c) Dense regular connective tissue (from tendon) (500×)

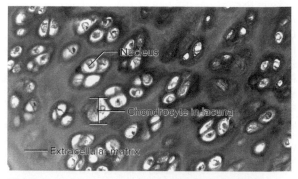

(d) Hyaline cartilage (from costal cartilage of ribs) (160×)

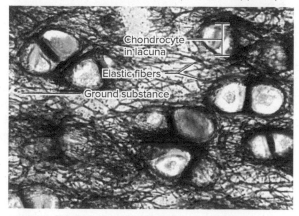

(e) Elastic cartilage (from ear) (200×)

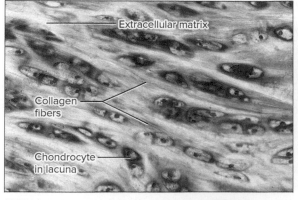

(f) Fibrocartilage (from intervertebral discs) (100×)

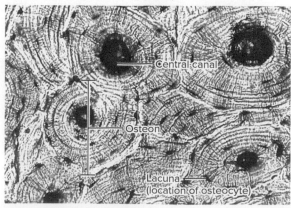

(g) Compact bone (from skeleton) (200×)

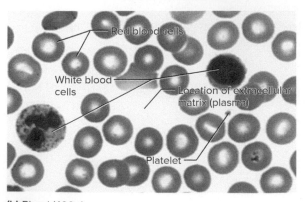

(h) Blood (400×)

Name _____

Date _____

Section _____

The Ⓐ corresponds to the indicated Learning Outcome(s) found at the beginning of the laboratory exercise.

LABORATORY
REPORT

9

Connective Tissues

PART A ASSESSMENTS

Match the tissues in column A with the characteristics in column B. Place the letter of your choice in the space provided. (Some answers may be used more than once.) Ⓐ1 Ⓐ3

Column A	Column B
a. Adipose tissue	_____ **1.** Forms framework of outer ear
b. Areolar tissue	_____ **2.** Functions as heat insulator beneath skin
c. Blood	_____ **3.** Contains large amounts of fluid and transports nutrients, waste, and gases
d. Bone (compact)	_____ **4.** Cells in solid matrix arranged around central canal
e. Dense connective tissue (regular)	_____ **5.** Binds skin to underlying organs
f. Elastic cartilage	_____ **6.** Main tissue of tendons and ligaments
g. Fibrocartilage	_____ **7.** Provides stored energy supply in fat droplets in cytoplasm
h. Hyaline cartilage	_____ **8.** Forms the ends of many long bones
	_____ **9.** Pads between vertebrae that are shock absorbers
	_____ **10.** Forms supporting rings of respiratory passages
	_____ **11.** Cells greatly enlarged have nuclei pushed close to cell membranes.
	_____ **12.** Forms delicate, thin layers between muscles

PART B ASSESSMENTS

In the space provided, sketch a small section of each of the types of connective tissues you observed. For each sketch, label the major characteristics, indicate the magnification used, write an example of a location in the body, and provide a function. Ⓐ Ⓐ Ⓐ

Areolar tissue (_____ ×) Location example: _____ Function: _____	Adipose tissue (_____ ×) Location example: _____ Function: _____
Dense connective tissue (regular type) (_____ ×) Location example: _____ Function: _____	Hyaline cartilage (_____ ×) Location example: _____ Function: _____
Elastic cartilage (_____ ×) Location example: _____ Function: _____	Fibrocartilage (_____ ×) Location example: _____ Function: _____
Bone (compact) (_____ ×) Location example: _____ Function: _____	Blood (_____ ×) Location example: _____ Function: _____

LEARNING EXTENSION

Use colored pencils to differentiate various cellular structures in Part B. Select a different color for the cells, fibers, and ground substance whenever visible.

10

Muscle and Nervous Tissues

MATERIALS NEEDED

Textbook
Compound light microscope
Prepared slides of the following:
    Skeletal muscle tissue
    Smooth muscle tissue
    Cardiac muscle tissue
    Nervous tissue (spinal cord smear and/or
        cerebellum)

For Learning Extension:
Colored pencils

PURPOSE OF THE EXERCISE

To review the characteristics of muscle and nervous tissues and to observe examples of these tissues.

LEARNING OUTCOMES APR

After completing this exercise, you should be able to

1. Differentiate the special characteristics of each type of muscle tissue and nervous tissue.
2. Sketch and label the characteristics of the different muscle and nervous tissues that you observed.
3. Indicate an example location and function of each type of muscle tissue and nervous tissue.
4. Identify three types of muscle tissues and nervous tissue on microscope slides.

Muscle tissues are characterized by the presence of elongated cells or muscle fibers that can contract in response to specific stimuli. As they shorten, these fibers pull at their attached ends and cause body parts to move. The three types of muscle tissues are *skeletal, smooth,* and *cardiac.*

Nervous tissues occur in the brain, spinal cord, and peripheral nerves. They consist of *neurons* (nerve cells), the impulse-conducting cells of the nervous system, and *neuroglia,* which perform supportive and protective functions for neurons.

EXPLORE

PROCEDURE—Muscle and Nervous Tissues APR

1. Review the sections 5.5 and 5.6 entitled "Muscle Tissues" and "Nervous Tissue" in chapter 5 of the textbook.
2. Complete Part A of Laboratory Report 10.
3. Using the microscope, observe each of the types of muscle tissues on the prepared slides. Look for the special features of each type, as described in the textbook. Compare your prepared slides of muscle tissues to the micrographs in figure 10.1 and the muscle tissue characteristics in table 10.1. As you observe each type of muscle tissue, prepare a labeled sketch of a representative portion of the tissue in Part B of the laboratory report.
4. Observe the prepared slide of nervous tissue and identify neurons (nerve cells), neuron cellular processes, and neuroglia. Compare your prepared slide of nervous tissue to the micrograph in figure 10.1. Prepare a labeled sketch of a representative portion of the tissue in Part B of the laboratory report.
5. Complete Part B of the laboratory report.
6. Test your ability to recognize each of these muscle and nervous tissues by having a laboratory partner select a slide, cover its label, and focus the microscope on this tissue. Then see if you correctly identify the tissue. 4

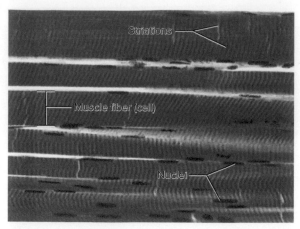

(a) Skeletal muscle (400x)

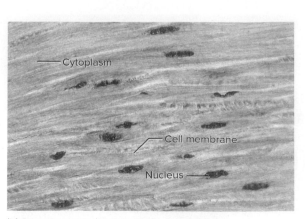

(b) Smooth muscle (from small intestine) (400x)

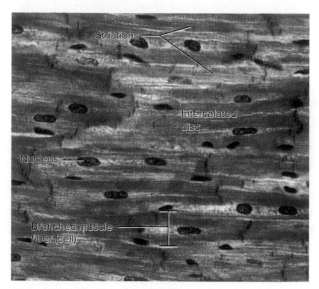

(c) Cardiac muscle (from heart) (400x)

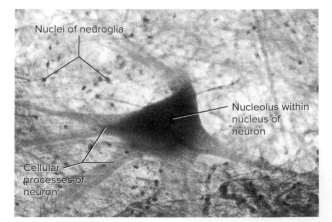

(d) Nervous tissue (350x)

FIGURE 10.1 Micrographs of muscle and nervous tissues *(a–d)*. **APR** **(a–c):** Al Telser/McGraw-Hill Education; **(d):** Alvin Telser/McGraw-Hill Education

TABLE 10.1 Muscle Tissue Characteristics

Characteristic	Skeletal Muscle	Smooth Muscle	Cardiac Muscle
Appearance of cells	Unbranched and relatively parallel	Spindle-shaped	Branched and connected in complex networks
Striations	Present and obvious	Absent	Present but faint
Nucleus	Multinucleated	Uninucleated	Uninucleated (usually)
Intercalated discs	Absent	Absent	Present
Control	Voluntary (usually)	Involuntary	Involuntary

The corresponds to the indicated Learning Outcome(s) found at the beginning of the laboratory exercise.

Muscle and Nervous Tissues

PART A ASSESSMENTS

Match the tissues in column A with the characteristics in column B. Place the letter of your choice in the space provided. (Some answers may be used more than once.) Ⓐ Ⓑ

Column A	Column B
a. Cardiac muscle	_____ **1.** Coordinates, regulates, and integrates body functions
b. Nervous tissue	_____ **2.** Contains intercalated discs
c. Skeletal muscle	_____ **3.** Muscle that lacks striations
d. Smooth muscle	_____ **4.** Striated and involuntary
	_____ **5.** Striated and voluntary
	_____ **6.** Contains neurons and neuroglia
	_____ **7.** Muscle attached to bones
	_____ **8.** Muscle that composes heart
	_____ **9.** Moves food through the digestive tract
	_____ **10.** Conducts impulses along cellular processes
	_____ **11.** Muscle under conscious control
	_____ **12.** Muscle of blood vessels and urinary bladder

PART B ASSESSMENTS

In the space that follows, sketch a few cells or fibers of each of the three types of muscle tissues and of nervous tissue as they appear through the microscope. For each sketch, label the major structures of the cells or fibers, indicate the magnification used, write an example of a location in the body, and provide a function. Ⓐ Ⓐ Ⓐ

Skeletal muscle tissue (_____x)
Location example:_____
Function:_____

Smooth muscle tissue (_____x)
Location example:_____
Function:_____

Cardiac muscle tissue (_____x)
Location example:_____
Function:_____

Nervous tissue (_____x)
Location example:_____
Function:_____

LEARNING EXTENSION

Use colored pencils to differentiate various cellular structures in Part B.

11

Integumentary System

MATERIALS NEEDED

Textbook
Skin model
Hand magnifier or dissecting microscope
Compound light microscope
Prepared microscope slide of human scalp or axilla
Forceps
Microscope slide and coverslip

For Learning Extension:
Prepared slide of dark (heavily pigmented) human skin

PURPOSE OF THE EXERCISE

To observe the structures and tissues of the integumentary system and to review the functions of these parts.

LEARNING OUTCOMES APR

After completing this exercise, you should be able to

1 Locate and name the structures of the integumentary system.

2 Describe the major functions of these structures.

3 Distinguish the locations and tissues among the epidermis, dermis, and subcutaneous layers.

4 Identify and sketch the layers of the skin and associated structures observed on the prepared slide.

The integumentary system includes the skin, hair, nails, sebaceous glands, and sweat glands. These structures provide a protective covering for deeper tissues, aid in regulating body temperature, retard water loss, house sensory receptors, synthesize various chemicals, and excrete small quantities of wastes.

The skin consists of two distinct layers. The outer layer, the *epidermis,* consists of keratinized stratified squamous epithelium. The inner layer, the *dermis,* consists of a thicker layer of mainly dense connective tissue. Beneath the dermis is the *subcutaneous layer* (not considered a true layer of the skin), composed of adipose and areolar connective tissues.

EXPLORE

PROCEDURE—Integumentary System

1. Review sections 6.2 and 6.3 entitled "Layers of the Skin" and "Accessory Structures of the Skin: Epidermal Derivatives" in chapter 6 of your textbook.
2. As a review activity, label figures 11.1 and 11.2. Locate as many of these structures as possible on a skin model.
3. Complete Part A of Laboratory Report 11.
4. Use the hand magnifier or dissecting microscope to do the following:
   a. Observe the skin, hair, and nails of your hand.
   b. Compare the type and distribution of hairs on the front and back of your forearm.
5. Pull out a single hair with the forceps and mount it on a microscope slide under a coverslip. Use the hand magnifier or dissecting microscope to observe the root and shaft of the hair. Note the scalelike parts that make up the shaft.
6. Complete Part B of the laboratory report.
7. As vertical sections of human skin are observed, remember that the lenses of the microscope invert and reverse images. It is important to orient the position of the epidermis, dermis, and subcutaneous (hypodermis) layers using scan magnification before continuing with additional observations. Compare all of your skin observations to figure 11.3. Use low-power magnification of the compound light microscope and proceed as follows:
   a. Observe the prepared slide of human scalp or axilla.
   b. Locate the epidermis, dermis, and subcutaneous layer; a hair follicle; an arrector pili muscle; a sebaceous gland; and a sweat gland.
   c. Focus on the epidermis with high power and locate the stratum corneum and stratum basale. Note how the shapes of the cells in these two layers differ.
   d. Observe the dense connective tissue (irregular type) that makes up the bulk of the dermis along with some areolar connective tissue.
   e. Observe the adipose tissue that composes most of the subcutaneous layer along with some areolar connective tissue.

8. Complete Part C of the laboratory report.
9. Using low-power magnification, locate a hair follicle sectioned longitudinally through its bulblike base, a sebaceous gland close to the follicle, and a sweat gland (fig. 11.3). Observe the detailed structure of these parts with high-power magnification.
10. Complete Parts D and E of the laboratory report.

LEARNING EXTENSION

Observe the prepared slide of dark (heavily pigmented) skin with low-power magnification. The pigment is most abundant in the epidermis. Focus on this region with the high-power objective. The pigment-producing cells, or melanocytes, are located among the deeper layers of epidermal cells. Differences in skin color are primarily due to the amount of pigment (melanin) produced by these cells (fig. 11.3a).

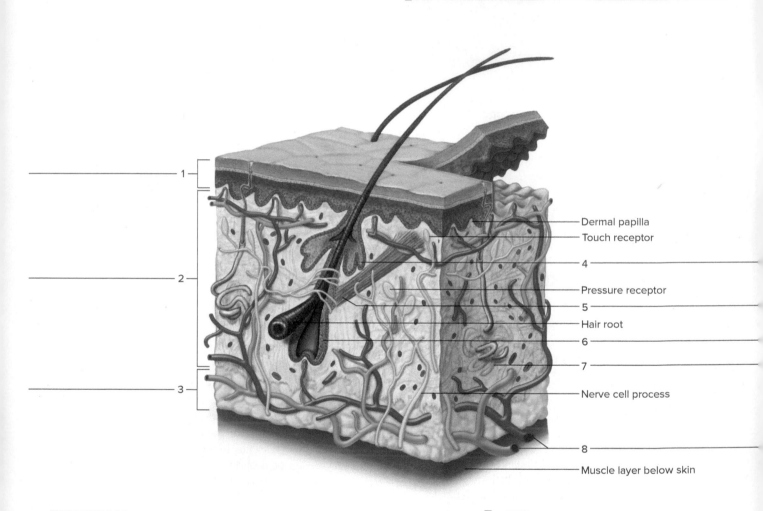

- Dermal papilla
- Touch receptor
- 4
- Pressure receptor
- 5
- Hair root
- 6
- 7
- Nerve cell process
- 8
- Muscle layer below skin

1
2
3

FIGURE 11.1   Label this vertical section of the skin and subcutaneous layer. ![APR icon]

![brain icon] **Critical Thinking Application**

Explain the advantage for melanin granules being located in the deep layer of the epidermis.

_____

_____

_____

_____

FIGURE 11.2
Label the features associated with this hair follicle. 🅰

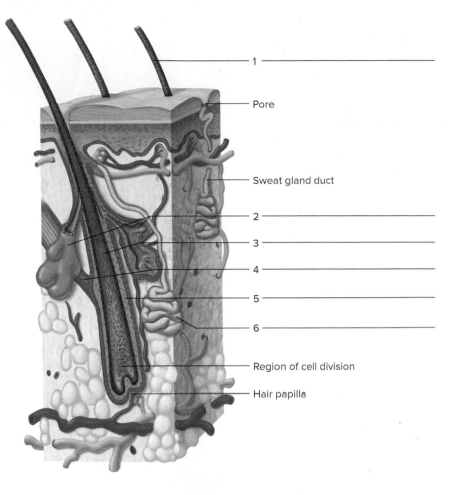

Pore

Sweat gland duct

1

2

3

4

5

6

Region of cell division

Hair papilla

FIGURE 11.3
Features of human skin are indicated in these micrographs (a–e). **APR**
(a): © Victor B. Eichler;
(b), (d): Al Telser/McGraw-Hill Education; **(c):** Image Source/Getty Images; **(e):** doc-stock/Alamy Stock Photo

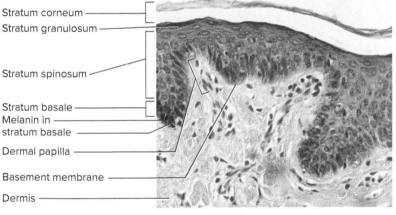

Stratum corneum
Stratum granulosum
Stratum spinosum
Stratum basale
Melanin in stratum basale
Dermal papilla
Basement membrane
Dermis

(a) Epidermis with melanin (400×)

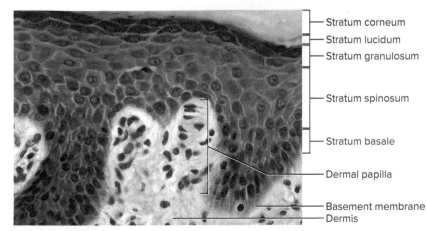

Stratum corneum
Stratum lucidum
Stratum granulosum
Stratum spinosum
Stratum basale
Dermal papilla
Basement membrane
Dermis

(b) Strata of thick skin (500×)

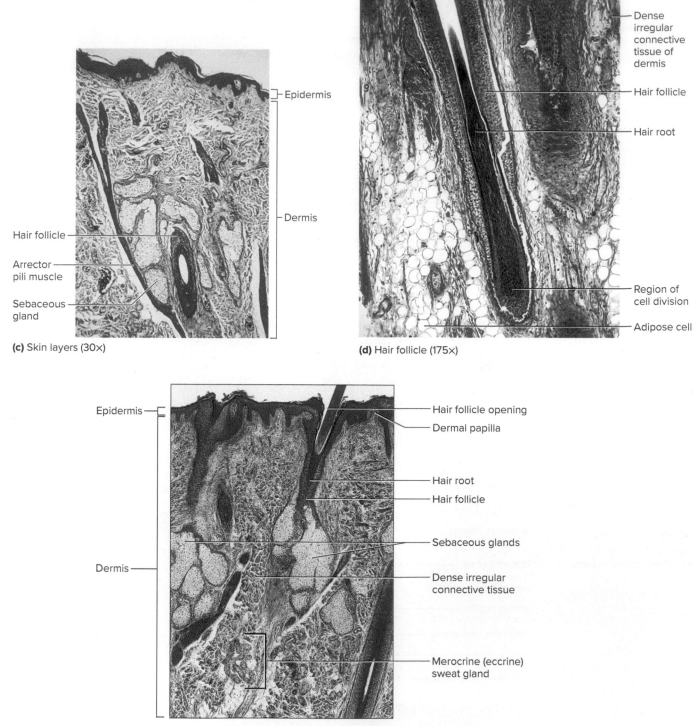

(c) Skin layers (30×)

(d) Hair follicle (175×)

(e) Skin layers with glands (30×)

FIGURE 11.3   *Continued.*

Name _____

Date _____

Section _____

The 🄰 corresponds to the indicated Learning Outcome(s) found at the beginning of the laboratory exercise.

Integumentary System

♻ PART A ASSESSMENTS

Match the structures in column A with the descriptions and functions in column B. Place the letter of your choice in the space provided. 🄰 🄐

Column A	Column B
a. Apocrine sweat gland	_____ 1. Oily secretion that helps to waterproof body surface
b. Arrector pili muscle	_____ 2. Outermost layer of epidermis
c. Dermis	_____ 3. Becomes active at puberty in axillary and groin regions
d. Epidermis	_____ 4. Epidermal pigment
e. Hair follicle	_____ 5. Inner layer of skin
f. Keratin	_____ 6. Responds to elevated body temperature
g. Melanin	_____ 7. General name of entire superficial layer of the skin
h. Merocrine sweat gland	_____ 8. Gland that secretes an oily mixture
i. Sebaceous gland	_____ 9. Tough protein of nails and hair
j. Sebum	_____ 10. Cell division and deepest layer of epidermis
k. Stratum basale	_____ 11. Tubelike part that contains the root of the hair
l. Stratum corneum	_____ 12. Causes hair to stand erect and goose bumps to appear

♻ PART B ASSESSMENTS

Complete the following:

1. How does the skin on your palm differ from that on the back (posterior) of your hand? 🄐 _____

_____

_____

2. Describe the differences you observed in the type and distribution of hair on the front (anterior) and back (posterior) of your forearm. 🄐 _____

_____

_____

3. Explain how the hair is formed. 🄐 _____

_____

_____

4. What cells produce the pigment in hair? 🄐 _____

_____

_____

PART C ASSESSMENTS

Complete the following:

1. Distinguish the locations and tissues among the epidermis, dermis, and subcutaneous layer. 🄰 _____

_____

_____

_____

2. How do the cells of stratum corneum and stratum basale differ? 🄰 _____

_____

_____

_____

3. What special qualities, due to the presence of fibers, does the connective tissue of the dermis have? 🄰 _____

_____

_____

_____

PART D ASSESSMENTS

Complete the following:

1. In which layer of skin are follicles usually found? 🄰 _____

2. How are sebaceous glands associated with hair follicles and what do they secrete? 🄰 _____

_____

3. The ducts of apocrine sweat glands open into _____. 🄰

PART E ASSESSMENTS

Using the scanning objective, sketch a vertical section of human skin. Label the skin layers and a hair follicle, a sebaceous gland, and a sweat gland. 🄰

12

Bone Structure

PURPOSE OF THE EXERCISE

To examine the structure of a long bone.

🔄 LEARNING OUTCOMES APR

After completing this exercise, you should be able to

(1) Locate the major structures of a long bone.

(2) Distinguish between compact and spongy bone.

(3) Differentiate the special characteristics of compact
    bone tissue.

(4) Describe the functions of various structures of a bone.

A bone represents an organ of the skeleton system. As
such, it is composed of a variety of tissues, including
bone tissue, cartilage, dense connective tissue, blood, and
nervous tissue. Bones are not only alive but also multi-
functional. They support and protect softer tissues, provide
points of attachment for muscles, house blood-producing
cells, and store inorganic salts.

Although bones of the skeleton vary greatly in size
and shape, they have much in common structurally and
functionally.

🔄 EXPLORE

PROCEDURE—Bone Structure

1. Review section 7.2 entitled "Bone Structure" in chapter 7
   of the textbook.
2. As a review activity, label figures 12.1 and 12.2.
3. Examine the sectioned bones and locate the following:

   epiphysis
       proximal—nearest limb attachment to torso
       distal—farthest from limb attachment to torso
   epiphyseal plate—growth zone of hyaline cartilage
   articular cartilage—on ends of epiphyses
   diaphysis—shaft between epiphyses
   periosteum—strong membrane around bone (except
       articular cartilage) of dense irregular connective
       tissue
   compact bone—forms diaphysis and epiphyseal
       surfaces
   spongy bone—within epiphyses
       trabeculae—a structural lattice in spongy bone
   medullary cavity—hollow chamber
   endosteum—thin membrane of reticular connective
       tissue that lines the medullary cavity
   yellow marrow—occupies medullary cavity and
       stores adipose tissue
   red marrow—occupies spongy bone in some
       epiphyses and flat bones and produces blood cells

4. Use the dissecting microscope to observe the compact
   bone and spongy bone of the sectioned specimens. Also
   examine the marrow in the medullary cavity and the
   spaces within the spongy bone of the fresh specimen.

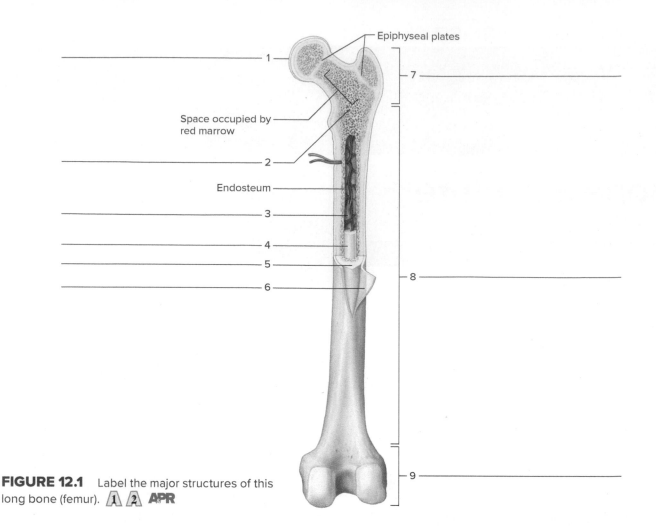

Epiphyseal plates

1

7

Space occupied by
red marrow

2

Endosteum

3

4

5

6

8

9

FIGURE 12.1 Label the major structures of this
long bone (femur). ⚠ ⚠ **APR**

5. Reexamine the microscopic structure of bone tissue by
   observing a prepared microscope slide of ground compact bone. Use figure 12.3, a micrograph of bone tissue,
   to locate the following features:

   osteon (Haversian system)—cylinder-shaped unit
   central canal (Haversian canal)—contains blood
     vessels and nerves
   lacuna—small chamber for an osteocyte
   bone extracellular matrix—collagen and calcium
     phosphate
   lamella—concentric ring of matrix around central
     canal
   canaliculus—minute tube containing cellular process

   ### Critical Thinking Application

   Explain how bone cells embedded in a solid ground substance obtain nutrients and eliminate wastes. ⚠

6. Complete Parts A and B of Laboratory Report 12.

DEMONSTRATION

Examine a fresh chicken bone and a chicken bone
that has been soaked for several days in vinegar or
exposed overnight in dilute hydrochloric acid. Wear
disposable gloves for handling these bones. This acid
treatment removes the inorganic salts from the bone
extracellular matrix. Rinse the bones in water and
note the texture and flexibility of each (fig. 12.4a).
The bone becomes soft and flexible without the support of the inorganic salts with calcium.

Examine the specimen of chicken bone that has
been exposed to high temperature (baked at 121°C
[250°F] for 2 hours). This treatment removes the
protein and other organic substances from the bone
extracellular matrix (fig. 12.4b). The bone becomes
brittle and fragile without the benefit of the collagen
fibers. A living bone with a combination of the qualities of inorganic and organic substances possesses
tensile strength.

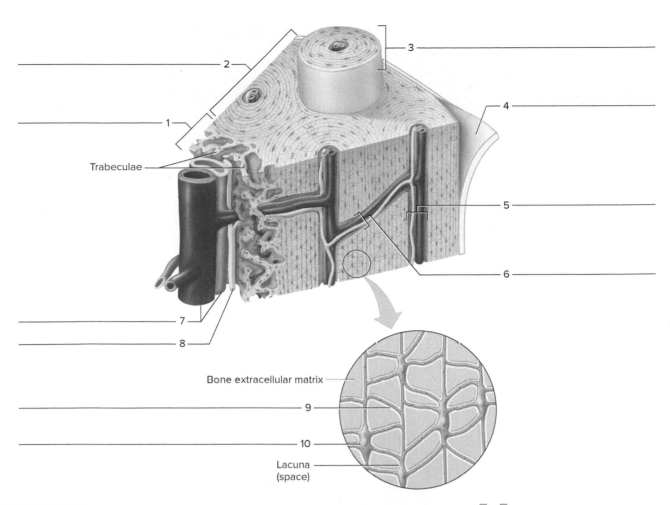

Trabeculae

Bone extracellular matrix

Lacuna
(space)

FIGURE 12.2 Label the features associated with the microscopic structure of bone. 🅐🅐

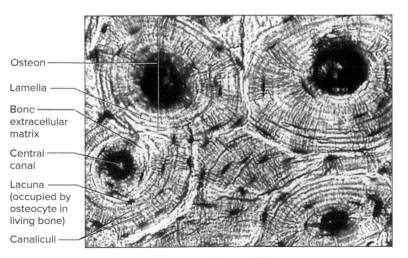

Osteon

Lamella

Bone
extracellular
matrix

Central
canal

Lacuna
(occupied by
osteocyte in
living bone)

Canaliculi

FIGURE 12.3 Micrograph of ground compact bone tissue (200×). **APR**

Dennis Strete/McGraw-Hill Education

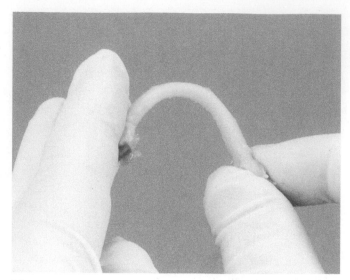

(a)

(b)

FIGURE 12.4 Results of fresh chicken bone demonstration: (*a*) soaked in vinegar; (*b*) baked in oven.
(a–b): ©J and J Photography

Name _____

Date _____

Section _____

The ⟁ corresponds to the indicated Learning Outcome(s) found at the beginning of the laboratory exercise.

LABORATORY
REPORT

12

Bone Structure

⟳ PART A ASSESSMENTS

Complete the following:

1. Where in the human skeleton are long bones found? ⟁

   _____

2. Distinguish between the epiphysis and the diaphysis of a long bone. ⟁

   _____

   _____

3. Describe where cartilage is found on the surface of a long bone. ⟁

   _____

   _____

4. Describe where the periosteum is found on the surface of a long bone. ⟁

   _____

5. In general, what is the function of bony processes? ⟁

   _____

   _____

6. Distinguish between the locations and tissues of the periosteum and those of the endosteum. ⟁

   _____

   _____

   _____

7. What structural differences did you note between the compact bone and the spongy bone? ⟁

   _____

   _____

   _____

8. How are these structural differences related to the locations and functions of these two types of bone? ⟁ ⟁

   _____

   _____

   _____

9. From your observations, how does the marrow in the medullary cavity compare with the marrow in the spaces of the spongy bone?

_____

_____

_____

PART B ASSESSMENTS

Identify the structures indicated in figure 12.5.

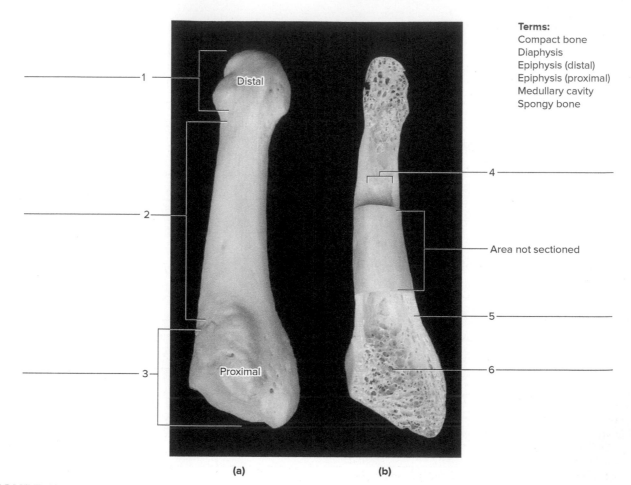

FIGURE 12.5 Identify the structures indicated in (*a*) the unsectioned long bone (fifth metatarsal) and (b) the partially sectioned long bone, using the terms provided. 🔺1 🔺2
©J and J Photography

13

Organization of the Skeleton

MATERIALS NEEDED

Textbook
Human skeleton, articulated
Human skeleton, disarticulated

For Demonstration:
Radiographs (X rays) of skeletal structures

PURPOSE OF THE EXERCISE

To review the organization of the skeleton, the major bones of the skeleton, and the terms used to describe skeletal structures.

 **LEARNING OUTCOMES APR**

After completing this exercise, you should be able to

1. Distinguish between the axial skeleton and the appendicular skeleton.
2. Locate and label the major bones of the human skeleton.
3. Associate the terms used to describe skeletal structures and locate examples of such structures on the human skeleton.

The skeleton can be separated into two major portions: (1) the *axial skeleton*, which consists of the bones and cartilages of the head, neck, and trunk; and (2) the *appendicular skeleton*, which consists of the bones of the limbs and those that anchor the limbs to the axial skeleton. The bones that anchor the limbs include the pectoral and pelvic girdles. Men and women, although variations can exist, possess the same total bone number of 206.

EXPLORE

PROCEDURE—Organization of the Skeleton

1. Review section 7.5 entitled "Skeletal Organization" in chapter 7 of the textbook. (Pronunciations of the names for major skeletal structures are included within the narrative of chapter 7.)

2. As a review activity, label figure 13.1.
3. Examine the articulated human skeleton and locate the following parts. As you locate the following bones, note the number of each in the skeleton. Palpate as many of the corresponding bones in your skeleton as possible.

axial skeleton	
skull	
cranium	(8)
face	(14)
middle ear bone	(6)
hyoid bone	(1)
vertebral column	
vertebra	(24)
sacrum	(1)
coccyx	(1)
thoracic cage	
rib	(24)
sternum	(1)
appendicular skeleton	
pectoral (shoulder) girdle	
scapula	(2)
clavicle	(2)
upper limbs	
humerus	(2)
radius	(2)
ulna	(2)
carpal	(16)
metacarpal	(10)
phalanx	(28)
pelvic girdle	
coxal bone	(2)
lower limbs	
femur	(2)
tibia	(2)
fibula	(2)
patella	(2)
tarsal	(14)
metatarsal	(10)
phalanx	(28)
Total	**206 bones**

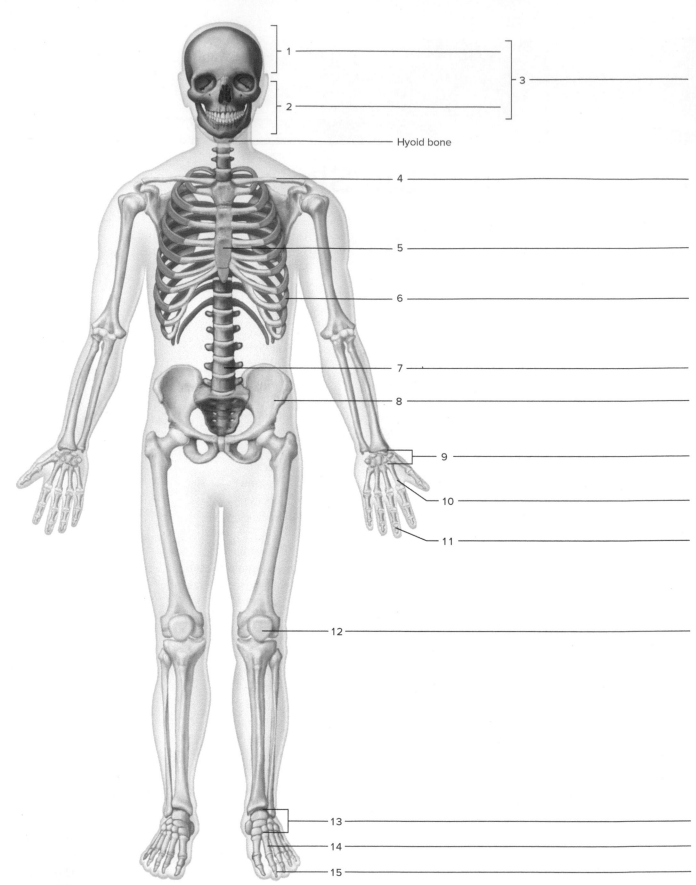

1 _____

2 _____

3 _____

Hyoid bone

4 _____

5 _____

6 _____

7 _____

8 _____

9 _____

10 _____

11 _____

12 _____

13 _____

14 _____

15 _____

(a)

FIGURE 13.1   Label the major bones of the skeleton: (*a*) anterior view; (*b*) posterior view. The axial portion is shown in orange; the appendicular portions are shown in yellow. A1 A2 APR

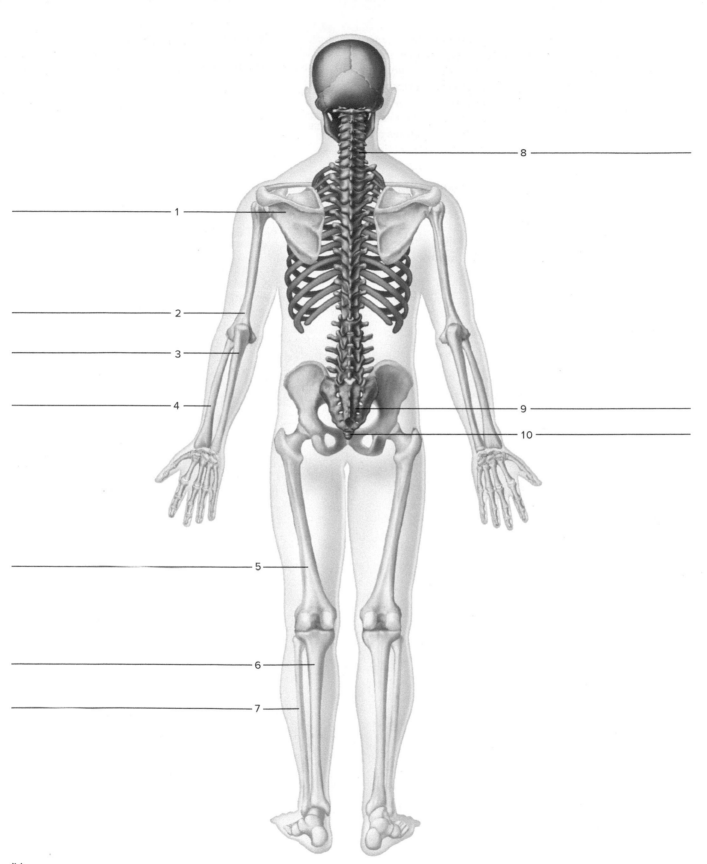

(b)

FIGURE 13.1 *Continued.*

4. Study table 7.2 in chapter 7 of the textbook. Bone features (bone markings) can be grouped together in a category of *projections*, *articulations*, *depressions*, or *openings*. Within each category, more specific examples occur. Each of the bones listed in this section represents only one example of a location in the human body. Locate an example of each of the following features on the example bone from a disarticulated skeleton, noting the size, shape, and location in the human skeleton:

Projections: sites for a tendon and ligament attachment

    crest (ridgelike)—hip bone

    epicondyle (superior to condyle)—femur

    line (linea) (slightly raised ridge)—femur

    process (prominent)—vertebra

    protuberance (outgrowth)—skull (occipital)

    spine (thornlike)—scapula

    trochanter (large)—femur

    tubercle (small, knoblike)—humerus

    tuberosity (rough elevation)—radius

Articulations: where bones connect at a joint

    condyle (rounded process)—skull (occipital)

    facet (nearly flat)—vertebra

    head (expanded end)—femur

Depressions: recessed areas in bones

    fossa (shallow basin)—humerus

    fovea (tiny pit)—femur

Openings: open spaces in bones

    fissure (slit)—skull (orbit)

    foramen (hole)—vertebra

    meatus (tubelike passageway)—skull (temporal)

    sinus (cavity)—skull (maxilla)

 Critical Thinking Application

Locate and name the largest foramen in the skull.

_____

_____

Locate and name the largest foramen in the skeleton.

_____

_____

5. Complete Parts A, B, C, and D of Laboratory Report 13.

DEMONSTRATION

Images on radiographs (X rays) are produced by allowing X rays from an X-ray tube to pass through a body part and to expose photographic film positioned on the opposite side of the part. The image that appears on the film after it is developed reveals the presence of parts with different densities. Bone, for example, is very dense tissue and is a good absorber of X rays. Thus, bone generally appears light on the film. Air-filled spaces, on the other hand, absorb almost no X rays and appear as dark areas on the film. Liquids and soft tissues absorb intermediate quantities of X rays, so they usually appear in various shades of gray.

Examine the available radiographs of skeletal structures by holding each film in front of a light source. Identify as many of the bones and features as you can.

Name _____

Date _____

Section _____

The corresponds to the indicated Learning Outcome(s) found at the beginning of the laboratory exercise.

Organization of the Skeleton

PART A ASSESSMENTS

Complete the following statements:

1. The two divisions of the skeleton are the _____ skeleton and the appendicular skeleton. ⚠1

2. The _____ bone supports the tongue. ⚠2

3. The _____ at the inferior end of the sacrum is composed of four fused vertebrae. ⚠2

4. The ribs are attached posteriorly to the _____ vertebrae. ⚠2

5. The thoracic cage is composed of _____ pairs of ribs. ⚠2

6. The scapulae and clavicles together form the _____ girdle. ⚠1

7. The humerus, radius, and _____ articulate to form the elbow joint. ⚠2

8. The wrist is composed of eight bones called _____. ⚠2

9. The hip bones are attached posteriorly to the _____. ⚠2

10. The _____ covers the anterior surface of the knee. ⚠2

11. The bones that articulate with the distal ends of the tibia and fibula are called _____. ⚠2

12. All finger and toe bones are called _____. ⚠2

PART B ASSESSMENTS

Match the terms in column A with the definitions in column B. Place the letter of your choice in the space provided. ⚠3

Column A
- a. Condyle
- b. Crest
- c. Facet
- d. Foramen
- e. Fossa
- f. Line
- g. Tuberosity

Column B

_____ 1. Small, nearly flat articular surface

_____ 2. Shallow basin

_____ 3. Rounded process

_____ 4. Opening or hole

_____ 5. Knoblike rough elevation

_____ 6. Ridgelike projection

_____ 7. Slightly raised ridge

Match the terms in column A with the definitions in column B. Place the letter of your choice in the space provided. ⒊

Column A	Column B
a. Fovea	_____ **1.** Tubelike passageway
b. Head	_____ **2.** Tiny pit or depression
c. Meatus	_____ **3.** Small, knoblike process
d. Sinus	_____ **4.** Rounded enlargement at end of bone
e. Spine	_____ **5.** Air-filled cavity within bone
f. Trochanter	_____ **6.** Relatively large process
g. Tubercle	_____ **7.** Thornlike projection

PART D ASSESSMENTS

1. Identify the bones indicated in figure 13.2.

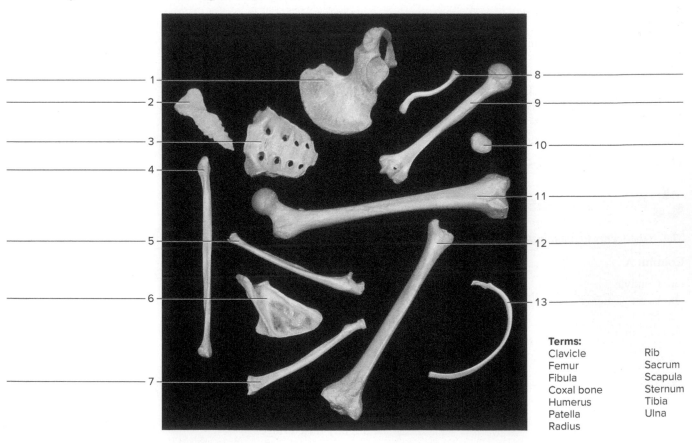

Terms:

Clavicle	Rib
Femur	Sacrum
Fibula	Scapula
Coxal bone	Sternum
Humerus	Tibia
Patella	Ulna
Radius	

FIGURE 13.2   Using the terms provided, identify the bones in this random arrangement. ⒉
©J and J Photography

2. List any of the bones shown in figure 13.2 that are included as part of the axial skeleton. ⒈ _____

_____

14

Skull

MATERIALS NEEDED

Textbook
Human skull, articulated
Human skull, disarticulated (Beauchene)
Human skull, sagittal section

For Learning Extension:
Colored pencils

For Demonstration:
Fetal skull

PURPOSE OF THE EXERCISE

To examine the structure of the human skull and to identify the bones and major features of the skull.

LEARNING OUTCOMES APR

After completing this exercise, you should be able to

1. Distinguish between the cranium and the facial skeleton.
2. Locate and label the bones of the skull and their major features.
3. Locate and label the major sutures of the cranium.
4. Locate and label the sinuses of the skull.

A human skull consists of twenty-two bones that, except for the lower jaw, are firmly interlocked along sutures. Eight of these immovable bones make up the braincase, or cranium, and thirteen more immovable bones and mandible form the facial skeleton.

EXPLORE

PROCEDURE—SKULL APR

1. Review section 7.6 entitled "Skull" in chapter 7 of the textbook.
2. As a review activity, label figures 14.1, 14.2, 14.3, 14.4, and 14.5.

3. Study the bones with paranasal sinuses (figure 14.6).
4. Examine the **cranial bones** of the articulated human skull and the sectioned skull. Also observe the corresponding disarticulated bones. Locate the following bones and features in the laboratory specimens and palpate as many of these bones and features in your skull as possible.

frontal bone (1)
    supraorbital foramen
    frontal sinus

parietal bone (2)
    sagittal suture
    coronal suture

occipital bone (1)
    lambdoid suture
    foramen magnum
    occipital condyle

temporal bone (2)
    squamous suture
    external acoustic meatus
    mandibular fossa
    mastoid process
    styloid process
    zygomatic process

sphenoid bone (1)
    sella turcica
    sphenoidal sinus

ethmoid bone (1)
    cribriform plate
    perpendicular plate
    superior nasal concha
    middle nasal concha
    ethmoidal sinus
    crista galli

5. Complete Parts A and B of Laboratory Report 14.

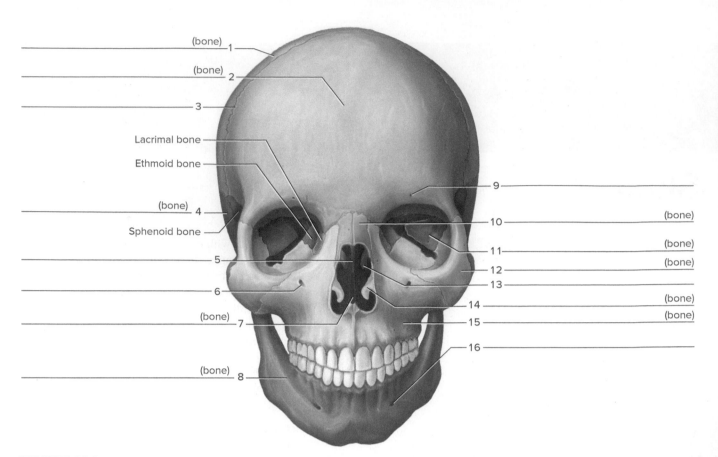

_____ (bone) 1
_____ (bone) 2
_____ 3
Lacrimal bone
Ethmoid bone
_____ (bone) 4
Sphenoid bone
_____ 5
_____ 6
_____ (bone) 7
_____ (bone) 8

9 _____
10 _____ (bone)
11 _____ (bone)
12 _____ (bone)
13
14 _____ (bone)
15 _____ (bone)
16 _____

FIGURE 14.1 Label the anterior bones and features of the skull. (If the line lacks the word *bone*, label the particular feature of that bone.) 🄰 **APR**

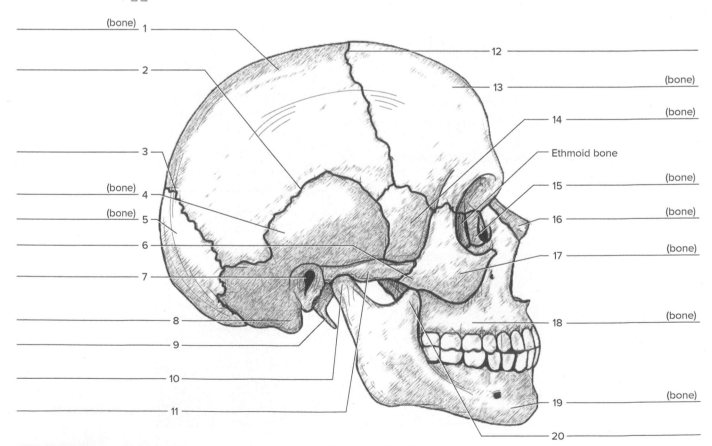

_____ (bone) 1
_____ 2
_____ 3
_____ (bone) 4
_____ (bone) 5
_____ 6
_____ 7
_____ 8
_____ 9
_____ 10
_____ 11

12 _____
13 _____ (bone)
14 _____ (bone)
Ethmoid bone
15 _____ (bone)
16 _____ (bone)
17 _____ (bone)
18 _____ (bone)
19 _____ (bone)
20 _____

FIGURE 14.2 Label the lateral bones and features of the skull. 🄰 🄱 **APR**

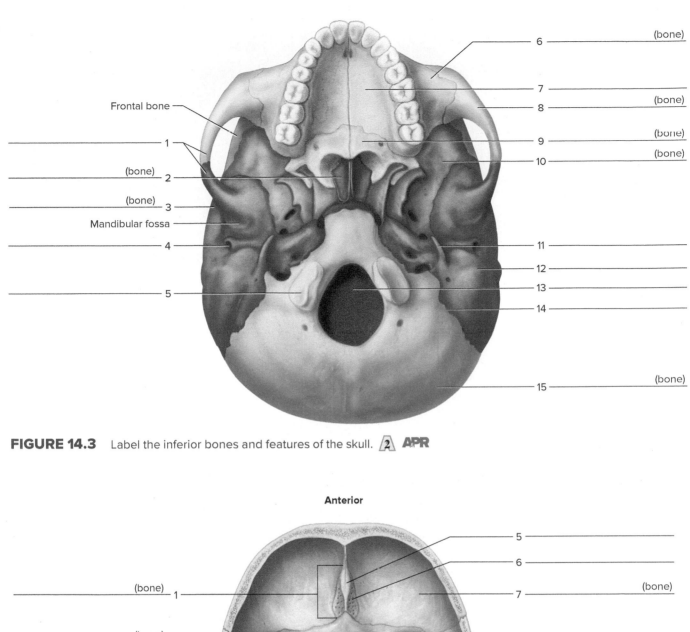

6 ——————————— (bone)

7 ———————————

8 ——————————— (bone)

9 ——————————— (bone)

10 ——————————— (bone)

Frontal bone ——

1

(bone) 2

(bone) 3

Mandibular fossa ——

4

5

11 ———————————

12 ———————————

13 ———————————

14 ———————————

15 ——————————— (bone)

FIGURE 14.3   Label the inferior bones and features of the skull.   2   APR

Anterior

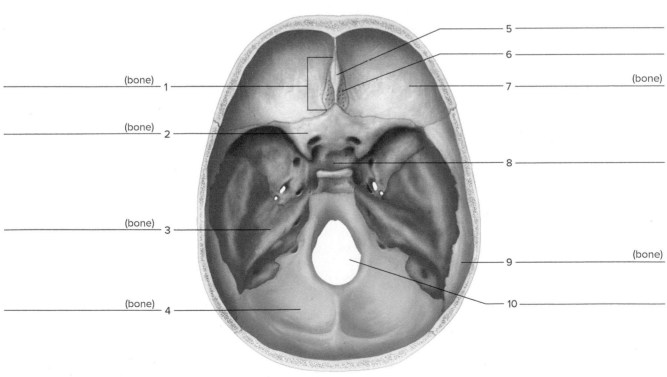

5 ———————————

6 ———————————

(bone) 1

7 ——————————— (bone)

(bone) 2

8 ———————————

(bone) 3

9 ——————————— (bone)

(bone) 4

10 ———————————

Posterior

FIGURE 14.4   Label the bones and features of the floor of the cranial cavity as viewed from above.   2   APR

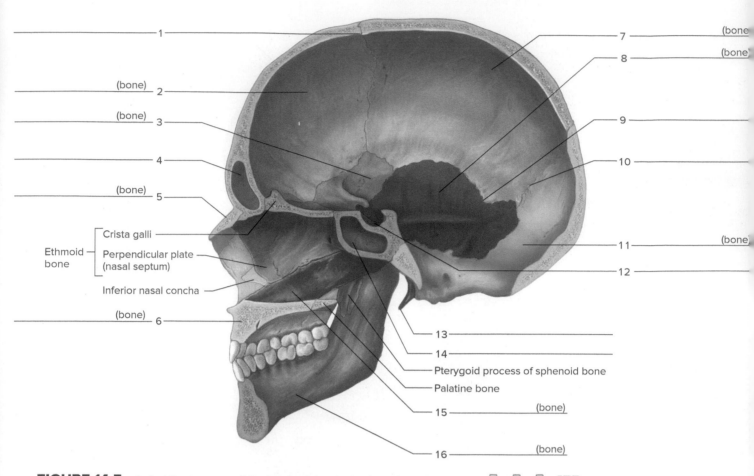

_____ 1

(bone) _____ 2

(bone) _____ 3

_____ 4

(bone) _____ 5

Ethmoid bone

   Crista galli _____

   Perpendicular plate (nasal septum) _____

   Inferior nasal concha _____

(bone) _____ 6

7 _____ (bone)

8 _____ (bone)

9 _____

10 _____

11 _____ (bone)

12 _____

13 _____

14 _____

Pterygoid process of sphenoid bone

Palatine bone

15 _____ (bone)

16 _____ (bone)

FIGURE 14.5 Label the bones and features of the sagittal section of the skull. 2 3 4 **APR**

6. Examine the **facial bones** of the articulated and sectioned skulls and the corresponding disarticulated bones. Locate the following:

   maxilla (2)

      maxillary sinus

      palatine process

      alveolar process

      alveolar arch

   palatine bone (2)

   zygomatic bone (2)

      temporal process

      zygomatic arch (formed by temporal and zygomatic processes)

   lacrimal bone (2)

   nasal bone (2)

   vomer (1)

   inferior nasal concha (2)

   mandible (1)

      mandibular condyle

      coronoid process

      alveolar arch

7. Study the skull bones of a disarticulated skull (fig. 14.7).

8. Complete Parts C and D of the laboratory report.

LEARNING EXTENSION

Use colored pencils to color the bones illustrated in figure 14.2. Select a different color for each bone. This activity should help you locate various bones shown in a lateral view of a skull. You can check your work by referring to the corresponding figure in the textbook, presented in full color.

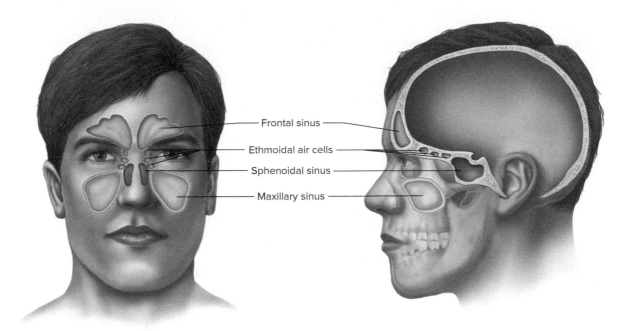

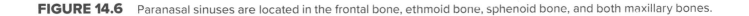

FIGURE 14.6 Paranasal sinuses are located in the frontal bone, ethmoid bone, sphenoid bone, and both maxillary bones.

Frontal sinus
Ethmoidal air cells
Sphenoidal sinus
Maxillary sinus

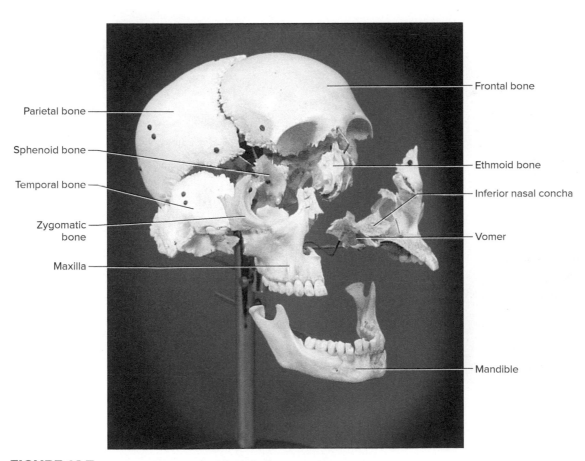

Parietal bone
Sphenoid bone
Temporal bone
Zygomatic bone
Maxilla

Frontal bone
Ethmoid bone
Inferior nasal concha
Vomer
Mandible

FIGURE 14.7 Bones of a disarticulated skull. ©J and J Photography

DEMONSTRATION

Examine the fetal skeleton and skull (figs. 14.8 and 14.9). The skull is incompletely developed and the cranial bones are separated by fibrous membranes. These membranous areas are called *fontanels,* or "soft spots." The fontanels close as the cranial bones grow together. The posterior and lateral fontanels usually close during the first year after birth, whereas the anterior fontanel may not close until the middle or end of the second year. What other features characterize the fetal skull? _____

_____

_____

_____

_____

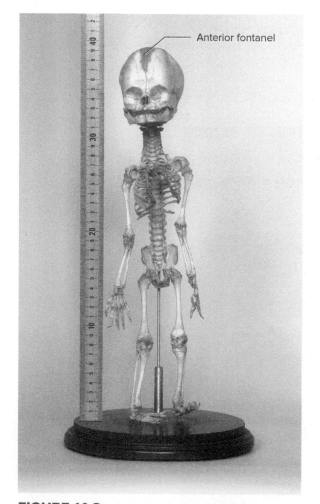

FIGURE 14.8 Anterior view of a fetal skeleton next to a metric scale. The gestational age of this skeleton is 7–8 months. ©J and J Photography

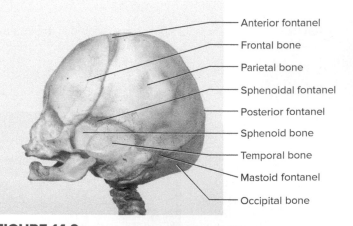

FIGURE 14.9 Lateral view of a fetal skull.
©J and J Photography

Name _____

Date _____

Section _____

The ⒶA corresponds to the indicated Learning Outcome(s) found at the beginning of the laboratory exercise.

LABORATORY
REPORT

14

Skull

↻ PART A ASSESSMENTS

Complete the following statements:

1. Name six cranial bones that are visible on a lateral view of a skull. Ⓐ1 _____

_____

2. The _____ suture joins the frontal bone to the parietal bones. Ⓐ3

3. The parietal bones are fused along the midline by the _____ suture. Ⓐ3

4. The _____ suture joins the parietal bones to the occipital bone. Ⓐ3

5. The temporal bones are joined to the parietal bones along the _____ sutures. Ⓐ3

6. Name the three cranial bones that contain sinuses. Ⓐ1 Ⓐ4 _____

7. Name a facial bone that contains a sinus. Ⓐ1 Ⓐ4 _____

↻ PART B ASSESSMENTS

Match the bones in column A with the features in column B. Place the letter of your choice in the space provided. (Some answers are used more than once.) Ⓐ2

Column A	Column B
a. Ethmoid bone	_____ **1.** Forms sagittal, coronal, squamous, and lambdoid sutures
b. Frontal bone	
c. Occipital bone	_____ **2.** Cribriform plate
d. Parietal bone	_____ **3.** Crista galli
e. Sphenoid bone	_____ **4.** External acoustic meatus
f. Temporal bone	_____ **5.** Foramen magnum
	_____ **6.** Mandibular fossa
	_____ **7.** Mastoid process
	_____ **8.** Middle nasal concha
	_____ **9.** Occipital condyle
	_____ **10.** Sella turcica
	_____ **11.** Styloid process
	_____ **12.** Supraorbital foramen

PART C ASSESSMENTS

Match the bones in column A with the characteristics in column B. Place the letter of your choice in the space provided. 🔺

Column A	Column B
a. Inferior nasal concha	_____ **1.** Forms bridge of nose
b. Lacrimal bone	_____ **2.** Only movable bone in the facial skeleton
c. Mandible	_____ **3.** Contains coronoid process
d. Maxilla	_____ **4.** Creates prominence of cheek inferior and lateral to the eye
e. Nasal bone	_____ **5.** Contains sockets of upper teeth
f. Palatine bone	_____ **6.** Forms inferior portion of nasal septum
g. Vomer	_____ **7.** Forms anterior portion of zygomatic arch
h. Zygomatic bone	_____ **8.** Scroll-shaped bone
	_____ **9.** Forms anterior roof of mouth
	_____ **10.** Forms posterior roof of mouth
	_____ **11.** Small medial bone of each orbit

PART D ASSESSMENTS

Identify the numbered bones and features of the skulls in figures 14.10, 14.11, 14.12, and 14.13. 🔺 🔺

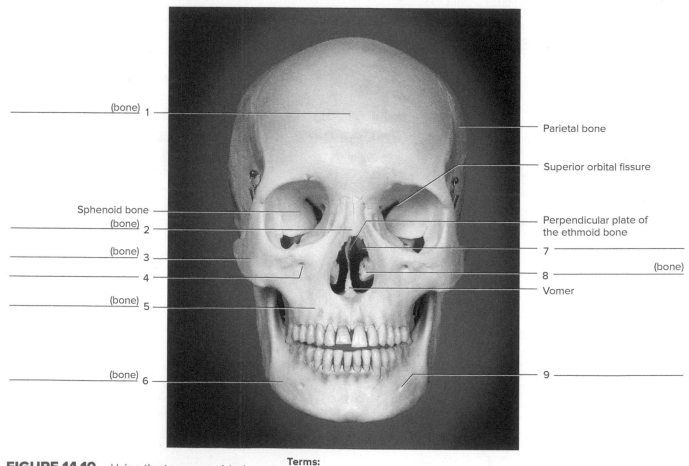

FIGURE 14.10 Using the terms provided, identify the bones and features indicated on this anterior view of the skull. ©J and J Photography

Terms:

Frontal bone	Mandible	Middle nasal concha (of ethmoid bone)
Inferior nasal concha	Maxilla	Nasal bone
Infraorbital foramen	Mental foramen	Zygomatic bone

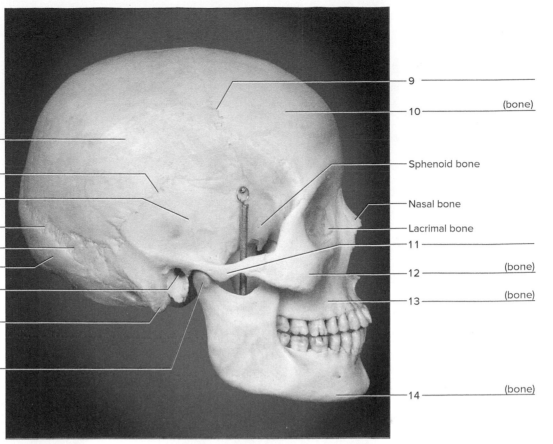

(bone) 1
2
(bone) 3
4
Sutural bone
(bone) 5
6
7
8

9
10 _____ (bone)
Sphenoid bone
Nasal bone
Lacrimal bone
11
12 _____ (bone)
13 _____ (bone)
14 _____ (bone)

FIGURE 14.11
Using the terms provided, identify the bones and features indicated on this lateral view of the skull.
©J and J Photography

Terms:

Coronal suture	Lambdoid suture	Mastoid process	Parietal bone	Zygomatic bone
External acoustic meatus	Mandible	Maxilla	Squamous suture	Zygomatic process
Frontal bone	Mandibular condyle	Occipital bone	Temporal bone	(of temporal bone)

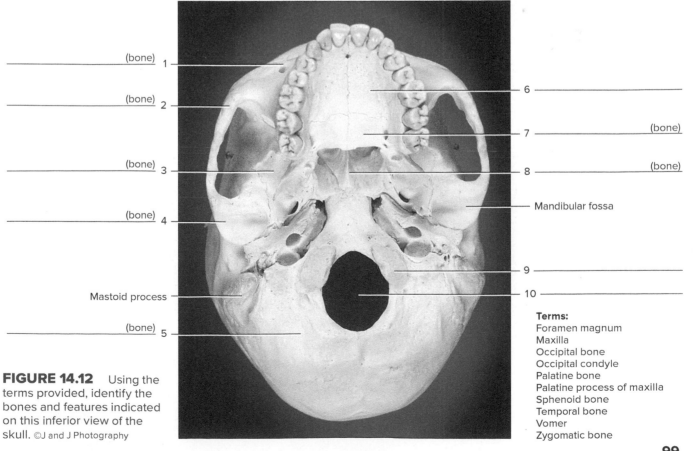

(bone) 1
(bone) 2
(bone) 3
(bone) 4
Mastoid process
(bone) 5

6
7 _____ (bone)
8 _____ (bone)
Mandibular fossa
9
10

FIGURE 14.12 Using the terms provided, identify the bones and features indicated on this inferior view of the skull. ©J and J Photography

Terms:
Foramen magnum
Maxilla
Occipital bone
Occipital condyle
Palatine bone
Palatine process of maxilla
Sphenoid bone
Temporal bone
Vomer
Zygomatic bone

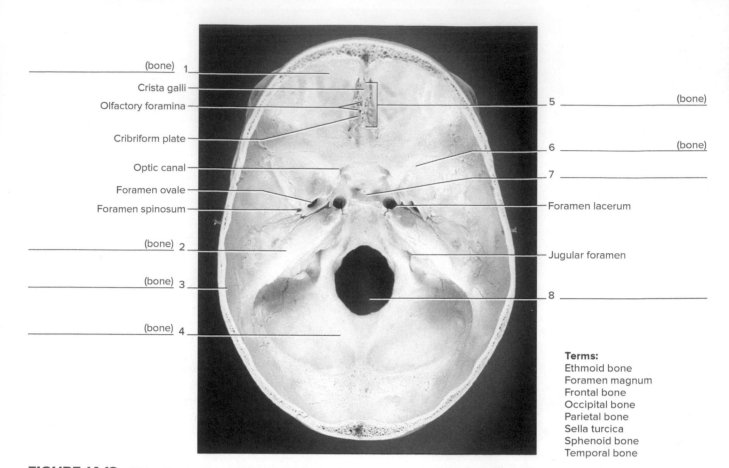

_____ (bone) 1

Crista galli —

Olfactory foramina —

Cribriform plate —

Optic canal —

Foramen ovale —

Foramen spinosum —

_____ (bone) 2

_____ (bone) 3

_____ (bone) 4

5 _____ (bone)

6 _____ (bone)

7 _____

Foramen lacerum

Jugular foramen

8 _____

Terms:
Ethmoid bone
Foramen magnum
Frontal bone
Occipital bone
Parietal bone
Sella turcica
Sphenoid bone
Temporal bone

FIGURE 14.13 Using the terms provided, identify the bones and features on this floor of the cranial cavity of a skull.
©J and J Photography

15

Vertebral Column and Thoracic Cage

MATERIALS NEEDED

Textbook
Human skeleton, articulated
Samples of cervical, thoracic, and lumbar vertebrae
Human skeleton, disarticulated

PURPOSE OF THE EXERCISE

To examine the vertebral column and the thoracic cage of the human skeleton and to identify the bones and major features of these parts.

LEARNING OUTCOMES APR

After completing this exercise, you should be able to

1. Identify the major features of the vertebral column.
2. Locate the features of a vertebra.
3. Distinguish the cervical, thoracic, and lumbar vertebrae and locate the sacrum and coccyx.
4. Identify the structures and functions of the thoracic cage.
5. Distinguish between true and false ribs.

The *vertebral column*, consisting of twenty-six bones, extends from the skull to the pelvis and forms the vertical axis of the human skeleton. The vertebral column includes seven cervical vertebrae, twelve thoracic vertebrae, five lumbar vertebrae, one sacrum of five fused vertebrae, and one coccyx of usually four fused vertebrae. To help remember the number of cervical, thoracic, and lumbar vertebrae from superior to inferior, consider this saying: breakfast at 7, lunch at 12, and dinner at 5. These vertebrae are separated from one another by cartilaginous intervertebral discs and are held together by ligaments.

The *thoracic cage* surrounds the thoracic and upper abdominal cavities. It includes the ribs, the thoracic vertebrae, the sternum, and the costal cartilages. The thoracic cage protects organs in the thoracic and upper abdominal cavities, supports the pectoral girdle and arm, and aids breathing.

EXPLORE

PROCEDURE A—Vertebral Column

1. Review section 7.7 entitled "Vertebral Column" in chapter 7 of the textbook.
2. As a review activity, label figures 15.1, 15.2, 15.3, and 15.4.
3. Examine the vertebral column of the human skeleton and locate the following bones and features. At the same time, locate as many of the corresponding bones and features in your skeleton as possible.

cervical vertebrae	**(7)**
atlas (C1)	(1)
axis (C2)	(1)
vertebra prominens (C7)	(1)
thoracic vertebrae	**(12)**
lumbar vertebrae	**(5)**
sacrum	**(1)**
coccyx	**(1)**

intervertebral discs (fibrocartilage)
vertebral canal (contains spinal cord)
cervical curvature
thoracic curvature
lumbar curvature
sacral curvature
intervertebral foramina (passageway for spinal nerves)

Critical Thinking Application

Note the four curvatures of the vertebral column (figure 15.1). What functional advantages exist with curvatures for skeletal structure instead of a straight vertebral column?

_____

_____

_____

_____

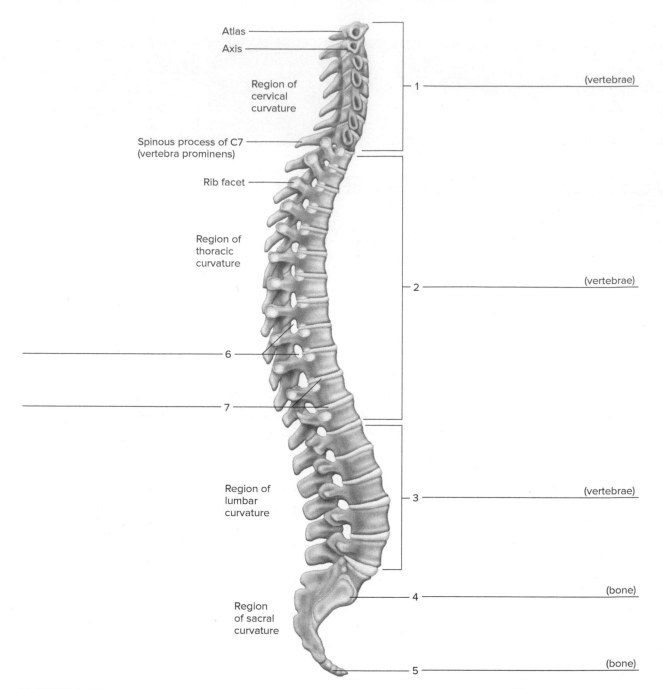

Atlas

Axis

Region of
cervical
curvature

Spinous process of C7
(vertebra prominens)

Rib facet

Region of
thoracic
curvature

6

7

Region of
lumbar
curvature

Region
of sacral
curvature

1 _____ (vertebrae)

2 _____ (vertebrae)

3 _____ (vertebrae)

4 _____ (bone)

5 _____ (bone)

FIGURE 15.1   Label the bones and features of the vertebral column (right lateral view).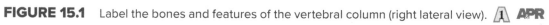

(a) Atlas (C1)

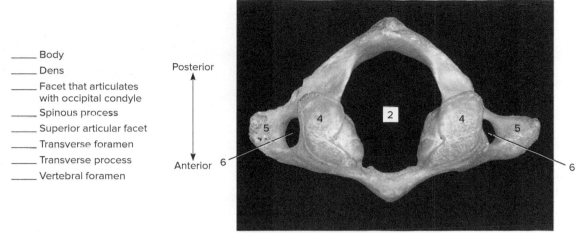

_____ Body
_____ Dens
_____ Facet that articulates
        with occipital condyle
_____ Spinous process
_____ Superior articular facet
_____ Transverse foramen
_____ Transverse process
_____ Vertebral foramen

Posterior

Anterior

Superior view

(b) Axis (C2)

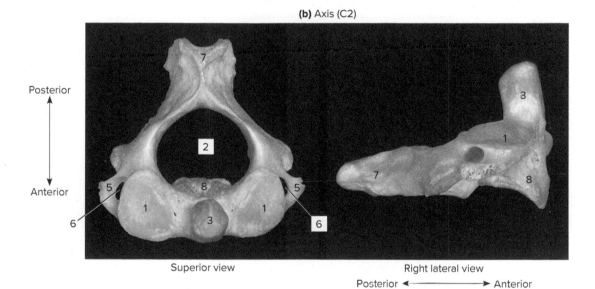

Posterior

Anterior

Superior view

Right lateral view

Posterior ◄────────► Anterior

FIGURE 15.2 Label the superior features of (*a*) the atlas and the superior and right lateral features of (*b*) the axis by placing the correct numbers in the spaces provided. ☑ ☑ APR (a, b) ©J and J Photography

_____ Body
_____ Inferior vertebral notch
_____ Lamina
_____ Pedicle
_____ Spinous process
_____ Superior articular process
_____ Transverse foramen
_____ Transverse process
_____ Vertebral foramen

Superior views   Right lateral views

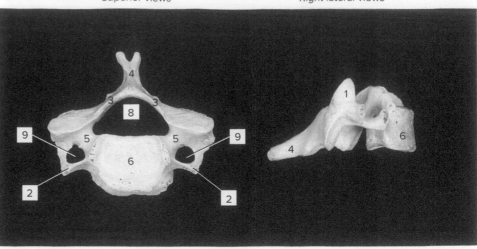

(a) Cervical vertebra

Posterior

Anterior

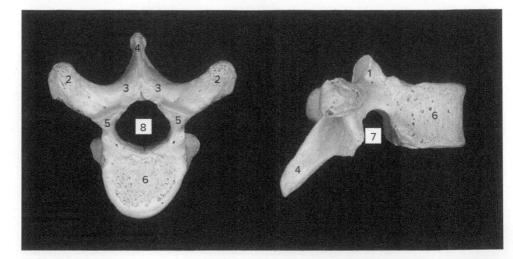

(b) Thoracic vertebra

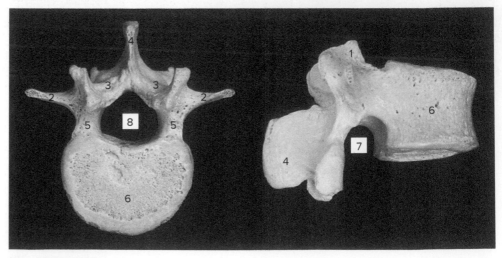

(c) Lumbar vertebra

Posterior ←——→ Anterior

FIGURE 15.3 Label the superior and right lateral features of the (_a_) cervical, (_b_) thoracic, and (_c_) lumbar vertebrae by placing the correct numbers in the spaces provided. 🄰 🄰 **APR** **(a–c)** ©J and J Photography

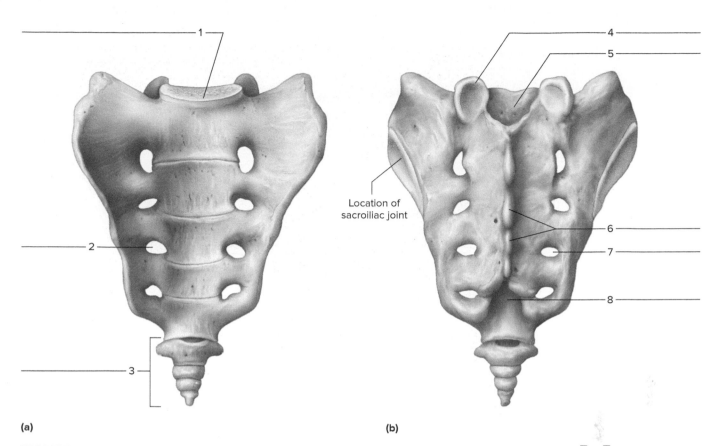

(a)

(b)

FIGURE 15.4 Label the coccyx and the features of the sacrum: (a) anterior view; (b) posterior view. 🅰 🅱 **APR**

4. Compare the available samples of cervical, thoracic, and lumbar vertebrae by noting differences in size and shape and by locating the following features:

vertebral foramen	superior articular
body	processes
pedicles	inferior articular
laminae	processes
spinous process	inferior vertebral
transverse processes	notch
dens of axis	transverse foramina
	facets

5. Examine the sacrum and coccyx. Locate the following features:

sacrum

    superior articular process

    posterior sacral foramen

    anterior sacral foramen

    sacral promontory

    sacral canal

    tubercles

    median sacral crest

    sacral hiatus

coccyx

6. Complete Parts A and B of Laboratory Report 15.

EXPLORE

PROCEDURE B—Thoracic Cage

1. Review section 7.8 entitled "Thoracic Cage" in chapter 7 of the textbook.
2. As a review activity, label figure 15.5.
3. Study a typical rib with articulation sites with a thoracic vertebra (fig. 15.6).
4. Examine the thoracic cage of the human skeleton and locate the following bones and features:

rib

    head

    neck

    shaft

    tubercle

    facets

    true ribs (pairs 1–7)

    false ribs (pairs 8–12; includes floating ribs)

        floating ribs (pairs 11–12)

costal cartilages

    (hyaline cartilage)

sternum

    jugular (suprasternal) notch

    manubrium

    sternal angle

    body

    xiphoid process

5. Complete Parts C and D of the laboratory report.

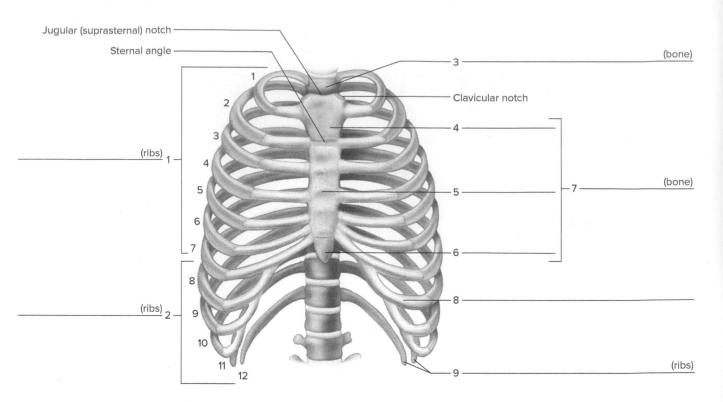

Jugular (suprasternal) notch

Sternal angle

Clavicular notch

3 _____ (bone)

4 _____

5 _____

6 _____

7 _____ (bone)

8 _____

9 _____ (ribs)

(ribs) 1

(ribs) 2

1
2
3
4
5
6
7
8
9
10
11
12

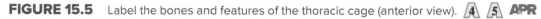

FIGURE 15.5 Label the bones and features of the thoracic cage (anterior view). 🄰 🄵 **APR**

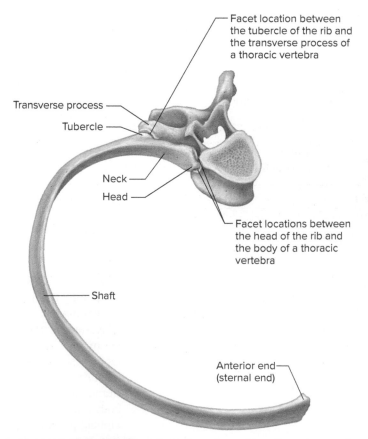

Facet location between the tubercle of the rib and the transverse process of a thoracic vertebra

Transverse process

Tubercle

Neck

Head

Facet locations between the head of the rib and the body of a thoracic vertebra

Shaft

Anterior end (sternal end)

FIGURE 15.6 Superior view of a typical rib with the articulation sites with a thoracic vertebra.

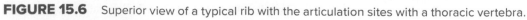

Name _____

Date _____

Section _____

The ⃝A corresponds to the indicated Learning Outcome(s) found at the beginning of the laboratory exercise.

Vertebral Column and Thoracic Cage

PART A ASSESSMENTS

Complete the following statements:

1. The vertebral column encloses and protects the _____. ⃝1

2. The _____ of the vertebrae support the weight of the head and trunk. ⃝1

3. The _____ separate adjacent vertebrae, and they soften the forces created by walking. ⃝1

4. The intervertebral foramina provide passageways for _____. ⃝1

5. Transverse foramina of cervical vertebrae serve as passageways for _____ leading to the brain. ⃝2

6. The first vertebra is also called the _____. ⃝3

7. When the head is moved from side to side, the first vertebra pivots around the _____ of the second vertebra. ⃝2

8. The _____ vertebrae have the largest and strongest bodies. ⃝3

9. The typical number of vertebrae that fuse in the adult to form the sacrum is _____. ⃝1

10. An opening called the _____ exists at the tip of the sacral canal. ⃝1

PART B ASSESSMENTS

Based on your observations, compare typical cervical, thoracic, and lumbar vertebrae in relation to the characteristics indicated in the table. The table is partly completed. For your responses, consider characteristics such as size, shape, presence or absence, and unique features. ⃝2 ⃝3

Vertebra	Number	Size	Body	Spinous Process	Transverse Foramina
Cervical	7		Smallest	C2 through C6 are forked (bifid)	
Thoracic		Intermediate			
Lumbar					Absent

PART C ASSESSMENTS

Complete the following statements:

1. The last two pairs of ribs that have no cartilaginous attachments to the sternum are sometimes called
   _____ ribs. ⑤

2. There are _____ pairs of true ribs. ⑤

3. The manubrium articulates with the _____ on its superior border. ④

4. List three general functions of the thoracic cage. ④
   a. _____
   b. _____
   c. _____

PART D ASSESSMENTS

Identify the bones and features indicated in the radiograph of the neck in figure 15.7. ① ②

Terms:
Atlas
Axis
Body
Intervertebral disc
Spinous process
Transverse process

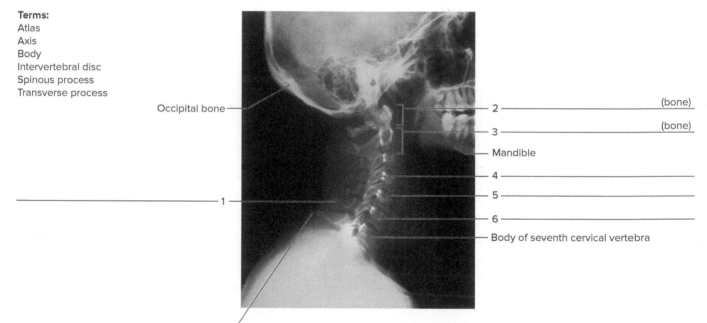

FIGURE 15.7 Using the terms provided, identify the bones and features indicated in this radiograph of the neck (lateral view). **APR** ©Stockbyte/PunchStock

16

Pectoral Girdle and Upper Limb

PURPOSE OF THE EXERCISE

To examine the bones of the pectoral girdle and upper limb and to identify the major features of these bones.

LEARNING OUTCOMES A&PR

After completing this exercise, you should be able to

1. Locate and identify the bones of the pectoral girdle and their major features.
2. Locate and identify the bones of the upper limb and their major features.
3. Recognize the anatomical terms pertaining to the pectoral girdle and upper limb.

The pectoral girdle (shoulder girdle) consists of two clavicles and two scapulae. These parts support the upper limbs and serve as attachments for various muscles that move these limbs.

Each upper limb includes a humerus, a radius, an ulna, eight carpals, five metacarpals, and fourteen phalanges. These bones form the framework of the arm, forearm, and hand. They also function as parts of levers when the limbs are moved.

EXPLORE

PROCEDURE A—Pectoral Girdle

1. Review section 7.9 entitled "Pectoral Girdle" in chapter 7 of the textbook.

2. As a review activity, label figures 16.1 and 16.2.
3. Examine the bones of the pectoral girdle and locate the following features. At the same time, locate as many of the corresponding surface bones and features of your own skeleton as possible.

clavicle
    medial (sternal) end
    lateral (acromial) end

scapula
    spine
    acromion process
    coracoid process
    glenoid cavity
    borders
        superior border
        medial (vertebral) border
        lateral (axillary) border
    fossae
        supraspinous fossa
        infraspinous fossa

4. Complete Part A of Laboratory Report 16.

Critical Thinking Application

Why is the clavicle a bone that can easily fracture?

_____

_____

_____

_____

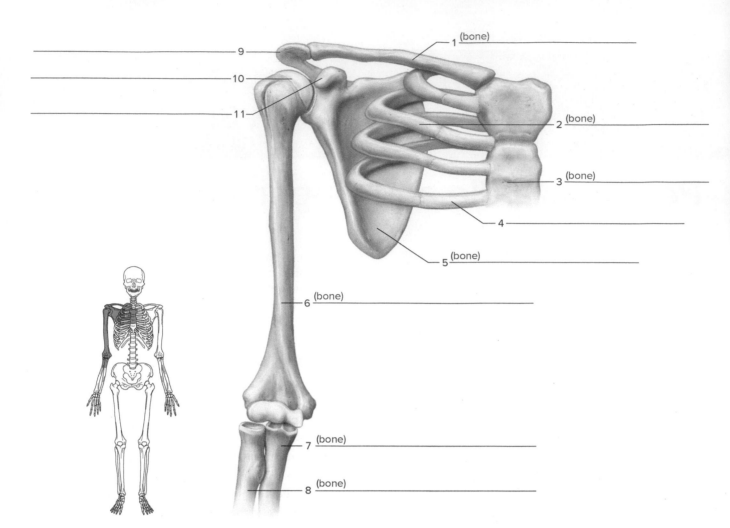

1 (bone) _____

2 (bone) _____

3 (bone) _____

4 _____

5 (bone) _____

6 (bone) _____

7 (bone) _____

8 (bone) _____

9 _____

10 _____

11 _____

FIGURE 16.1   Label the bones and features of the right shoulder and upper limb (anterior view). A̱ 2̱ APR

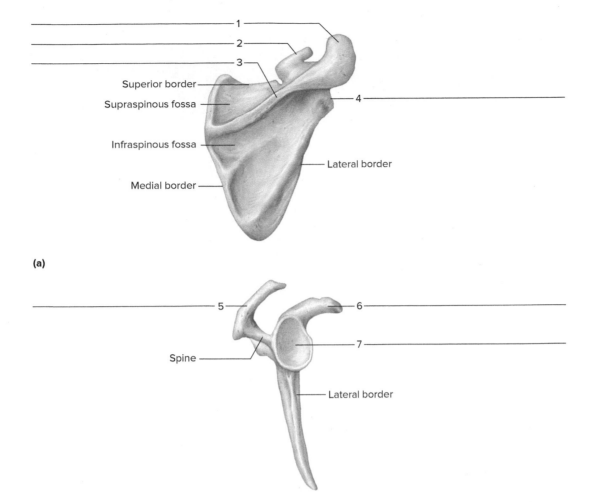

Superior border

Supraspinous fossa

Infraspinous fossa

Lateral border

Medial border

(a)

Spine

Lateral border

(b)

FIGURE 16.2   Label (*a*) the posterior surface of the right scapula and (*b*) the lateral aspect of the right scapula.

EXPLORE

PROCEDURE B—Upper Limb

1. Review section 7.10 entitled "Upper Limb" in chapter 7 of the textbook.
2. As a review activity, label figures 16.3, 16.4, and 16.5.
3. Examine the following bones and features of the upper limb:

humerus
    proximal features
        head
        greater tubercle
        lesser tubercle
        anatomical neck
        surgical neck
        intertubercular sulcus
    shaft
        deltoid tuberosity

    distal features
        capitulum
        trochlea
        medial epicondyle
        lateral epicondyle
        coronoid fossa
        olecranon fossa
radius
    head
    radial tuberosity
    styloid process
ulna
    trochlear notch (semilunar notch)
    olecranon process
    coronoid process
    styloid process
    head

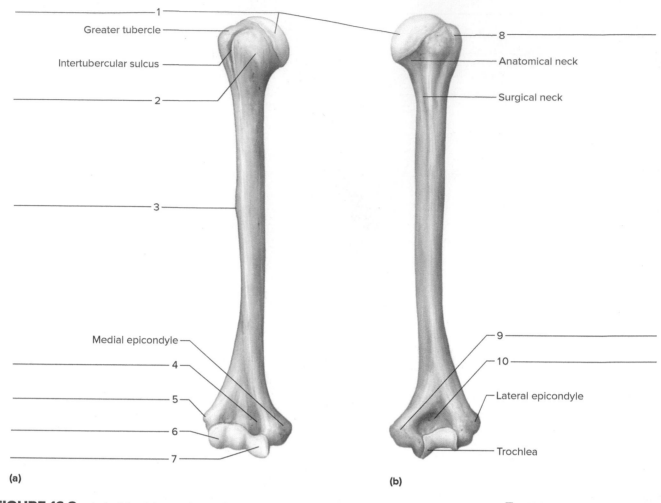

Greater tubercle ———— 1

Intertubercular sulcus ————

———— 2

———— 3

Medial epicondyle ————

———— 4

———— 5

———— 6

———— 7

(a)

8 ————

Anatomical neck

Surgical neck

9 ————

10 ————

Lateral epicondyle

Trochlea

(b)

FIGURE 16.3 Label the (*a*) anterior surface and (*b*) posterior surface of the right humerus. APR

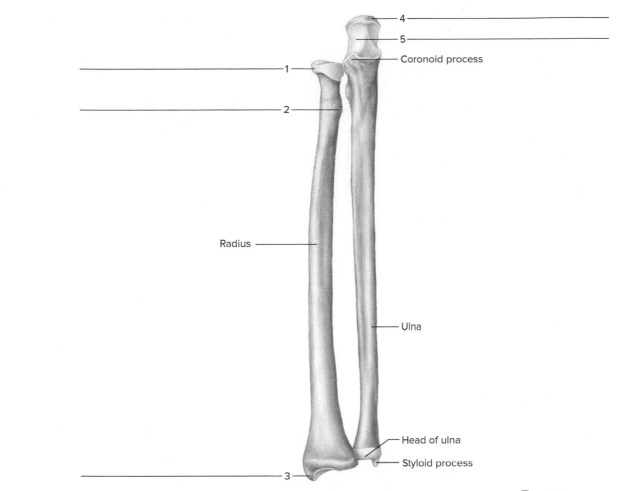

4 ———————————————

5 ———————————————

— Coronoid process

1 —

2 —

Radius ———————

— Ulna

— Head of ulna

— Styloid process

3 —

FIGURE 16.4 Label the major anterior features of the right radius and ulna (anterior view). APR

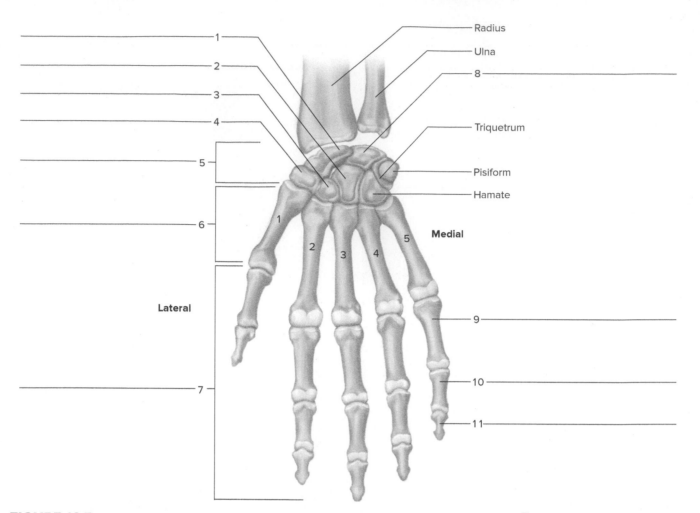

Radius

Ulna

8

Triquetrum

Pisiform

Hamate

Medial

Lateral

FIGURE 16.5 Label the bones and groups of bones in this anterior view of the right hand. **2** **APR**

carpal bones

    proximal row (listed lateral to medial)

        scaphoid

        lunate

        triquetrum

        pisiform

    distal row (listed medial to lateral)

        hamate

        capitate

        trapezoid

        trapezium

The following mnemonic device will help you learn the eight carpals:

So **L**ong **T**op **P**art
Here **C**omes **T**he **T**humb

The first letter of each word corresponds to the first letter of a carpal. This device arranges the carpals in order for

the proximal, transverse row of four bones from lateral to medial, followed by the distal, transverse row from medial to lateral, which ends nearest the thumb. This arrangement assumes the anatomical position of the hand.

    metacarpal bones

    phalanges

        proximal phalanx

        middle phalanx

        distal phalanx

4. Complete Parts B, C, D, and E of the laboratory report.

LEARNING EXTENSION

Use different colored pencils to distinguish the individual bones in figure 16.5.

Name _____

Date _____

Section _____

The  corresponds to the indicated Learning Outcome(s) found at the beginning of the laboratory exercise.

Pectoral Girdle and Upper Limb

PART A ASSESSMENTS

Complete the following statements:

1. The pectoral girdle is an incomplete ring because it is open in the back between the _____.

2. The medial end of a clavicle articulates with the _____ of the sternum.

3. The lateral end of a clavicle articulates with the _____ process of the scapula.

4. The _____ divides the posterior side of the scapula into unequal portions.

5. The lateral tip of the shoulder is the _____ process of the scapula.

6. Near the lateral end of the scapula, the _____ process curves anteriorly and inferiorly from the clavicle.

7. The glenoid cavity of the scapula articulates with the _____ of the humerus.

PART B ASSESSMENTS

Match the bones in column A with the features in column B. Place the letter of your choice in the space provided.

Column A	Column B
a. Carpals	_____ 1. Capitate
b. Humerus	_____ 2. Coronoid fossa
c. Metacarpals	
d. Phalanges	_____ 3. Deltoid tuberosity
e. Radius	_____ 4. Greater tubercle
f. Ulna	_____ 5. Intertubercular sulcus
	_____ 6. Lunate
	_____ 7. Olecranon fossa
	_____ 8. Five palmar bones
	_____ 9. Radial tuberosity
	_____ 10. Trapezium
	_____ 11. Trochlear notch
	_____ 12. Fourteen bones in digits

Identify the bones and features indicated in the radiographs of figures 16.6, 16.7, and 16.8. 🄰 🄰

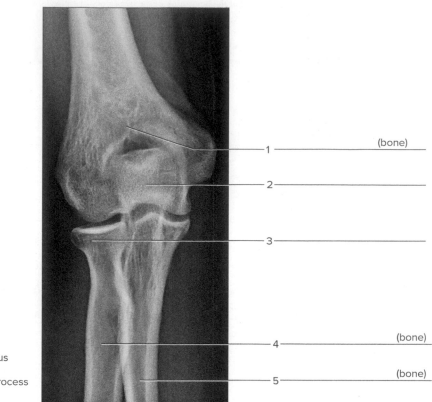

1 ——————————————————— (bone)

2 ———————————————————

3 ———————————————————

4 ——————————————————— (bone)

5 ——————————————————— (bone)

Terms:
Head of radius
Humerus
Olecranon process
Radius
Ulna

FIGURE 16.6 Using the terms provided, identify the bones and features indicated on this radiograph of the right elbow (anterior view). **APR** ©Dale Butler

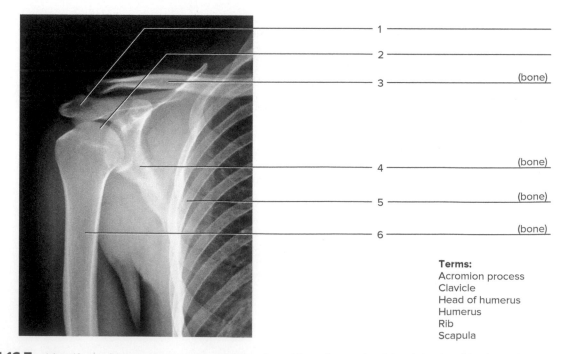

1 ———————————————————

2 ———————————————————

3 ——————————————————— (bone)

4 ——————————————————— (bone)

5 ——————————————————— (bone)

6 ——————————————————— (bone)

Terms:
Acromion process
Clavicle
Head of humerus
Humerus
Rib
Scapula

FIGURE 16.7 Identify the bones and features indicated on this radiograph of the right shoulder (anterior view), using the terms provided. **APR** sanyanwuji/Getty Images

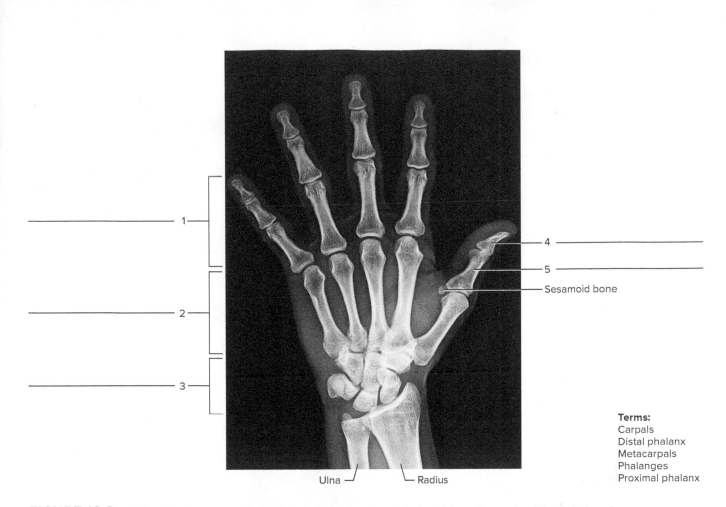

1 _____

2 _____

3 _____

4 _____

5 _____

Sesamoid bone

Ulna   Radius

Terms:
Carpals
Distal phalanx
Metacarpals
Phalanges
Proximal phalanx

FIGURE 16.8 Using the terms provided, identify the bones indicated on this radiograph of the right hand (anterior view). **APR** ©Dale Butler

Complete the labeling of the bones of the hand in figure 16.9.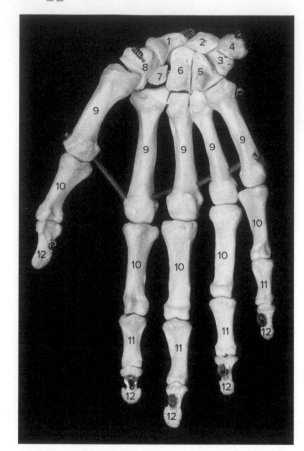

_____	Capitate
_____	Distal phalanges
_____	Hamate
_____	Lunate
_____	Metacarpals
_____	Middle phalanges
___4___	Pisiform
_____	Proximal phalanges
_____	Scaphoid
_____	Trapezium
_____	Trapezoid
___3___	Triquetrum

FIGURE 16.9 Label the bones numbered on this anterior view of the right hand by placing the rest of the correct numbers in the spaces provided.
©J and J Photography

PART E ASSESSMENTS

Locate the 21 anatomical terms pertaining to the "Pectoral Girdle and Upper Limb." 3

E	T	A	M	A	H	S	U	I	D	A	R
M	A	R	D	U	B	E	N	I	P	S	E
U	E	K	I	L	I	P	O	F	U	S	L
L	L	E	O	N	X	Z	H	C	Z	S	C
U	H	V	H	A	E	V	E	A	P	E	R
T	C	O	P	P	Q	F	B	P	H	C	E
I	O	O	A	S	S	O	F	I	A	O	B
P	R	R	C	D	A	E	H	T	L	R	U
A	T	G	S	H	A	F	T	A	A	P	T
C	O	R	A	C	O	I	D	T	N	G	K
H	U	M	E	R	U	S	T	E	X	J	M
P	N	E	C	K	A	L	U	P	A	C	S

Find and circle words pertaining to Laboratory Exercise 16. The words may be horizontal, vertical, diagonal, or backward.

10 bones of the pectoral girdle and upper limb

11 features of bones of the pectoral girdle and upper limb

17

Pelvic Girdle and Lower Limb

MATERIALS NEEDED

Textbook
Human skeleton, articulated
Human skeleton, disarticulated
Male and female pelves

For Learning Extension:
Colored pencils

PURPOSE OF THE EXERCISE

To examine the bones of the pelvic girdle and lower limb and to identify the major features of these bones.

LEARNING OUTCOMES A&PR

After completing this exercise, you should be able to

1. Locate and identify the bones of the pelvic girdle and their major features.
2. Locate and identify the bones of the lower limb and their major features.
3. Recognize the anatomical terms pertaining to the pelvic girdle and lower limb.

The pelvic girdle includes two coxal (hip) bones that articulate with each other anteriorly at the pubic symphysis and posteriorly with the sacrum. Together, the pelvic girdle, sacrum, and coccyx constitute the pelvis. The pelvis, in turn, provides support for the trunk of the body and provides attachments for the lower limbs.

The bones of the lower limb form the framework of the thigh, leg, and foot. Each limb includes a femur, a patella, a tibia, a fibula, seven tarsals, five metatarsals, and fourteen phalanges.

EXPLORE

PROCEDURE A—Pelvic Girdle

1. Review section 7.11 entitled "Pelvic Girdle" in chapter 7 of the textbook.

2. As a review activity, label figures 17.1 and 17.2.
3. Examine the bones of the pelvic girdle and locate the following:

coxal bone (hip bone; pelvic bone; innominate bone)
    acetabulum
    ilium
    iliac crest
    sacroiliac joint
    anterior superior iliac spine
    greater sciatic notch
    ischium
    ischial tuberosity
    ischial spine
    pubis
    pubic symphysis
    pubic arch
    obturator foramen

4. Complete Part A of Laboratory Report 17.

 Critical Thinking Application

Examine the male and female pelves. Look for major differences between them. Note especially the flare of the iliac bones, the angle of the pubic arch, the distance between the ischial spines and ischial tuberosities, and the curve and width of the sacrum. In what ways are the differences you observed related to the function of the female pelvis as a birth canal?

_____
_____
_____
_____
_____

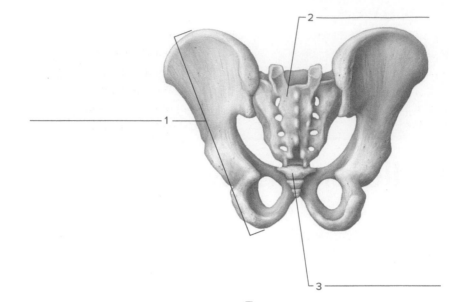

FIGURE 17.1 Label the bones of the pelvis (posterior view). ⒜

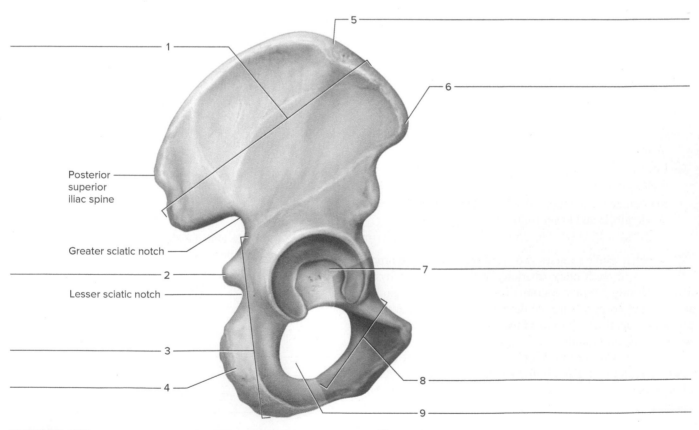

Posterior
superior
iliac spine

Greater sciatic notch

Lesser sciatic notch

FIGURE 17.2 Label the lateral features of the right coxal bone. ⒜ **APR**

 **EXPLORE**

PROCEDURE B—Lower Limb

1. Review section 7.12 entitled "Lower Limb" in chapter 7 of the textbook.
2. As a review activity, label figures 17.3, 17.4, and 17.5.
3. Examine the bones of the lower limb and locate each of the following:

femur
    proximal features
        head
        fovea capitis
        neck
        greater trochanter
        lesser trochanter
    shaft
        gluteal tuberosity
        linea aspera
    distal features
        lateral epicondyle
        medial epicondyle
        lateral condyle
        medial condyle

patella
tibia
    medial condyle
    lateral condyle
    tibial tuberosity
    anterior crest (border)
    medial malleolus
fibula
    head
    lateral malleolus
tarsal bones
    talus
    calcaneus
    navicular
    cuboid
    lateral cuneiform
    intermediate cuneiform
    medial cuneiform
metatarsal bones
phalanges
    proximal phalanx
    middle phalanx
    distal phalanx

4. Complete Parts B, C, D, and E of the laboratory report.

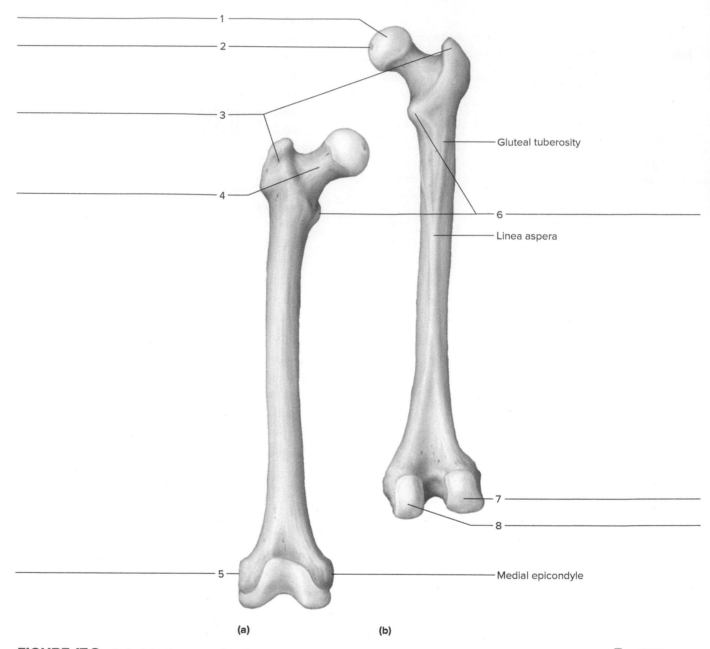

1 _____

2 _____

3 _____

4 _____

Gluteal tuberosity

6 _____

Linea aspera

7 _____

8 _____

5 _____

Medial epicondyle

(a)   (b)

FIGURE 17.3   Label the features of (*a*) the anterior surface and (*b*) the posterior surface of the right femur.   APR

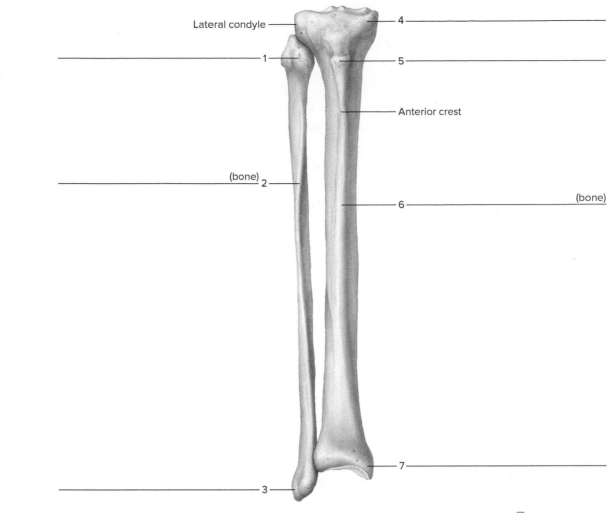

Lateral condyle

(bone)

Anterior crest

(bone)

1

2

3

4

5

6

7

FIGURE 17.4 Label the bones and features of the right tibia and fibula in this anterior view.

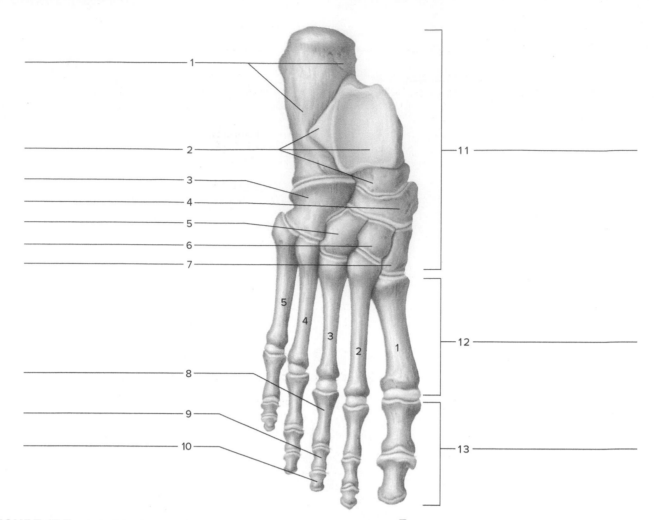

FIGURE 17.5 Label the bones of the superior surface of the right foot. ⚠2

LEARNING EXTENSION

Use different colored pencils to distinguish the individual bones in figure 17.5.

The ⚠ corresponds to the indicated Learning Outcome(s) found at the beginning of the laboratory exercise.

Pelvic Girdle and Lower Limb

♻ PART A ASSESSMENTS

Complete the following statements:

1. The pelvic girdle consists of two _____. ⚠

2. The head of the femur articulates with the _____ of the coxal bone. ⚠

3. The _____ is the largest portion of the coxal bone. ⚠

4. The pubic bones come together anteriorly to form a cartilaginous joint called the _____. ⚠

5. The _____ is the portion of the ilium that causes the prominence of the hip. ⚠

6. When a person sits, the _____ of the ischium supports the weight of the body. ⚠

7. The angle formed by the pubic bones below the pubic symphysis is called the _____. ⚠

8. The _____ is the largest foramen in the skeleton. ⚠

9. The ilium joins the sacrum at the _____ joint. ⚠

♻ PART B ASSESSMENTS

Match the bones in column A with the features in column B. Place the letter of your choice in the space provided. ⚠

Column A	Column B
a. Femur	_____ 1. Middle phalanx
b. Fibula	_____ 2. Lesser trochanter
c. Metatarsals	_____ 3. Medial malleolus
d. Patella	_____ 4. Fovea capitis
e. Phalanges	_____ 5. Calcaneus
f. Tarsals	_____ 6. Lateral cuneiform
g. Tibia	_____ 7. Tibial tuberosity
	_____ 8. Talus
	_____ 9. Cuboid
	_____ 10. Lateral malleolus
	_____ 11. Located in a tendon over the knee
	_____ 12. Five bones that form the instep

Identify the bones and features indicated in the radiographs of figures 17.6, 17.7, and 17.8. ⓘ ②

Terms:
Head of femur
Ilium
Obturator foramen
Pubic symphysis
Pubis
Sacrum

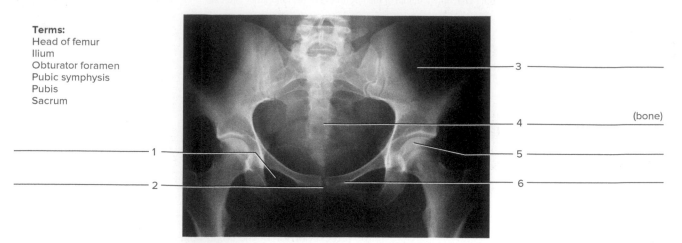

FIGURE 17.6 Using the terms provided, identify the bones and features indicated on this radiograph of the anterior view of the pelvic region. **APR** Courtesy Dale Butler

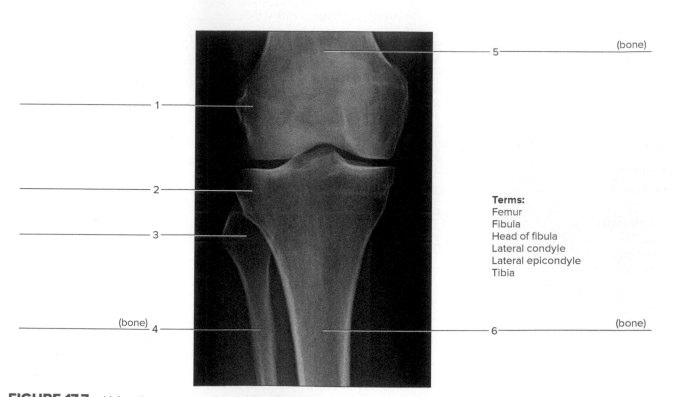

Terms:
Femur
Fibula
Head of fibula
Lateral condyle
Lateral epicondyle
Tibia

FIGURE 17.7 Using the terms provided, identify the bones and features indicated in this radiograph of the right knee (anterior view). **APR** Courtesy Dale Butler

Terms:
Calcaneus
Distal phalanx
Metatarsal
Proximal phalanx
Talus
Tibia

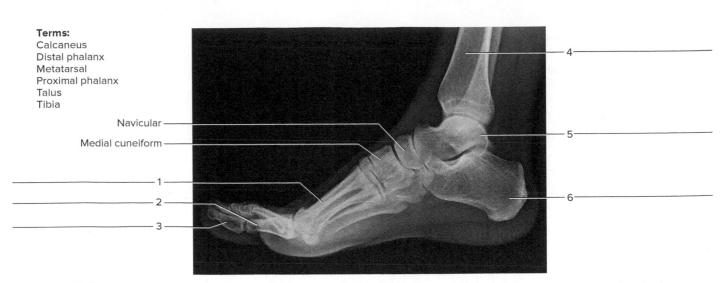

Navicular ───

Medial cuneiform ───

────── 1

────── 2

────── 3

─── 4

─── 5

─── 6

FIGURE 17.8 Using the terms provided, identify the bones indicated on this radiograph of the right foot (medial view).
APR Courtesy Dale Butler

PART D ASSESSMENTS

Identify the bones of the foot in figure 17.9.

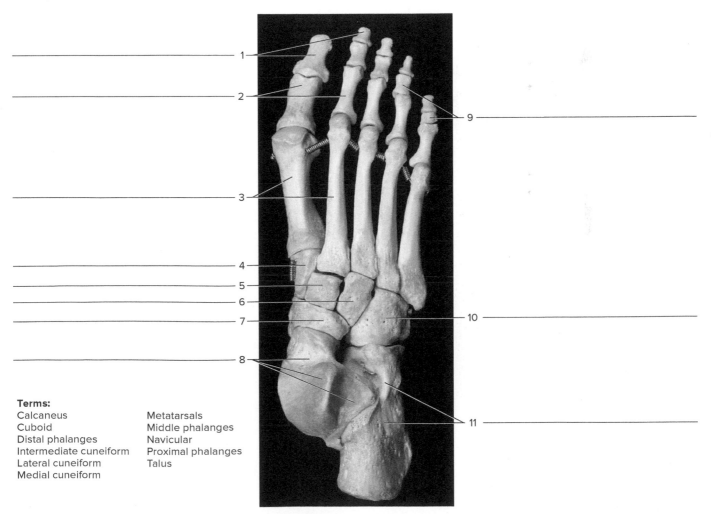

────── 1

────── 2

────── 3

────── 4

────── 5

────── 6

────── 7

────── 8

─── 9

─── 10

─── 11

Terms:

Calcaneus	Metatarsals
Cuboid	Middle phalanges
Distal phalanges	Navicular
Intermediate cuneiform	Proximal phalanges
Lateral cuneiform	Talus
Medial cuneiform	

FIGURE 17.9 Using the terms provided, identify the bones indicated on this superior view of the right foot.
©J and J Photography

127

Locate the 22 anatomical terms pertaining to the "Pelvic Girdle and Lower Limb." 📐

P	L	A	S	R	A	T	A	T	E	M	B
A	C	E	T	A	B	U	L	U	M	Q	H
T	E	L	Y	D	N	O	C	B	T	S	C
E	A	T	A	L	U	S	W	E	O	U	U
L	E	D	B	N	K	H	C	R	A	L	N
L	N	E	C	K	A	E	V	O	F	O	E
A	I	L	I	U	M	D	P	S	E	E	I
T	L	X	R	S	I	W	D	I	M	L	F
Z	I	U	I	O	W	H	A	T	U	L	O
H	V	B	B	M	Y	W	E	Y	R	A	R
I	U	U	I	S	C	H	I	U	M	M	
P	C	H	Y	A	F	O	R	A	M	E	N

Find and circle words pertaining to Laboratory Exercise 17. The words may be horizontal, vertical, diagonal, or backward.

9 bones of the pelvic girdle and lower limb

13 features of the pelvic girdle and lower limb

Joint Structure and Movements

PURPOSE OF THE EXERCISE

To examine examples of the three types of joints, to identify the major features of these joints, and to review the types of movements produced at synovial joints.

LEARNING OUTCOMES APR

After completing this exercise, you should be able to

1. Identify key structural and functional characteristics of fibrous, cartilaginous, and synovial joints.
2. Locate examples of each type of joint.
3. Demonstrate the types of movements that occur at synovial joints.
4. Examine the structure of the knee joint.

Joints are junctions between bones. Although they vary considerably in structure, they can be classified according to the type of tissue that binds the bones together. Thus, the three groups of *structural joints* can be identified as *fibrous joints, cartilaginous joints,* and *synovial joints.*

Fibrous joints are filled with dense fibrous connective tissue, cartilaginous joints are filled with a type of cartilage, and synovial joints contain synovial fluid inside a joint cavity.

Functional joints can also be classified by the degree of movement allowed: *synarthroses* are immovable, *amphiarthroses* allow slight movement, and *diarthroses* allow free movement. Movements occurring at freely movable synovial joints are due to the contractions of skeletal muscles. In each case, the type of movement depends on the type of joint involved and the way in which the muscles are attached to the bones on either side of the joint.

EXPLORE

PROCEDURE A—Types of Joints APR

1. Review section 7.13 entitled "Joints" in chapter 7 of the textbook.
2. Study the knee joint model and locate the structures indicated in figure 18.1.

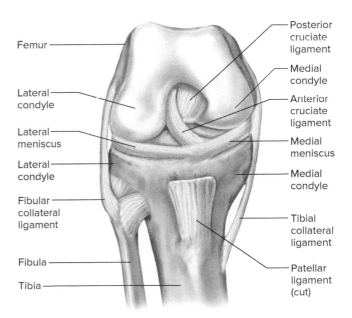

FIGURE 18.1 Anterior view of right bent knee (patella removed). APR

3. Examine the human skull and articulated skeleton and locate examples of the following types of structural and functional joints:

fibrous joints

> suture of skull—synarthrotic joint
>
> tibiofibular joint—amphiarthrotic joint

cartilaginous joints

> intervertebral disc—amphiarthrotic joint
>
> pubic symphysis—amphiarthrotic joint

4. Locate examples of the following types of synovial joints (diarthrotic joints) in the skeleton and in your own body. Experiment with each joint to experience its range of movements. ⒶⒶ

synovial joints

> ball-and-socket joint—hip; shoulder
>
> condylar (ellipsoidal) joint—radiocarpal; meta-carpophalangeal
>
> plane (gliding) joint—intercarpal; intertarsal
>
> hinge joint—elbow; knee; interphalangeal
>
> pivot joint—radioulnar at elbow; dens at atlas
>
> saddle joint—base of thumb with trapezium

5. Complete Parts A, B, and C of Laboratory Report 18.

DEMONSTRATION

Examine a longitudinal section of a fresh synovial animal knee joint. Locate the dense connective tissue that forms the joint capsule and the hyaline cartilage that forms the articular cartilage on the ends of the bones. Locate the synovial membrane on the inner surface of the joint capsule. Does the joint have any semilunar cartilages (menisci)? ⑤

_____

_____

What is the function of such cartilages?

_____

EXPLORE

PROCEDURE B—Joint Movements

1. Review the section entitled "Types of Joint Movements" in chapter 7 of the textbook and study figures 18.2 and 18.3.

2. Most joints and body parts are extended and/or adducted when the body is positioned in anatomical position. Skeletal muscle action involves the more movable end (*insertion*) being pulled toward the more stationary end (*origin*). In the limbs, the origin is usually proximal to the insertion; in the trunk, the origin is usually medial to the insertion. Use these concepts as reference points as you move joints. Move various parts of your body to demonstrate the synovial joint movements described in tables 18.1 and 18.2. Ⓐ

3. Have your laboratory partner do some of the movements and see if you can correctly identify the movements made. Ⓐ

Critical Thinking Application

Describe a body position that can exist when all major body parts are flexed.

_____

_____

_____

_____

4. Complete Part D of the laboratory report.

DEMONSTRATION

Study the available radiographs of joints by holding them in front of a light source. Identify the type of joint and the bones incorporated in the joint. Also identify other major visible features.

TABLE 18.1   Angular and Rotational Movements

Movement	Description
Flexion	Decrease of an angle (usually in the sagittal plane)
Extension	Increase of an angle (usually in the sagittal plane)
Hyperextension	Extension beyond anatomical position
Abduction	Movement of a body part away from the midline (usually in the frontal plane)
Adduction	Movement of a body part toward the midline (usually in the frontal plane)
Circumduction	Circular movement (combines flexion, abduction, extension, and adduction)
Rotation	Movement of part around its long axis
Medial (internal)	Inward rotation
Lateral (external)	Outward rotation

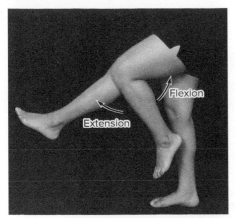

(a) Flexion and extension of the knee joint

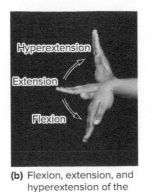

(b) Flexion, extension, and hyperextension of the hand at the wrist joint

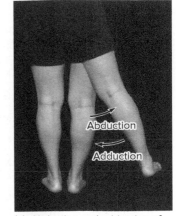

(c) Abduction and adduction of the hip joint

(d) Circumduction of the shoulder joint

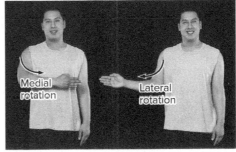

(e) Medial and lateral rotation of the arm at the shoulder joint

FIGURE 18.2 Examples of angular and rotational movements of synovial joints (*a–e*). **APR** **(a–e):** ©J and J Photography

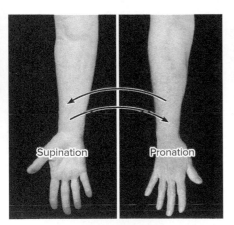

(a) Supination and pronation of the forearm and hand involving movement at the radioulnar joint

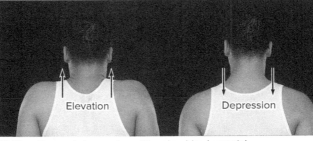

(b) Elevation and depression of the shoulder (scapula)

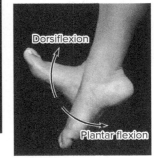

(c) Dorsiflexion and plantar flexion of the foot at the ankle joint

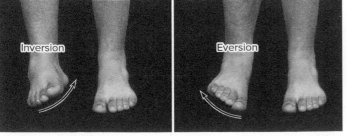

(d) Inversion and eversion of the right foot at the ankle joint (The left foot is unchanged for comparison.)

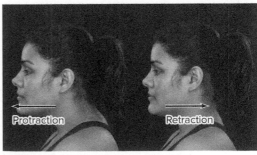

(e) Protraction and retraction of the head

FIGURE 18.3 Special movements of synovial joints (*a–e*). **APR** **(a–e):** ©J and J Photography

TABLE 18.2 Special Movements (pertaining to specific joints)

Movement	Description
Supination	Rotation of forearm and palm of hand anteriorly or upward
Pronation	Rotation of forearm and palm of hand posteriorly or downward
Elevation	Movement of body part upward
Depression	Movement of body part downward
Dorsiflexion	Movement of ankle joint so dorsum (superior) of foot becomes closer to anterior surface of leg (as standing on heels)
Plantar flexion	Movement of ankle joint so the plantar surface of foot becomes closer to the posterior surface of leg (as standing on toes)
Inversion (called supination in some health professions)	Medial movement of sole of foot at ankle joint
Eversion (called pronation in some health professions)	Lateral movement of sole of foot at ankle joint
Protraction	Anterior movement in the transverse plane
Retraction	Posterior movement in the transverse plane

Name _____

Date _____

Section _____

The  corresponds to the indicated Learning Outcome(s) found at the beginning of the laboratory exercise.

LABORATORY
REPORT

18

Joint Structure and Movements

PART A ASSESSMENTS

Complete the following statements:

1. A _____ of the skull is an example of an immovable (synarthrotic) fibrous joint. 🔼2

2. The structural type of joint at the distal tibia and fibula is _____ . 🔼2

3. Intervertebral discs are composed of _____ . 🔼1

4. _____ are the most common type of structural joints in the human body. 🔼1

5. _____ acts as a joint lubricant in synovial joints. 🔼1

6. The discs of fibrocartilage between the articulating surfaces of the knee are called _____ . 🔼5

7. The pubic symphysis is an example of a _____ structural joint. 🔼2

8. Articular cartilage is composed of _____ tissue. 🔼1

PART B ASSESSMENTS

Match the types of synovial joints in column A with the examples in column B. Place the letter of your choice in the space provided. 🔼3

Column A	Column B
a. Ball-and-socket	_____ 1. Hip joint
b. Condylar (ellipsoidal)	_____ 2. Metacarpal-phalanx
c. Hinge	_____ 3. Proximal radius-ulna
d. Pivot	_____ 4. Humerus-ulna of the elbow joint
e. Plane (gliding)	_____ 5. Phalanx-phalanx
f. Saddle	_____ 6. Shoulder joint
	_____ 7. Tarsal-tarsal
	_____ 8. Carpal-metacarpal of the thumb
	_____ 9. Carpal-carpal

PART C ASSESSMENTS

Identify the types of structural and functional joints numbered in figure 18.4. **1** **2**

Structural Classification	Functional Classification
1.	
2.	
3.	
4.	
5.	
6.	
7.	
8.	
9.	

PART D ASSESSMENTS

Complete the answers for the types of joint movements numbered in figure 18.5. **4**

1. _____ (of head)
2. _____ (of shoulder)
3. _____ (of shoulder)
4. Supination _____ (of hand at radioulnar joint)
5. _____ (of hand at radioulnar joint)
6. _____ (of arm at shoulder)
7. Adduction _____ (of arm at shoulder)
8. _____ (of wrist)
9. Extension _____ (of wrist)
10. _____ (of thigh at hip)
11. _____ (of thigh at hip)
12. _____ (of lower limb at hip)
13. _____ (of chin/mandible)
14. _____ (of chin/mandible)

15. _____ (of vertebral column)
16. _____ (of vertebral column)
17. _____ (of head and neck)
18. _____ (of head and neck)
19. Flexion _____ (of arm at shoulder)
20. _____ (of arm at shoulder)
21. Flexion _____ (of elbow)
22. _____ (of elbow)
23. _____ (of thigh at hip)
24. _____ (of thigh at hip)
25. _____ (of knee)
26. Extension _____ (of knee)
27. _____ (of foot at ankle)
28. _____ (of foot at ankle)

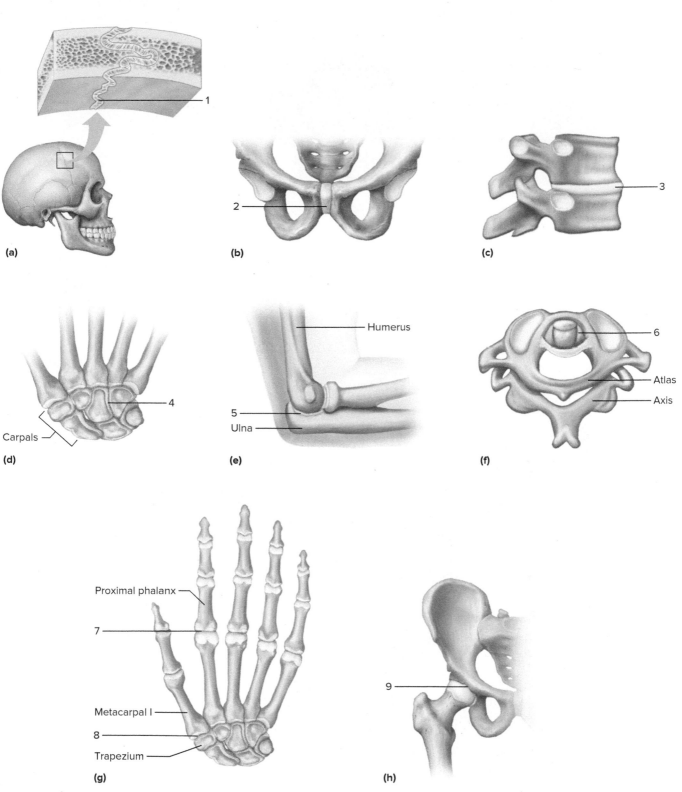

FIGURE 18.4 Identify the types of structural and functional joints numbered in these illustrations (*a–h*).

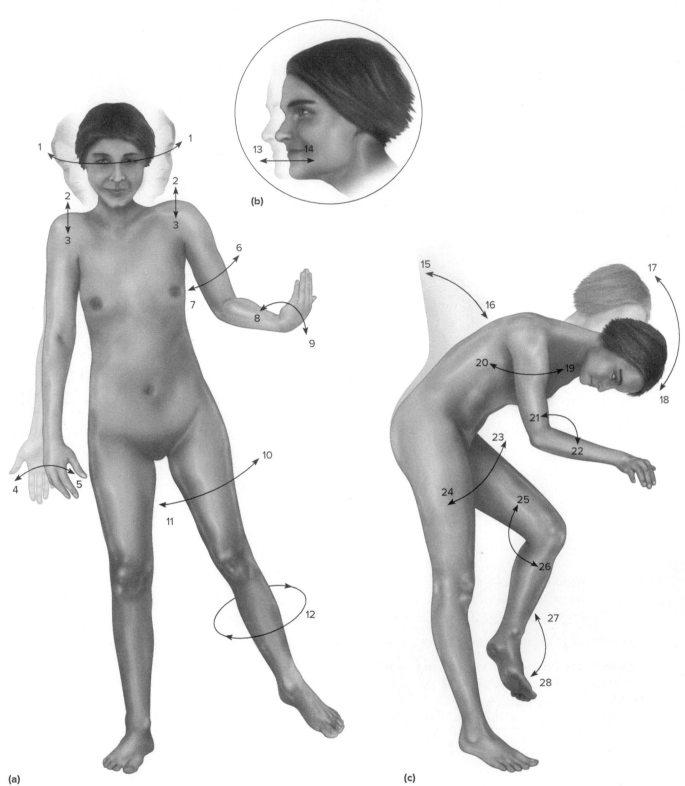

FIGURE 18.5 Identify each of the types of movements numbered and illustrated: (*a*) anterior view; (*b*) lateral view of head; (*c*) lateral view.

19

Skeletal Muscle Structure and Function

MATERIALS NEEDED

Textbook
Compound light microscope
Prepared microscope slide of skeletal muscle tissue
   (longitudinal section and cross section)
Human muscular model
Model of skeletal muscle fiber

For Demonstration:
Fresh round beefsteak

⚠ SAFETY

- Wear disposable gloves when handling the fresh beefsteak.
- Wash your hands before leaving the laboratory.

PURPOSE OF THE EXERCISE

To review the structure of a skeletal muscle.

LEARNING OUTCOMES APR

After completing this exercise, you should be able to

1. Locate the major structures of a skeletal muscle fiber on a microscope slide of skeletal muscle tissue and on a model.
2. Describe how connective tissue is associated with muscle tissue within a skeletal muscle.
3. Distinguish between the origin and insertion of a muscle.
4. Describe and demonstrate the general actions of agonists, synergists, and antagonists.

A skeletal muscle represents an organ of the muscular system and is composed of several types of tissues. These tissues include skeletal muscle tissue, nervous tissue, blood, and various connective tissues.

Each skeletal muscle is encased within and permeated by connective tissue sheaths. The connective tissue often projects beyond the end of a muscle, providing an attachment to other muscles or to bones as a tendon. The connective tissues also extend into the structure of a muscle and separate it into compartments. The outer layer, called *epimysium,* extends deeper into the muscle as *perimysium,* containing bundles of cells (fascicles), and farther inward extends around each muscle cell (fiber) as a thin *endomysium.* The connective tissues provide support and reinforcement during muscular contractions and allow portions of a muscle to contract somewhat independently. Some of the collagen fibers are continuous with the tendon, periosteum, and inward extensions as bone fibers, allowing a strong structural continuity.

Muscles are named according to their location, size, action, shape, attachments, number of origins, or the direction of the fibers. Examples of how muscles are named include these: gluteus maximus (location and size); adductor longus (action and size); sternocleidomastoid (attachments); serratus anterior (shape and location); biceps brachii (two origins and location); and orbicularis oculi (direction of fibers and location).

⟳ EXPLORE

PROCEDURE—Skeletal Muscle Structure and Function

1. Review the concept heading of "Skeletal muscle tissue" in section 5.5 in chapter 5 of the textbook.
2. Examine the microscopic structure of a longitudinal section of skeletal muscle by observing a prepared microscope slide of this tissue. Use figure 19.1 of skeletal muscle tissue to locate the following features:
   skeletal muscle fiber (cell)
   nuclei
   striations (alternating light and dark)
3. Examine the microscopic structure of a cross section of skeletal muscle by observing a prepared microscope slide of this tissue. Use figure 19.2 of skeletal muscle tissue to locate the following features:
   connective tissue
      perimysium
      endomysium

fascicle

muscle cell

    nucleus

    myofibrils

4. Review section 8.2 entitled "Structure of a Skeletal Muscle" in chapter 8 of the textbook.
5. As a review activity, label figure 19.3 and study figure 19.4.
6. Examine the human torso model and locate examples of fascia, tendons, and aponeuroses. An origin tendon is attached to a less movable end of the muscle, whereas the insertion tendon is attached to a more movable end of the muscle. Sheets of connective tissue, called aponeuroses, also serve for some muscle attachments. Locate examples of tendons in your body.

DEMONSTRATION

Examine the fresh round beefsteak. It represents a cross section through the beef thigh muscles. Note the white lines of connective tissue that separate the individual skeletal muscles. Also note how the connective tissue extends into the structure of a muscle and separates it into small compartments of muscle tissue. Locate the epimysium and the perimysium of an individual muscle.

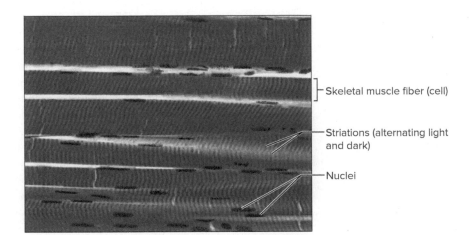

FIGURE 19.1   Micrograph of a longitudinal section of skeletal muscle fibers (cells) (400×). **APR**

Al Telser/McGraw-Hill Education

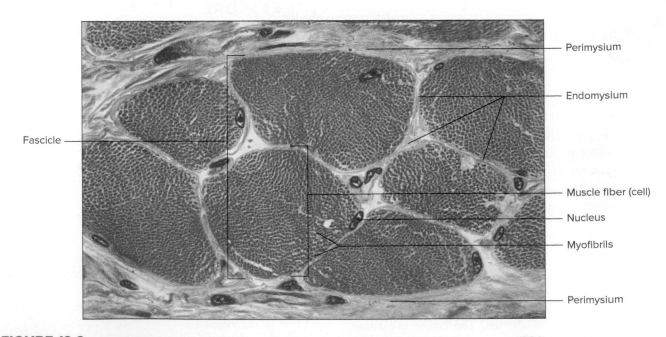

FIGURE 19.2   Micrograph of a cross section of a fascicle and associated connective tissues (800×). ©Ed Reschke

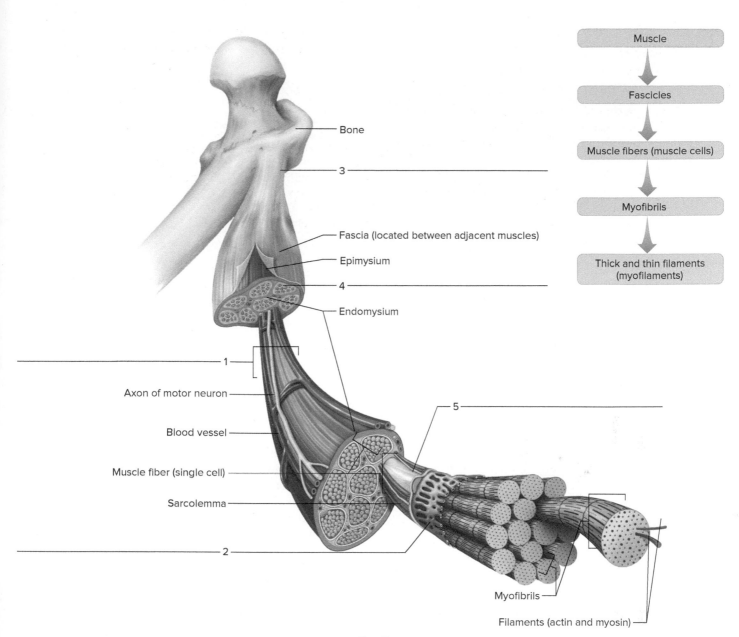

Muscle

↓

Fascicles

↓

Muscle fibers (muscle cells)

↓

Myofibrils

↓

Thick and thin filaments (myofilaments)

Bone

3

Fascia (located between adjacent muscles)

Epimysium

4

Endomysium

1

Axon of motor neuron

Blood vessel

Muscle fiber (single cell)

Sarcolemma

2

5

Myofibrils

Filaments (actin and myosin)

FIGURE 19.3 Label the structures of a skeletal muscle. 🔺1 🔺2

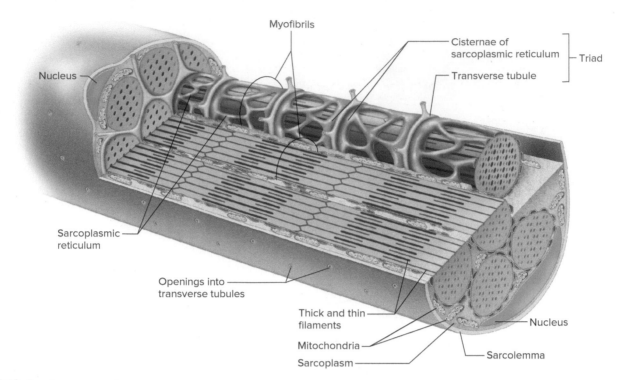

FIGURE 19.4 Structures of a skeletal muscle fiber.

7. Complete Part A of Laboratory Report 19.
8. Examine the model of the skeletal muscle fiber and locate the following:

   sarcolemma

   sarcoplasm

   myofibril

       thick (myosin) filament

       thin (actin) filament

   sarcomere (functional contractile unit)

       A band (dark)

       I band (light)

       Z line (disc)

   sarcoplasmic reticulum

       cisternae

   transverse tubules

9. Complete Part B of the laboratory report.
10. Review section 8.7 entitled "Skeletal Muscle Actions" in chapter 8 of the textbook.
11. Locate the biceps brachii muscle and its origin and insertion in the human torso model and in your body.
12. Make various movements with your upper limb at the shoulder and elbow. For each movement, determine the location of the muscles that function as *agonists* and as *antagonists*. An agonist causes an action, whereas an antagonist works against the action; however, the role of the muscle is dependent on the particular movement involved. Antagonistic muscle pairs pull from opposite sides of the same body region. *Synergistic* muscles often supplement the contraction force of an agonist or by acting to stabilize nearby joints. **The role of a muscle as an agonist, antagonist, or a synergist depends upon the movement under consideration, as the role can change**. These voluntary movements of complex actions are controlled by the nervous system. See table 19.1.
13. Complete Part C of the laboratory report.

TABLE 19.1 Various Roles of Muscles

Functional Category	Description
Agonist*	Muscle responsible for the action (movement)
Antagonist	Muscle responsible for action in the opposite direction of an agonist or for resistance to an agonist
Synergist	Muscle that assists an agonist, often by supplementing the contraction force

* The term *agonist* is often used interchangeably with *prime mover*. When the term *prime mover* is used, it usually refers to the agonist that provides most of the force of the movement of the action involved.

Name _____

Date _____

Section _____

The  corresponds to the indicated Learning Outcome(s) found at the beginning of the laboratory exercise.

Skeletal Muscle Structure and Function

PART A ASSESSMENTS

Match the terms in column A with the definitions in column B. Place the letter of your choice in the space provided. 🄰 🄰

Column A	Column B
a. Endomysium	_____ **1.** Membranous channel extending inward from muscle fiber membrane
b. Epimysium	
c. Fascia	_____ **2.** Cytoplasm of a muscle fiber
d. Fascicle	_____ **3.** Connective tissue located between adjacent muscles
e. Myosin	_____ **4.** Layer of connective tissue that separates a muscle into small bundles called fascicles
f. Perimysium	
g. Sarcolemma	
h. Sarcomere	_____ **5.** Cell membrane of a muscle fiber
i. Sarcoplasm	_____ **6.** Layer of connective tissue that surrounds a skeletal muscle
j. Sarcoplasmic reticulum	_____ **7.** Unit of alternating light and dark striations between Z lines
k. Tendon	_____ **8.** Layer of connective tissue that surrounds an individual muscle fiber
l. Transverse (T) tubule	
	_____ **9.** Cellular organelle in muscle fiber corresponding to the endoplasmic reticulum
	_____ **10.** Cordlike part that attaches a muscle to a bone
	_____ **11.** Protein found within thick filament
	_____ **12.** Small bundle of muscle fibers within a muscle

PART B ASSESSMENTS

Provide the labels for the electron micrograph in figure 19.5. 🄰

1. _____

2. _____

3. _____

4. _____

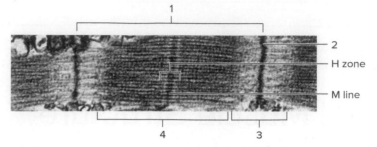

Terms:
A band (dark)
I band (light)
Sarcomere
Z line

1
2
H zone
M line
4   3

FIGURE 19.5 Using the terms provided, identify the bands and lines of the striations of a transmission electron micrograph of relaxed sarcomeres (16,000×). **APR** Biology Pics/Science Source

PART C ASSESSMENTS

Complete the following statements:

1. The _____ of a muscle is usually attached to a more fixed location. ⚠3

2. The _____ of a muscle is usually attached to a more movable location. ⚠3

3. The forearm is flexed at the elbow when the _____ muscle contracts. ⚠4

4. A muscle responsible for an action (movement) is called a(n) _____. ⚠4

5. Assisting muscles of an agonist are called _____. ⚠4

6. Antagonists are muscles that resist the actions of _____ and cause movement in the opposite direction. ⚠4

7. When the forearm is extended at the elbow joint, the _____ muscle acts as the agonist. ⚠4

8. An agonist that provides the most force of a specific action is sometimes referred to as a _____. ⚠4

20

Muscles of the Head and Neck

Textbook
Human torso model with musculature
Human skull
Human skeleton, articulated

For Learning Extension:
Long rubber band

PURPOSE OF THE EXERCISE

To review the locations, actions, origins, and insertions of
the muscles of the face, head, and neck.

LEARNING OUTCOMES A&PR

After completing this exercise, you should be able to

1. Locate and identify the muscles of facial expression,
   the muscles of mastication, and the muscles that move
   the head.
2. Describe and demonstrate the action of each of these
   muscles.
3. Locate the origin and insertion of each of these
   muscles in a human skeleton and the musculature of
   the human torso model.

The skeletal muscles of the face and head include the
muscles of facial expression, which lie just beneath the
skin; the muscles of mastication, attached to the mandible;
and the muscles that move the head, located in the neck.

EXPLORE

PROCEDURE—Muscles of the Head and Neck

1. Review the concept headings entitled "Muscles of Facial
   Expression," "Muscles of Mastication," and "Muscles That
   Move the Head" in section 8.8 in chapter 8 of the textbook.

2. As a review activity, label figures 20.1 and 20.2.
3. Locate the following muscles in the human torso model
   and in your body whenever possible:

   muscles of facial expression

   epicranius
     frontalis
     occipitalis
   orbicularis oculi
   orbicularis oris
   buccinator
   zygomaticus major
   zygomaticus minor
   platysma

   muscles of mastication

   masseter
   temporalis

   muscles that move the head and neck

   sternocleidomastoid
   splenius capitis
   semispinalis capitis
   scalene muscles

4. Demonstrate the action of these muscles in your body.
5. Locate the origins and insertions of these muscles in
   the human skull and skeleton.
6. Complete Parts A, B, and C of Laboratory Report 20.

LEARNING EXTENSION

A long rubber band can be used to simulate muscle
locations, origins, insertions, and actions on the human
torso model, the skeleton, or a laboratory partner. Hold
one end of the rubber band firmly on the origin location
of a muscle; then slightly stretch the rubber band and
hold the other end on the insertion site. Allow the insertion end to slowly move toward the origin end to simulate
the contraction and action of the muscle. 1 2 3

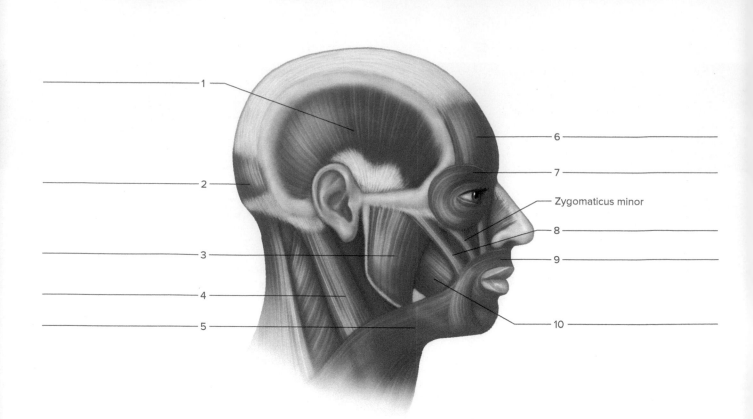

FIGURE 20.1 Label the muscles of expression and mastication. 🔺 **APR**

Zygomaticus minor

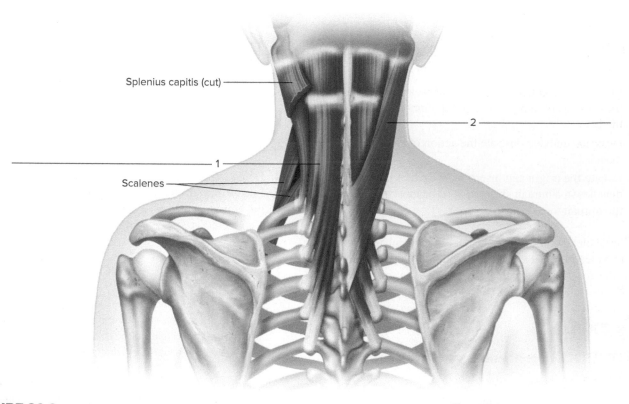

Splenius capitis (cut)

Scalenes

FIGURE 20.2 Label these deep muscles of the posterior neck (trapezius removed). 🔺 **APR**

The Ⓐ corresponds to the indicated Learning Outcome(s) found at the beginning of the laboratory exercise.

Muscles of the Head and Neck

 PART A ASSESSMENTS

Complete the following statements:

1. When the _____ contracts, the corner of the mouth is elevated. 🔼2

2. The _____ acts to compress the wall of the cheeks as when air is blown out of the mouth. 🔼2

3. The _____ causes the lips to close and pucker during kissing, whistling, and speaking. 🔼2

4. The _____ acts to elevate and retract the mandible. 🔼2

5. The _____ can close the eye, as in blinking. 🔼2

6. The _____ can flex the head and neck toward the chest. 🔼2

7. The muscle used to pout and to express horror by pulling the lower lip downward is the _____. 🔼2

 PART B ASSESSMENTS

Name the muscle indicated by the following combinations of origin and insertion. 🔼3

Origin	Insertion	Muscle
1. Occipital bone	Skin around eye	_____
2. Zygomatic bone	Corner of mouth skin and muscle	_____
3. Zygomatic arch	Lateral surface of mandible	_____
4. Manubrium of sternum and clavicle	Mastoid process of temporal bone	_____
5. Alveolar processes of mandible and maxilla	Orbicularis oris	_____
6. Fascia in upper chest	Mandible and skin below mouth	_____
7. Temporal bone	Coronoid process of mandible	_____
8. Transverse processes of cervical vertebrae	Surfaces of ribs 1-2	_____
9. Transverse processes of upper thoracic vertebrae	Occipital bone	_____

Critical Thinking Application

Using the terms provided, identify the muscles of facial expression being contracted in each of these photographs. ⚠ ⚠

Terms:
Epicranius/frontalis
Orbicularis oculi
Orbicularis oris
Zygomaticus

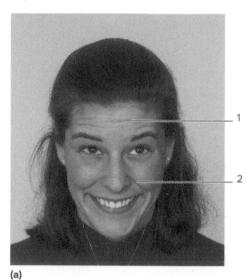

(a)

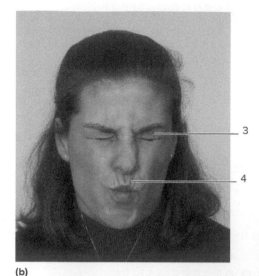

(b)

FIGURE 20.3   The muscles of facial expression being contracted (a and b).   **(a and b)**: ©J and J Photography

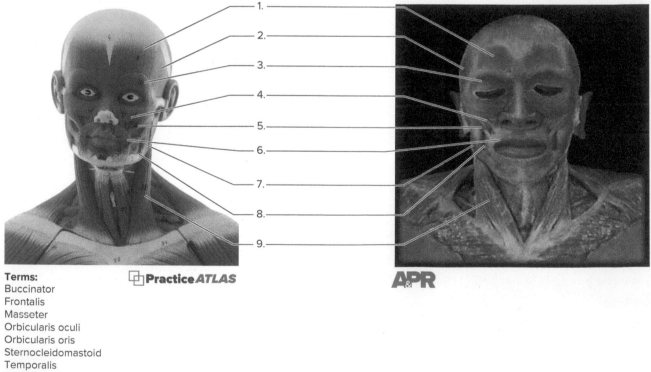

Terms:
Buccinator
Frontalis
Masseter
Orbicularis oculi
Orbicularis oris
Sternocleidomastoid
Temporalis
Zygomaticus major
Zygomaticus minor

PracticeATLAS

APR

FIGURE 20.4   Using the terms provided identify the muscles of the heads and neck on the model image and the cadaver image.

(left): B59 [1000212] 1/3 Life-Size Muscle Figure, 2-part © 3B Scientific GmbH, Germany, 2017 www.3bscientific.com; (right): McGraw-Hill Education/APR

21

Muscles of the Chest, Shoulder, and Upper Limb

MATERIALS NEEDED

Textbook
Human torso model
Human skeleton, articulated
Muscular models of the upper limb

For Learning Extension:
Long rubber band

PURPOSE OF THE EXERCISE

To review the locations, actions, origins, and insertions of the muscles in the chest, shoulder, and upper limb.

LEARNING OUTCOMES APR

After completing this exercise, you should be able to

1. Locate and identify the muscles of the chest, shoulder, and upper limb.
2. Describe and demonstrate the action of each of these muscles.
3. Locate the origin and insertion of each of these muscles in a human skeleton and on the muscular models.

The muscles of the chest and shoulder are responsible for moving the scapula and arm, whereas those within the arm and forearm move joints in the elbow and hand.

EXPLORE

PROCEDURE—Muscles of the Chest, Shoulder, and Upper Limb

1. Review the concept headings entitled "Muscles That Move the Pectoral Girdle," "Muscles That Move the Arm," "Muscles That Move the Forearm," and "Muscles That Move the Hand" of section 8.8 in chapter 8 of the textbook.
2. As a review activity, label figures 21.1, 21.2, 21.3, and 21.4.

3. Locate the following muscles in the human torso model and models of the upper limb. Also locate in your body as many of the muscles as you can.

muscles that move the pectoral girdle
   anterior muscles
      pectoralis minor
      serratus anterior
   posterior muscles
      trapezius
      rhomboid major
      rhomboid minor
      levator scapulae

muscles that move the arm
   origins on axial skeleton
      pectoralis major
      latissimus dorsi
   origins on scapula
      deltoid
      teres major
      coracobrachialis
   rotator cuff (SITS) muscles (origins also on scapula)
      supraspinatus
      infraspinatus
      teres minor
      subscapularis

muscles that move the forearm
   muscle bellies in arm
      biceps brachii
      brachialis
      triceps brachii
   muscle bellies in forearm
      brachioradialis
      supinator
      pronator teres
      pronator quadratus

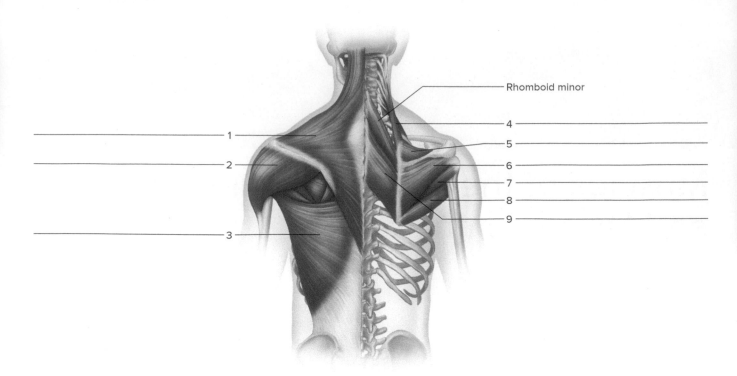

Rhomboid minor

1

2

3

4

5

6

7

8

9

FIGURE 21.1 Label the muscles of the posterior shoulder. Superficial muscles are illustrated on the left side and deep muscles on the right side. Ⓐ **APR**

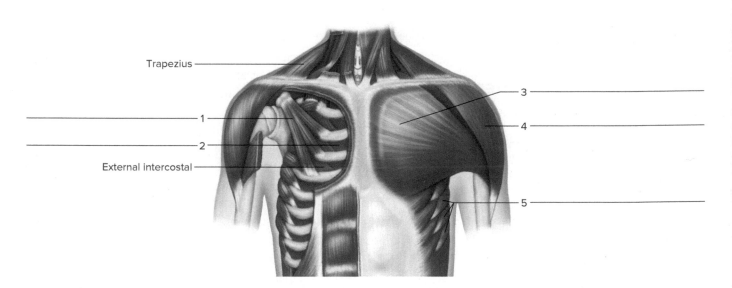

Trapezius

1

2

External intercostal

3

4

5

FIGURE 21.2 Label the muscles of the anterior chest. Superficial muscles are illustrated on the left side and deep muscles on the right side. Ⓐ **APR**

muscles that move the hand
  anterior flexor muscles
    flexor carpi radialis
    flexor carpi ulnaris
    palmaris longus
    flexor digitorum profundus

posterior extensor muscles
  extensor carpi radialis longus
  extensor carpi radialis brevis
  extensor carpi ulnaris
  extensor digitorum

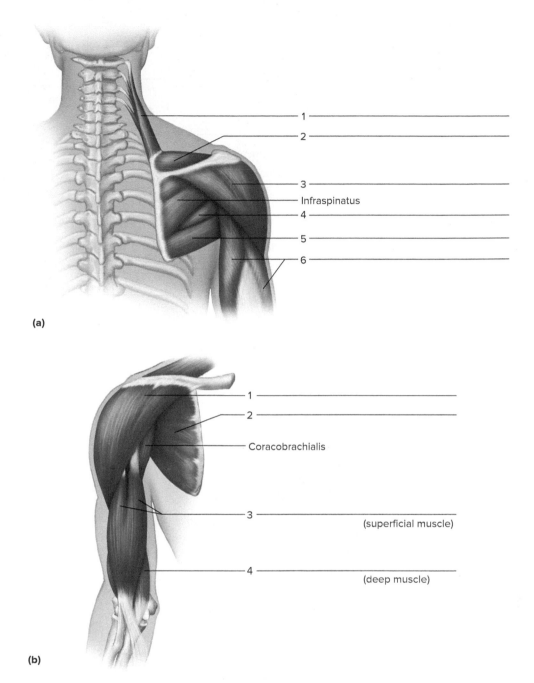

(a)

1 ——————————

2 ——————————

3 ——————————

Infraspinatus

4 ——————————

5 ——————————

6 ——————————

(b)

1 ——————————

2 ——————————

Coracobrachialis

3 ——————————

(superficial muscle)

4 ——————————

(deep muscle)

FIGURE 21.3 Label (*a*) the muscles of the posterior shoulder and arm and (*b*) the muscles of the anterior shoulder and arm, with the rib cage removed. ⚠ **APR**

LEARNING EXTENSION

A long rubber band can be used to simulate muscle locations, origins, insertions, and actions on muscular models, the skeleton, or a laboratory partner. Hold one end of the rubber band firmly on the origin location of a muscle; then slightly stretch the rubber band and hold the other end on the insertion site. Allow the insertion end to slowly move toward the origin end to simulate the contraction and action of the muscle. ⚠ ⚠ ⚠

4. Demonstrate the action of these muscles in your body.
5. Locate the origins and insertions of these muscles in the human skeleton.
6. Complete Parts A, B, and C of Laboratory Report 21.

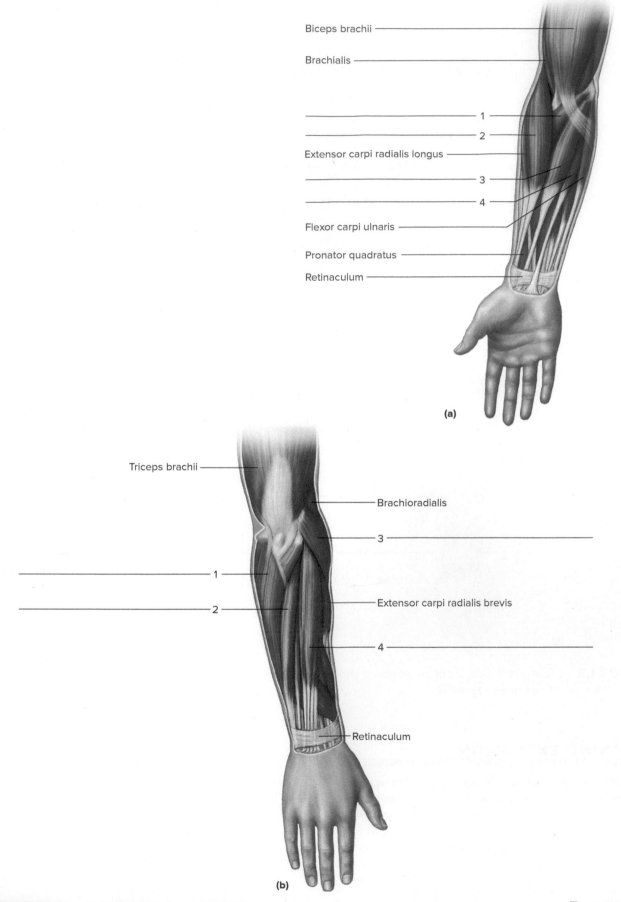

Biceps brachii

Brachialis

1

2

Extensor carpi radialis longus

3

4

Flexor carpi ulnaris

Pronator quadratus

Retinaculum

(a)

Triceps brachii

Brachioradialis

3

1

2

Extensor carpi radialis brevis

4

Retinaculum

(b)

FIGURE 21.4 Label (*a*) the muscles of the anterior forearm and (*b*) the muscles of the posterior forearm.

Name _____

Date _____

Section _____

The corresponds to the indicated Learning Outcome(s) found at the beginning of the laboratory exercise.

LABORATORY
REPORT

21

Muscles of the Chest, Shoulder, and Upper Limb

PART A ASSESSMENTS

Match the muscles in column A with the actions in column B. Place the letter of your choice in the space provided. 🄰

Column A	**Column B**
a. Brachialis	_____ **1.** Abducts arm and flexes and extends arm at shoulder
b. Coracobrachialis	_____ **2.** Flexes arm at shoulder, adducts and medially rotates arm
c. Deltoid	
d. Extensor carpi ulnaris	_____ **3.** Flexes wrist and adducts hand
e. Flexor carpi ulnaris	_____ **4.** Elevates and retracts scapula
f. Infraspinatus	_____ **5.** Depresses and protracts scapula
g. Pectoralis major	_____ **6.** Used to thrust shoulder anteriorly (protraction), as when pushing something
h. Pectoralis minor	
i. Rhomboid major	
j. Serratus anterior	_____ **7.** Flexes elbow
k. Teres major	_____ **8.** Flexes shoulder and adducts arm
l. Triceps brachii	_____ **9.** Extends elbow
	_____ **10.** Extends shoulder and adducts and medially rotates arm
	_____ **11.** Extends wrist and adducts hand
	_____ **12.** Laterally rotates arm

Name the muscle indicated by the following combinations of origin and insertion. ⒊

Origin	Insertion	Muscle
1. Spinous processes of upper thoracic vertebrae	Medial border of scapula	_____
2. Anterior surfaces of most ribs	Medial border of scapula	_____
3. Anterior surfaces of ribs 3-5	Coracoid process of scapula	_____
4. Coracoid process of scapula	Shaft of humerus	_____
5. Lateral border of scapula	Intertubercular sulcus of humerus	_____
6. Anterior surface of scapula	Lesser tubercle of humerus	_____
7. Lateral border of scapula	Greater tubercle of humerus	_____
8. Anterior shaft of humerus	Coronoid process of ulna	_____
9. Medial epicondyle of humerus and coronoid process of ulna	Lateral surface of radius	_____
10. Distal lateral end of humerus	Lateral surface of radius above styloid process	_____
11. Medial epicondyle of humerus	Base of second and third metacarpals	_____
12. Medial epicondyle of humerus	Fascia of palm	_____

Critical Thinking Application

FIGURE 21.5a   Using the terms provided label the indicated muscles on the model and cadaver images in **figure 21.5a.**

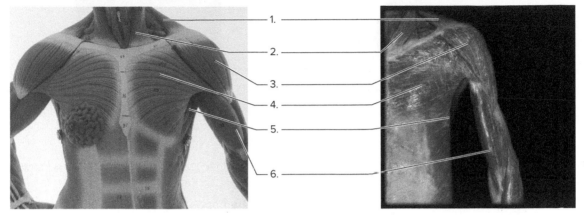

1. _____
2. _____
3. _____
4. _____
5. _____
6. _____

Terms:
Biceps brachii
Deltoid
Pectoralis major
Serratus anterior
Sternocleidomastoid
Trapezius

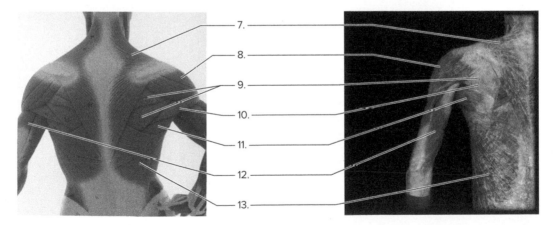

7. _____
8. _____
9. _____
10. _____
11. _____
12. _____
13. _____

Terms:
Deltoid
Infraspinatus
Latissimus dorsi
Teres major
Teres minor
Trapezius
Triceps brachii

FIGURE 21.5b   Using the terms provided label the indicated muscles on the model and cadaver images in figure 21.5b.
(*a, b,* and *c*). a(left): B59 [1000212] 1/3 Life-Size Muscle Figure, 2-part © 3B Scientific GmbH, Germany, 2017 www.3bscientific.com; a(right): ©McGraw-Hill
Education/APR; b(left): B59 [1000212] 1/3 Life-Size Muscle Figure, 2-part © 3B Scientific GmbH, Germany, 2017 www.3bscientific.com; b(right): ©McGraw-Hill
Education/APR; c(left): B59 [1000212] 1/3 Life-Size Muscle Figure, 2-part © 3B Scientific GmbH, Germany, 2017 www.3bscientific.com; c(right): ©McGraw-Hill
Education/APR

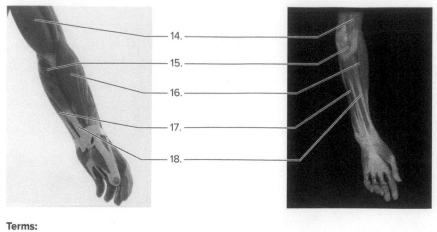

Terms:
Biceps brachii
Brachioradialis
Flexor carpi radialis
Palmaris longus
Pronator teres

FIGURE 21.5c Using the terms provided label the indicated muscles on the model and cadaver images in figure 21.5c.

22

Muscles of the Abdominal Wall and Pelvic Floor

MATERIALS NEEDED

Textbook
Human torso model with musculature
Human skeleton, articulated
Muscular models of male and female pelves

PURPOSE OF THE EXERCISE

To review the actions, origins, and insertions of the muscles of the abdominal wall and pelvic floor.

LEARNING OUTCOMES **A&PR**

After completing this exercise, you should be able to

1. Locate and identify the muscles of the abdominal wall and pelvic floor.
2. Describe the action of each of these muscles.
3. Locate the origin and insertion of each of these muscles in a human skeleton or on muscular models.

The anterior and lateral walls of the abdomen contain broad, flattened muscles arranged in layers. These muscles connect the rib cage and vertebral column to the pelvic girdle. The abdominal wall muscles compress the abdominal visceral organs, help maintain posture, assist in forceful exhalation, and contribute in trunk flexion and waist rotation.

The muscles of the pelvic floor are arranged in two muscular sheets: (1) a pelvic diaphragm that spans the outlet of the pelvic cavity and (2) a urogenital diaphragm that fills the space within the pubic arch. The pelvic floor is penetrated by the urethra, vagina, and anus in a female; thus, pelvic floor muscles are important in obstetrics.

EXPLORE

PROCEDURE A—Muscles of the Abdominal Wall

1. Review the concept heading entitled "Muscles of the Abdominal Wall" of section 8.8 in chapter 8 of the textbook.
2. Study figure 22.1.
3. Locate the following muscles in the human torso model:

   external oblique
   internal oblique
   transversus abdominis
   rectus abdominis

4. Demonstrate the actions of these muscles in your body. 2
5. Locate the origin and insertion of each of these muscles in the human skeleton. 3
6. Complete Parts A and B of Laboratory Report 22.

Critical Thinking Application

List the muscles from superficial to deep for an appendectomy incision.

_____
_____
_____
_____
_____

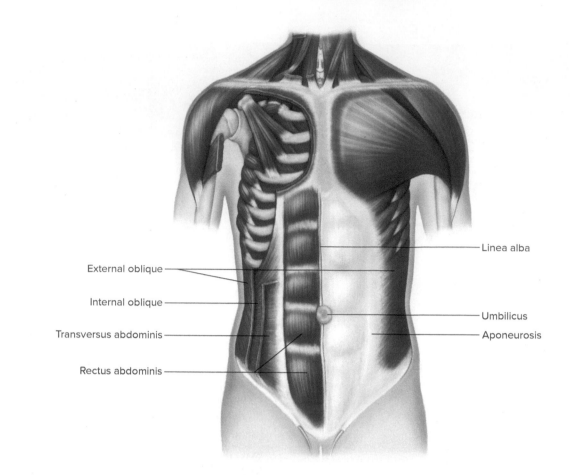

External oblique

Internal oblique

Transversus abdominis

Rectus abdominis

Linea alba

Umbilicus

Aponeurosis

FIGURE 22.1 Muscles of the abdominal wall. **APR**

EXPLORE

PROCEDURE B—Muscles of the Pelvic Floor

1. Review the concept heading entitled "Muscles of the Pelvic Floor" of section 8.8 in chapter 8 of the textbook.
2. As a review activity, label figures 22.2 and 22.3.
3. Study figure 22.4 as it relates to the pelvic muscles indicated in figures 22.2 and 22.3.

4. Locate the following muscles in the models of the male and female pelves:

   levator ani
   superficial transversus perinei
   bulbospongiosus
   ischiocavernosus

5. Locate the origin and insertion of each of these muscles in the human skeleton. 🔺
6. Complete Part C of the laboratory report.

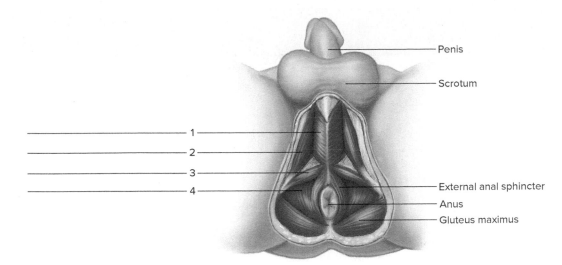

Penis

Scrotum

1

2

3

4

External anal sphincter

Anus

Gluteus maximus

FIGURE 22.2 Label the muscles of the male pelvic floor.

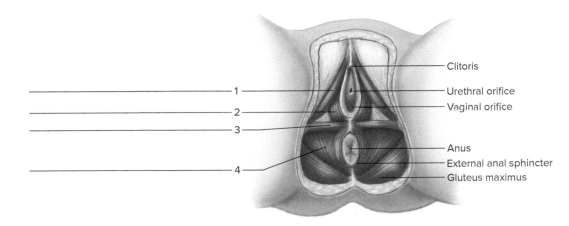

Clitoris

Urethral orifice

Vaginal orifice

1

2

3

4

Anus

External anal sphincter

Gluteus maximus

FIGURE 22.3 Label the muscles of the female pelvic floor.

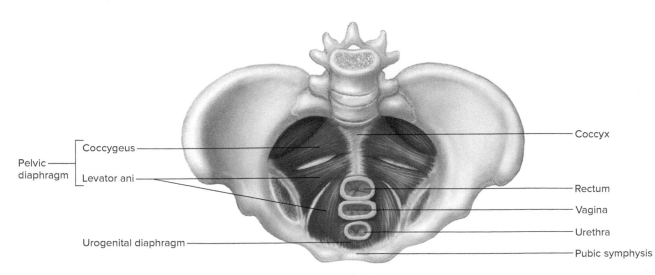

Coccyx

Pelvic diaphragm

Coccygeus

Levator ani

Rectum

Vagina

Urethra

Urogenital diaphragm

Pubic symphysis

FIGURE 22.4 Internal view of the female pelvic and urogenital diaphragms.

Notes

158

Name _____

Date _____

Section _____

The corresponds to the indicated Learning Outcome(s) found at the beginning of the laboratory exercise.

Muscles of the Abdominal Wall and Pelvic Floor

PART A ASSESSMENTS

Complete the following statements:

1. A band of tough connective tissue in the midline of the anterior abdominal wall called the _____ serves as a muscle attachment.

2. The _____ muscle spans from the costal cartilage of ribs and the xiphoid process of the sternum to the pubic bones. 3

3. The _____ forms the third layer (deepest layer) of the abdominal wall muscles. 1

4. The external oblique muscles have an origin on the _____. 3

5. The action of the rectus abdominis is to _____. 2

PART B ASSESSMENTS

Label the muscles of the abdominal wall in figure 22.5. 1

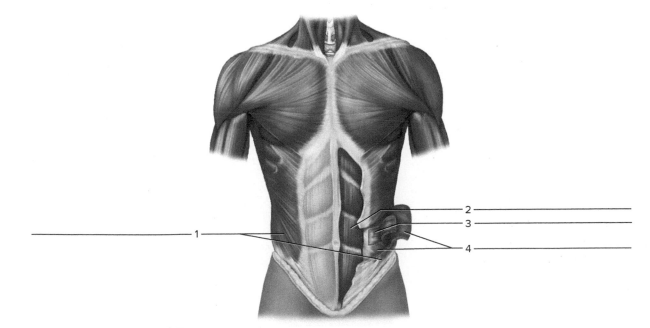

FIGURE 22.5 Label the muscles of the abdominal wall.

Complete the following statements:

1. The levator ani forms the _____ diaphragm. △1

2. The action of the levator ani is to compress the _____ and support pelvic viscera. △2

3. The action of the superficial transversus perinei is to _____ . △2

4. The _____ assists in emptying the male urethra. △2

5. In females, the bulbospongiosus acts to constrict the _____ . △2

6. The ischiocavernosus extends from the penis or clitoris of the pubic arch to the _____ . △3

23

Muscles of the Hip and Lower Limb

MATERIALS NEEDED

Textbook
Human torso model with musculature
Human skeleton, articulated
Muscular models of the lower limb

For Learning Extension:
Long rubber band

PURPOSE OF THE EXERCISE

To review the actions, origins, and insertions of the muscles that move the thigh, leg, and foot.

LEARNING OUTCOMES A&PR

After completing this exercise, you should be able to

① Locate and identify the muscles that move the thigh, leg, and foot.

② Describe and demonstrate the actions of each of these muscles.

③ Locate the origin and insertion of each of these muscles in a human skeleton and on muscular models.

The muscles that move the thigh are attached to the femur and to some part of the pelvic girdle. Those attached anteriorly primarily act to flex the thigh at the hip, whereas those attached posteriorly act to extend, abduct, or rotate the thigh.

The muscles that move the leg connect the tibia or fibula to the femur or to the pelvic girdle. They function to flex or extend the leg at the knee. Other muscles, located in the leg, act to move the foot.

EXPLORE

PROCEDURE—Muscles of the Hip and Lower Limb

1. Review the concept headings entitled "Muscles That Move the Thigh," "Muscles That Move the Leg," and "Muscles That Move the Foot" of section 8.8 in chapter 8 of the textbook.

2. As a review activity, label figures 23.1, 23.2, 23.3, 23.4, 23.5, and 23.6.

3. Locate the following muscles in the human torso model and in the lower limb models. Also locate as many of them as possible in your body.

muscles that move the thigh
  anterior hip muscles
    iliopsoas group
      psoas major
      iliacus
  posterior and lateral hip muscles
    gluteus maximus
    gluteus medius
    gluteus minimus
    tensor fasciae latae
  medial adductor muscles
    adductor longus
    adductor magnus
    gracilis
muscles that move the leg
  anterior thigh muscles
    sartorius
    quadriceps femoris group
      rectus femoris
      vastus lateralis
      vastus medialis
      vastus intermedius
  posterior thigh muscles
    hamstring group
      biceps femoris
      semitendinosus
      semimembranosus
muscles that move the foot
  anterior leg muscles
    tibialis anterior
    fibularis (peroneus) tertius
    extensor digitorum longus
  posterior leg muscles
    gastrocnemius
    soleus
    flexor digitorum longus
  lateral leg muscle
    fibularis (peroneus) longus
    fibularis (peroneus) brevis

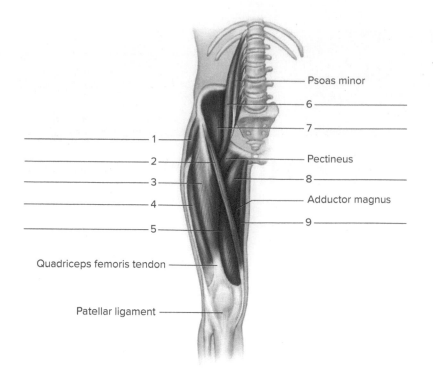

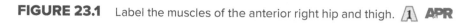

Psoas minor

6

7

1

2

Pectineus

3

8

4

Adductor magnus

5

9

Quadriceps femoris tendon

Patellar ligament

FIGURE 23.1   Label the muscles of the anterior right hip and thigh.   1   APR

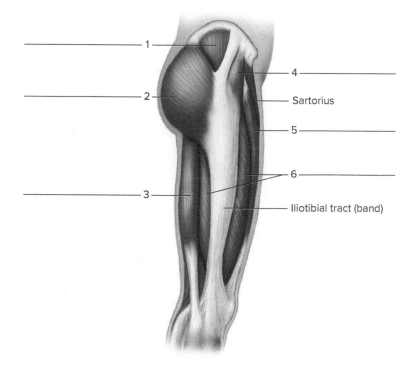

1

2

4

Sartorius

5

6

3

Iliotibial tract (band)

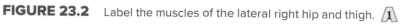

FIGURE 23.2   Label the muscles of the lateral right hip and thigh.   1

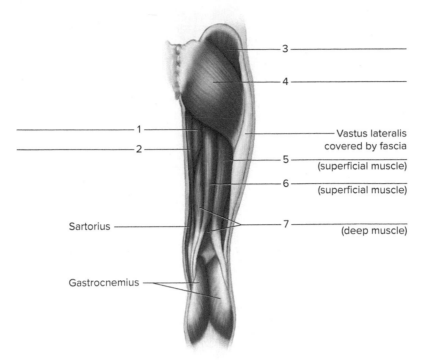
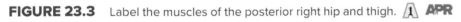

3

4

1 ————————————————————— Vastus lateralis
covered by fascia

2 ————————————————— 5 ——————————— (superficial muscle)

6 ——————————— (superficial muscle)

Sartorius ———————————— 7 ——————————— (deep muscle)

Gastrocnemius ————————

FIGURE 23.3 Label the muscles of the posterior right hip and thigh. 🅐 **APR**

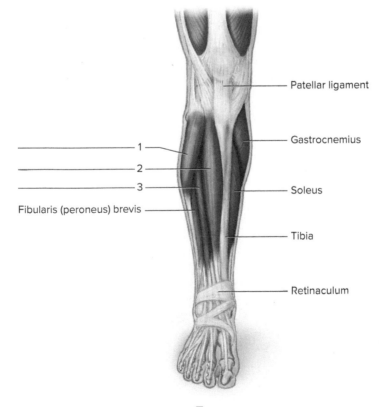

Patellar ligament

Gastrocnemius

1

2

Soleus

3

Fibularis (peroneus) brevis ————

Tibia

Retinaculum

FIGURE 23.4 Label the muscles of the anterior right leg. 🅐 **APR**

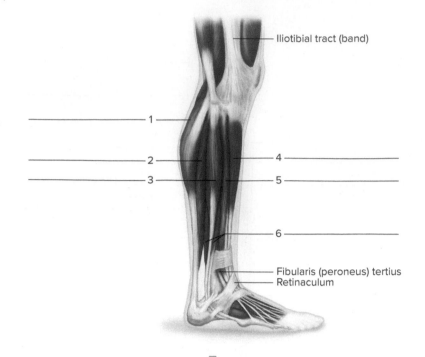

Iliotibial tract (band)

1

2 — 4

3 — 5

6

Fibularis (peroneus) tertius
Retinaculum

FIGURE 23.5 Label the muscles of the lateral right leg.

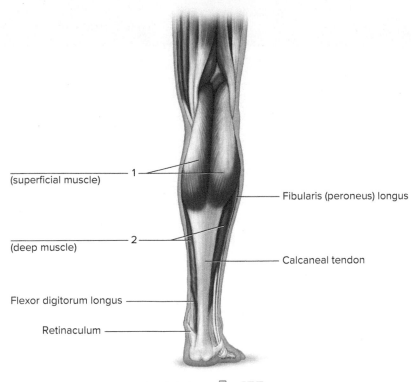

_____ 1
(superficial muscle)

_____ Fibularis (peroneus) longus

_____ 2
(deep muscle)

_____ Calcaneal tendon

Flexor digitorum longus _____

Retinaculum _____

FIGURE 23.6   Label the muscles of the posterior right leg. ⚠ APR

4. Demonstrate the action of each of these muscles in your body.
5. Locate the origin and insertion of each of these muscles in the human skeleton.
6. Complete Parts A, B, and C of Laboratory Report 23.

LEARNING EXTENSION

A long rubber band can be used to simulate muscle locations, origins, insertions, and actions on muscular models, the skeleton, or a laboratory partner. Hold one end of the rubber band firmly on the origin location of a muscle; then slightly stretch the rubber band and hold the other end on the insertion site. Allow the insertion end to slowly move toward the origin end to simulate the contraction and action of the muscle. ⚠ ⚠ ⚠

The Ⓐ corresponds to the indicated Learning Outcome(s) found at the beginning of the laboratory exercise.

Muscles of the Hip and Lower Limb

PART A ASSESSMENTS

Match the muscles in column A with the actions in column B. Place the letter of your choice in the space provided. Ⓐ2

Column A	Column B
a. Biceps femoris	_____ **1.** Adducts thigh and flexes thigh at hip
b. Fibularis (peroneus) longus	_____ **2.** Plantar flexion and eversion of foot
c. Gluteus medius	_____ **3.** Flexes thigh at the hip
d. Gracilis	_____ **4.** Abducts and laterally rotates thigh; flexes leg at knee and thigh at hip
e. Psoas major and iliacus	
f. Quadriceps femoris group	_____ **5.** Abducts thigh and medially rotates it
g. Sartorius	_____ **6.** Flexes leg at the knee and extends thigh at hip
h. Tibialis anterior	_____ **7.** Extends leg at the knee
	_____ **8.** Dorsiflexion and inversion of foot

PART B ASSESSMENTS

Name the muscle indicated by the following combinations of origin and insertion. Ⓐ3

Origin	Insertion	Muscle
1. Lateral surface of ilium	Greater trochanter of femur	_____
2. Ischial tuberosity and pubis	Posterior surface of femur	_____
3. Anterior superior iliac spine	Medial surface of proximal tibia	_____
4. Lateral and medial condyles of femur	Posterior surface of calcaneus	_____
5. Anterior iliac crest	Fascia (iliotibial tract) of the thigh	_____
6. Greater trochanter and posterior surface of femur	Patella and to tibial tuberosity	_____
7. Ischial tuberosity	Medial surface of proximal tibia	_____
8. Medial surface of femur	Patella and to tibial tuberosity	_____
9. Posterior surface of tibia	Distal phalanges of four lateral toes	_____
10. Lateral condyle and lateral surface of tibia	Medial cuneiform and first metatarsal	_____

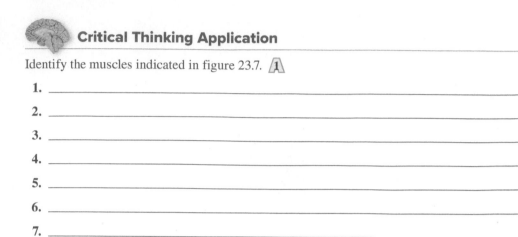

Critical Thinking Application

Identify the muscles indicated in figure 23.7.

1. _____

2. _____

3. _____

4. _____

5. _____

6. _____

7. _____

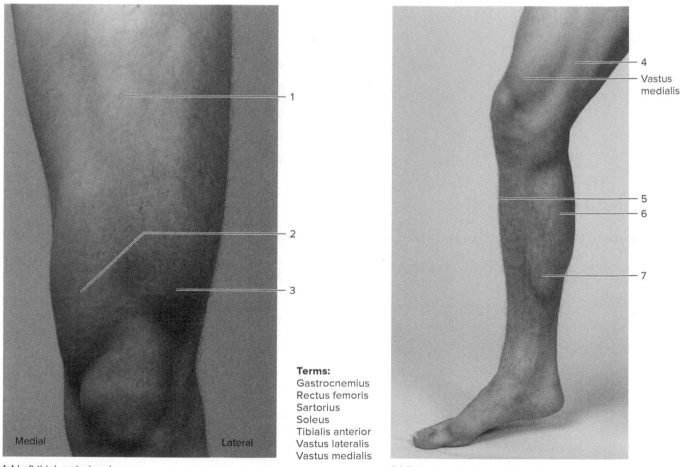

Terms:
Gastrocnemius
Rectus femoris
Sartorius
Soleus
Tibialis anterior
Vastus lateralis
Vastus medialis

Medial   Lateral

(a) Left thigh, anterior view

4
Vastus medialis
5
6
7

(b) Right lower limb, medial view

FIGURE 23.7 Using the terms provided, identify the muscles that appear as lower limb surface features in these photographs (*a* and *b*). **(a, b):** ©J and J Photography

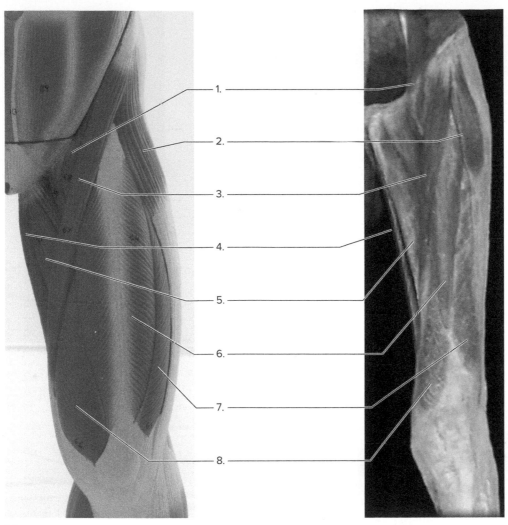

1. _____

2. _____

3. _____

4. _____

5. _____

6. _____

7. _____

8. _____

Terms:
Adductor magnus
Gracilis
Iliopsoas
Rectus femoris
Sartorius
Tensor fasciae latae
Vastus lateralis
Vastus mediais

(a)

FIGURE 23.8a   Using the terms provided, label the indicated muscles of the lower limb on the model and cadaver images.   **(a, b left):** M20 Muscle Leg © 3B Scientific GmbH, Germany, 2017 www.3bscientific.com; **(a, b right):** ©McGraw-Hill Education/APR

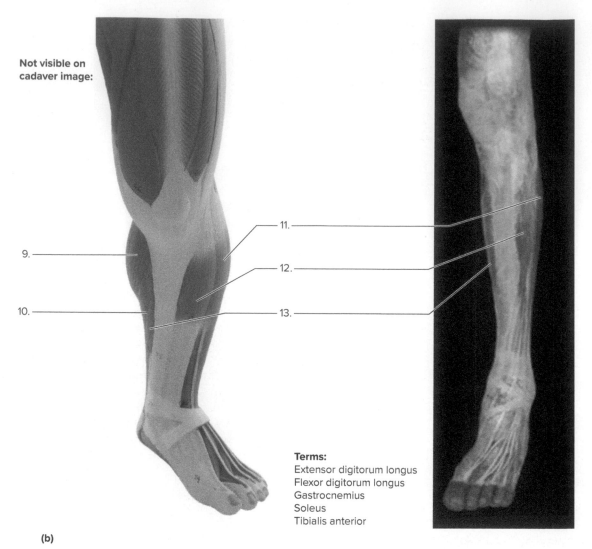

9. _____

10. _____

11. _____

12. _____

13. _____

Terms:
Extensor digitorum longus
Flexor digitorum longus
Gastrocnemius
Soleus
Tibialis anterior

(b)

FIGURE 23.8b   Using the terms provided, label the indicated muscles of the lower limb on the model and cadaver image.

24

Surface Anatomy

PURPOSE OF THE EXERCISE

To examine the surface features of the human body and the terms used to describe them.

LEARNING OUTCOMES

After completing this exercise you should be able to

1. Locate and identify major surface features of the human body.
2. Arrange surface features by body region.
3. Distinguish between surface features that are bony landmarks and soft tissue.

External landmarks, called *surface anatomy*, located on the human body provide an opportunity to examine surface features and help locate other internal structures. This exercise will focus on bony landmarks and superficial soft tissue structures, primarily related to the skeletal and muscular systems. The technique used for the examination of surface features is called *palpation* and is accomplished by touching with the hands or fingers. This enables us to determine the location, size, and texture of the underlying anatomical structure. Some of the surface features can be visible without the need for palpation, especially on body builders and very thin individuals. If the individual involved has considerable subcutaneous adipose tissue, additional pressure might be necessary to palpate some of the structures.

These surface features will help us to locate additional surface features and deeper structures during the study of the organ systems. Surface features will help us to determine the pulse locations of superficial arteries and enable proper assessments of injection sites; descriptions of pain and injury sites; insertion of examination tubes; and heart, lung, and bowel sounds heard with a stethoscope. For those who choose careers in emergency medical services, nursing, medicine, physical education, physical therapy, chiropractic, and massage therapy, surface anatomy is especially valuable.

EXPLORE

PROCEDURE — Surface Anatomy

1. Review textbook figures on body regions, human organism reference plates, the skeletal system, and the muscular system. These figures are located in chapters 1, 7, and 8.
2. Use the textbook figures and the other figures provided to complete figures 24.1 through 24.8. Many features have already been labeled. Features in **boldface** are *bony features*. Other labeled features (not boldface) are *soft tissue features*. Bony features are a part of a bone and the skeletal system. Soft tissue features are composed of tissue other than bone, such as a muscle, tendon, ligament, cartilage, or skin. The missing features are listed by each figure.

Terms:
Hyoid bone
Mandible
Sternocleidomastoid
Trapezius (upper)

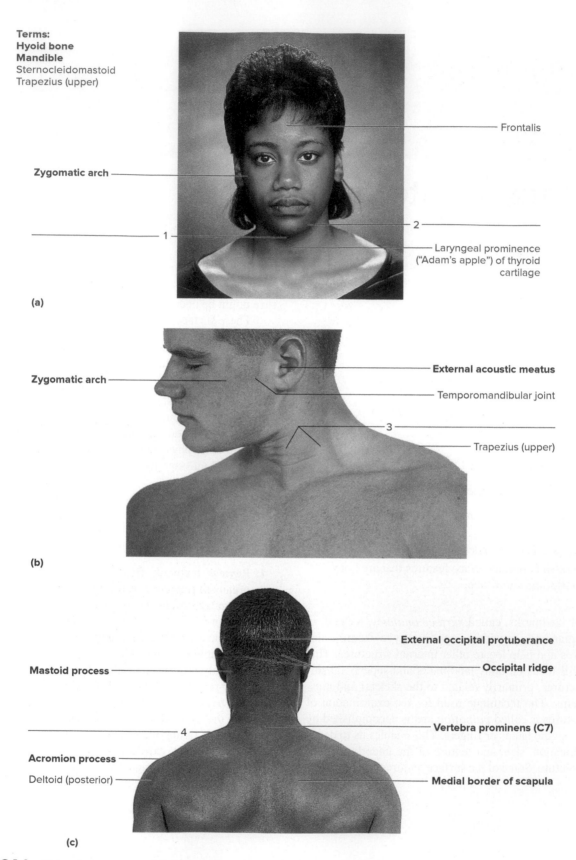

Frontalis

Zygomatic arch

1

2

Laryngeal prominence
("Adam's apple") of thyroid
cartilage

(a)

External acoustic meatus

Zygomatic arch

Temporomandibular joint

3

Trapezius (upper)

(b)

External occipital protuberance

Occipital ridge

Mastoid process

Vertebra prominens (C7)

4

Acromion process

Deltoid (posterior)

Medial border of scapula

(c)

FIGURE 24.1   Using the terms provided, label the surface features of (a) the anterior head and neck, (b) the lateral head and neck, and (c) the posterior head and neck. (**Boldface** indicates bony features; not boldface indicates soft tissue features.)

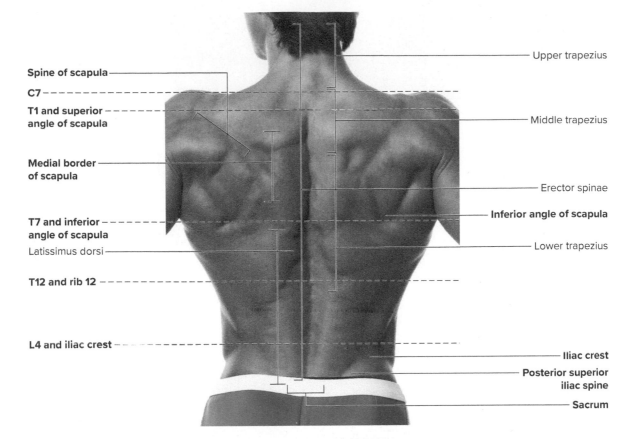

Spine of scapula

C7

T1 and superior angle of scapula

Medial border of scapula

T7 and inferior angle of scapula

Latissimus dorsi

T12 and rib 12

L4 and iliac crest

Upper trapezius

Middle trapezius

Erector spinae

Inferior angle of scapula

Lower trapezius

Iliac crest

Posterior superior iliac spine

Sacrum

FIGURE 24.2 Surface features of the posterior shoulder and torso. (**Boldface** indicates bony features; not boldface indicates soft tissue features.) BLACKDAY/Shutterstock

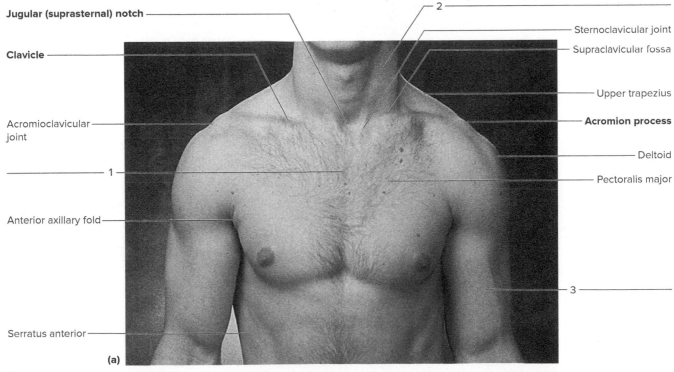

Jugular (suprasternal) notch

Clavicle

Acromioclavicular joint

1

Anterior axillary fold

Serratus anterior

(a)

2

Sternoclavicular joint

Supraclavicular fossa

Upper trapezius

Acromion process

Deltoid

Pectoralis major

3

Terms:
Biceps brachii
Serratus anterior
Sternocleidomastoid
Sternum

FIGURE 24.3 Using the terms provided, label the anterior surface features of (*a*) upper torso and (*b*) lower torso. (**Boldface** indicates bony features; not boldface indicates soft tissue features.) 3
(a): Joe DeGrandis/McGraw-Hill Education; (b): Juice Images/Alamy Stock Photo

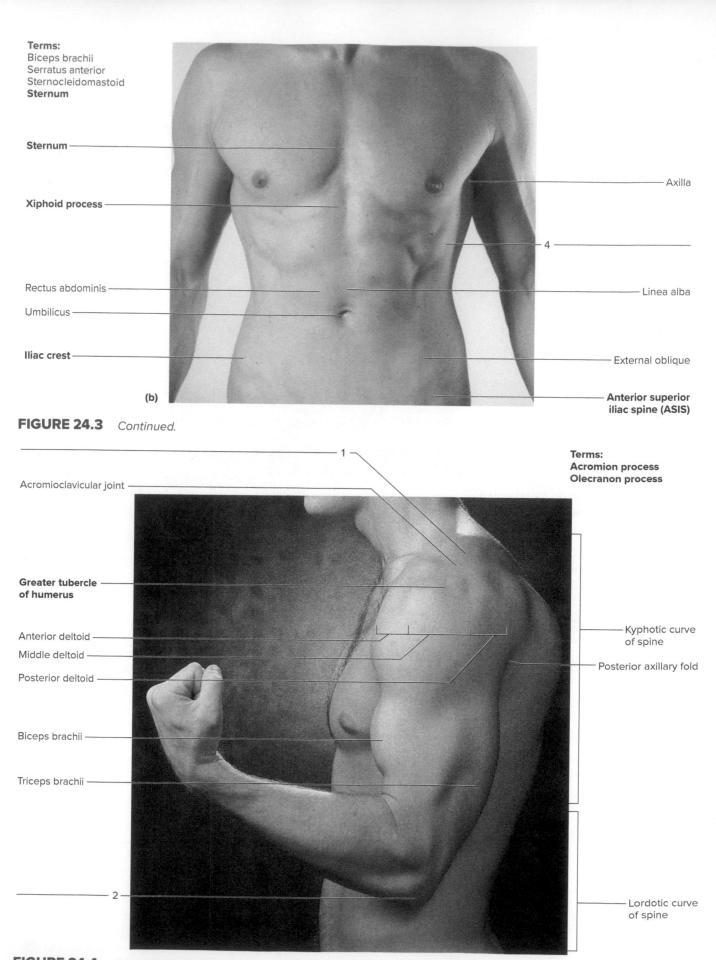

Terms:
Biceps brachii
Serratus anterior
Sternocleidomastoid
Sternum

Sternum —————

Xiphoid process —————

Rectus abdominis —————

Umbilicus —————

Iliac crest —————

(b)

————— Axilla

————— 4

————— Linea alba

————— External oblique

————— **Anterior superior iliac spine (ASIS)**

FIGURE 24.3   *Continued.*

————— 1

Acromioclavicular joint —————

Terms:
Acromion process
Olecranon process

Greater tubercle of humerus —————

Anterior deltoid —————
Middle deltoid —————
Posterior deltoid —————

Biceps brachii —————

Triceps brachii —————

————— 2

————— Kyphotic curve of spine

————— Posterior axillary fold

————— Lordotic curve of spine

FIGURE 24.4   Using the terms provided, label the surface features of the lateral shoulder and upper limb. (**Boldface** indicates bony features; not boldface indicates soft tissue features.)   Joe DeGrandis/McGraw-Hill Education

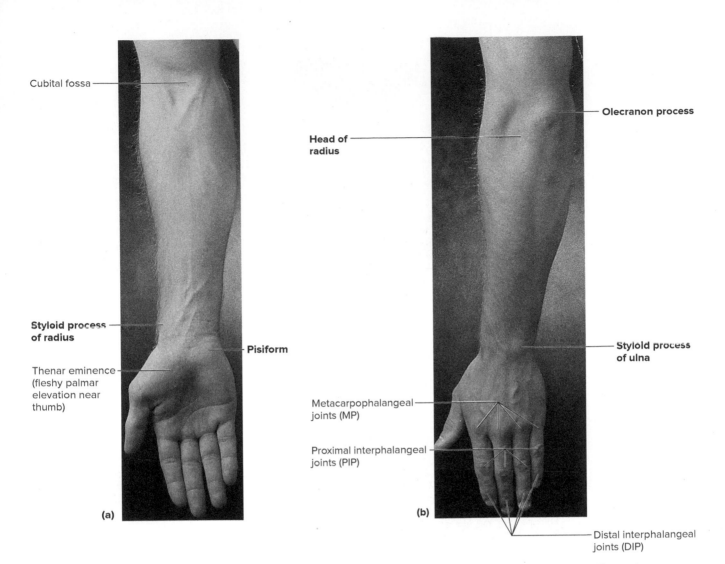

Cubital fossa

Olecranon process

Head of radius

Styloid process of radius

Pisiform

Thenar eminence (fleshy palmar elevation near thumb)

Styloid process of ulna

Metacarpophalangeal joints (MP)

Proximal interphalangeal joints (PIP)

Distal interphalangeal joints (DIP)

(a)

(b)

FIGURE 24.5 Surface features of the upper limb (*a*) anterior view and (*b*) posterior view. (**Boldface** indicates bony features; not boldface indicates soft tissue features.) **(a, b):** Joe DeGrandis/McGraw-Hill Education

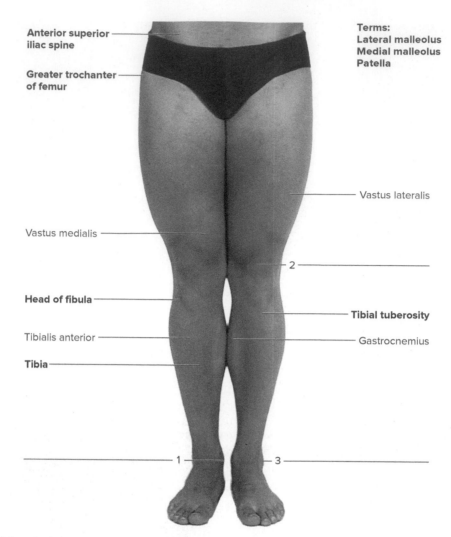

Anterior superior
iliac spine

Greater trochanter
of femur

Terms:
Lateral malleolus
Medial malleolus
Patella

Vastus lateralis

Vastus medialis

2

Head of fibula

Tibial tuberosity

Tibialis anterior

Gastrocnemius

Tibia

1   3

FIGURE 24.6 Using the terms provided, label the surface features of the anterior view of the body. (**Boldface** indicates bony features; not boldface indicates soft tissue features.) ⚠**1** ©Eric Wise

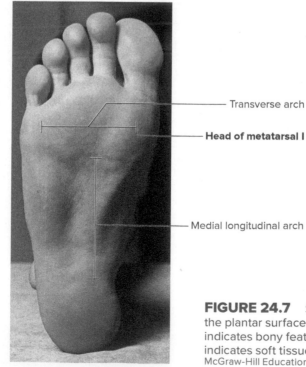

Transverse arch

Head of metatarsal I

Medial longitudinal arch

FIGURE 24.7 Surface features of the plantar surface of the foot. (**Boldface** indicates bony features; not boldface indicates soft tissue features.) Joe DeGrandis/ McGraw-Hill Education

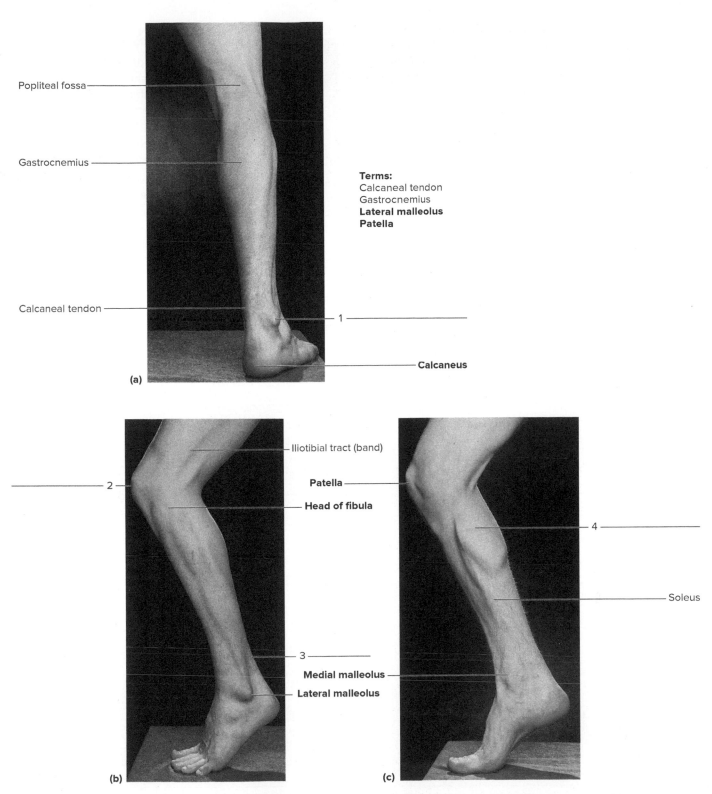

Popliteal fossa

Gastrocnemius

Calcaneal tendon

Terms:
Calcaneal tendon
Gastrocnemius
Lateral malleolus
Patella

1

Calcaneus

(a)

Iliotibial tract (band)

Patella

Head of fibula

2

3

Medial malleolus

Lateral malleolus

(b)

4

Soleus

(c)

FIGURE 24.8 Using the terms provided, label the surface features of the (*a*) posterior lower limb, (*b*) lateral lower limb, and (c) medial lower limb. (**Boldface** indicates bony features; not boldface indicates soft tissue features.) (a–c): Joe DeGrandis/McGraw-Hill Education

3. Complete Parts A and B of Laboratory Report 24.
4. Use figures 24.1 through 24.8 to determine which items listed in table 24.1 are bony features and which items are soft tissue features. If the feature listed is a bony surface feature, place an "X" in the second column; if the feature listed is a soft tissue feature, place an "X" in the third column. △3
5. Complete Part C of the laboratory report.

TABLE 24.1 Representative Bony and Soft Tissue Surface Features

Bony and Soft Tissues	Bony Features	Soft Tissue Features
Biceps brachii		
Calcaneal tendon		
Deltoid		
Gastrocnemius		
Greater trochanter		
Head of fibula		
Iliac crest		
Iliotibial tract		
Lateral malleolus		
Mandible		
Mastoid process		
Medial border of scapula		
Rectus abdominis		
Sacrum		
Serratus anterior		
Sternocleidomastoid		
Sternum		
Thenar eminence		
Tibialis anterior		
Zygomatic arch		

(*Note:* After indicating your "X" in the appropriate column, use the extra space in columns 2 and 3 to add personal descriptions of the nature of the body feature or the soft tissue feature palpatation.)

Name _____

Date _____

Section _____

The A corresponds to the indicated Learning Outcome(s) found at the beginning of the laboratory exercise.

Surface Anatomy

PART A ASSESSMENTS

Match the body regions in column A with the surface features in column B. Place the letter of your choice in the space provided. 1 2

Column A	Column B
a. Head	_____ _____ **1.** Umbilicus
b. Trunk (torso, including shoulder, and hip)	_____ **2.** Medial malleolus
c. Upper limb	_____ **3.** Iliac crest
d. Lower limb	_____ **4.** Transverse arch
	_____ **5.** Spine of scapula
	_____ **6.** External occipital protuberance
	_____ **7.** Metacarpophalangeal joints
	_____ **8.** Sternum
	_____ **9.** Olecranon process
	_____ **10.** Zygomatic arch
	_____ **11.** Mastoid process
	_____ **12.** Thenar eminence
	_____ **13.** Popliteal fossa
	_____ **14.** Cupital fossa

Label figure 24.9 with the surface features provided.

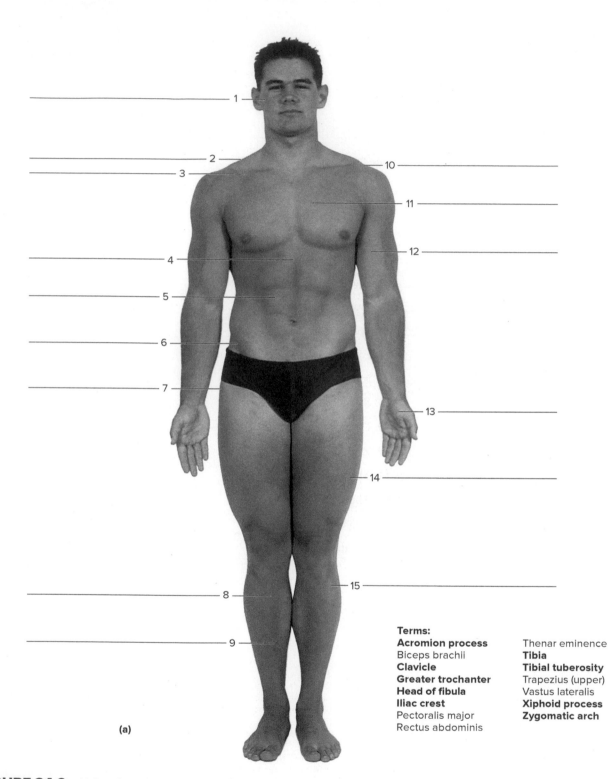

Terms:

Acromion process
Biceps brachii
Clavicle
Greater trochanter
Head of fibula
Iliac crest
Pectoralis major
Rectus abdominis

Thenar eminence
Tibia
Tibial tuberosity
Trapezius (upper)
Vastus lateralis
Xiphoid process
Zygomatic arch

(a)

FIGURE 24.9 Using the terms provided, label the surface features of the (a) anterior view of the body and (b) posterior view of the body. (**Boldface** indicates bony features; not boldface indicates soft tissue features.) (a, b): ©Eric Wis

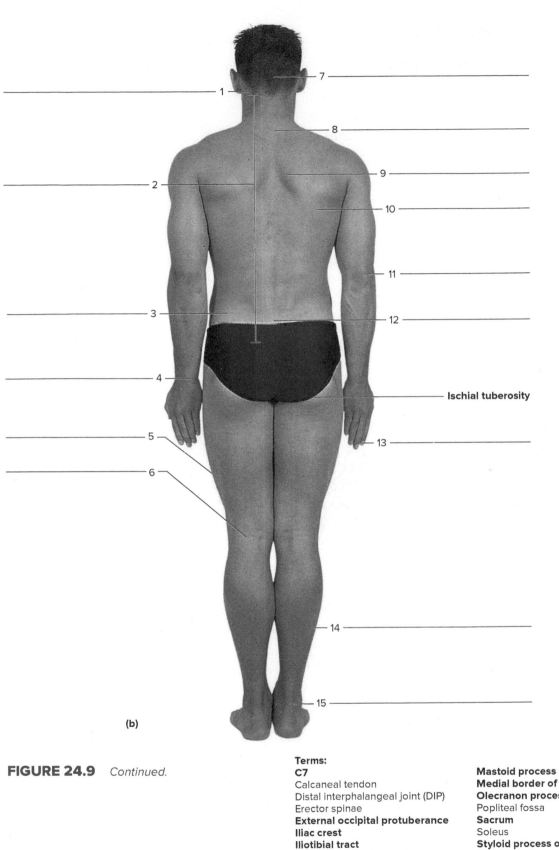

(b)

FIGURE 24.9 *Continued.*

Ischial tuberosity

Terms:

C7	**Mastoid process**
Calcaneal tendon	**Medial border of scapula**
Distal interphalangeal joint (DIP)	**Olecranon process**
Erector spinae	Popliteal fossa
External occipital protuberance	**Sacrum**
Iliac crest	Soleus
Iliotibial tract	**Styloid process of ulna**
Inferior angle of scapula	

Using table 24.1, indicate the locations of the surface features with an "X" on figure 24.10. Use a black "X" for the bony features and a red "X" for the soft tissue features.

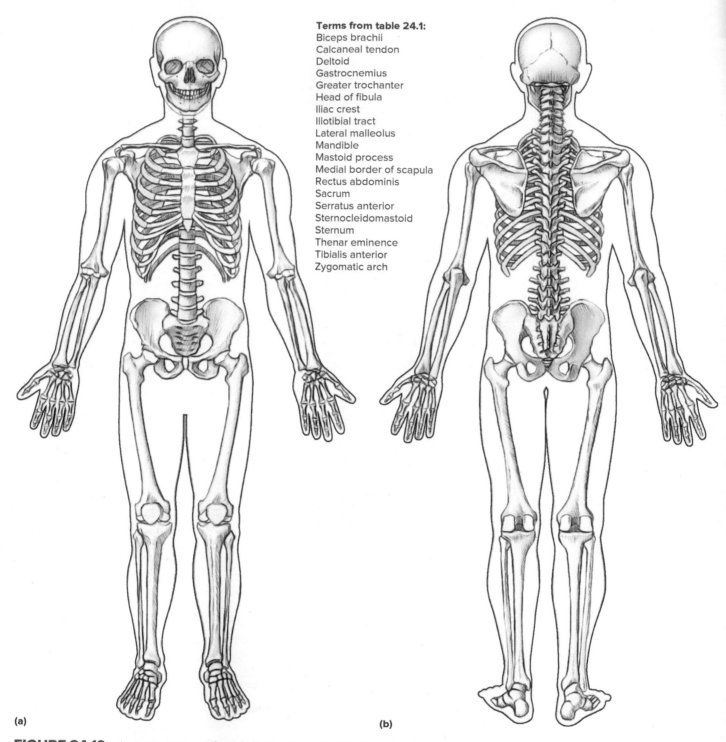

Terms from table 24.1:
Biceps brachii
Calcaneal tendon
Deltoid
Gastrocnemius
Greater trochanter
Head of fibula
Iliac crest
Iliotibial tract
Lateral malleolus
Mandible
Mastoid process
Medial border of scapula
Rectus abdominis
Sacrum
Serratus anterior
Sternocleidomastoid
Sternum
Thenar eminence
Tibialis anterior
Zygomatic arch

(a)   (b)

FIGURE 24.10   Indicate bony surface features with a black "X" and soft tissue surface features with a red "X" using the terms from table 24.1 on the (a) anterior and (b) posterior diagrams of the body. ⒊

25

Nervous Tissue and Nerves

MATERIALS NEEDED

Textbook
Compound light microscope
Prepared microscope slides of the following:
    Spinal cord (smear)
    Dorsal root ganglion (section)
    Neuroglia (astrocytes)
    Peripheral nerve (cross section and longitudinal
      section)
Neuron model

For Learning Extension:
Prepared microscope slide of Purkinje cells from
    cerebellum

PURPOSE OF THE EXERCISE

To review the characteristics of nervous tissue and to observe neurons, neuroglia, and various features of a nerve.

LEARNING OUTCOMES APR

After completing this exercise, you should be able to

① Describe the location and characteristics of nervous tissue.

② Distinguish structural and functional characteristics of neurons and neuroglia.

③ Identify and sketch the major structures of a neuron and a nerve.

Nervous tissue, which occurs in the brain, spinal cord, and nerves, contains neurons and neuroglia. The neurons are the basic structural and functional units of the nervous system involved in making decisions, detecting stimuli, and conducting impulses. The neuroglia, also called glial cells, have quite diverse functions. These include protection, insulation, and general support.

EXPLORE

PROCEDURE—Nervous Tissue and Nerves APR

1. Review sections 9.3 and 9.4 entitled "Neurons" and "Neuroglia" in chapter 9 of the textbook.
2. As a review activity, label figures 25.1 and 25.2.
3. Complete Part A of Laboratory Report 25.
4. Obtain a prepared microscope slide of a spinal cord smear. Using low-power magnification, search the slide and locate the relatively large, deeply stained cell bodies of motor neurons (multipolar neurons).
5. Observe a single, multipolar motor neuron using high-power magnification, and note the following features:

   cell body
       nucleus
       nucleolus
       neurofibrils (threadlike structures extending into the nerve fibers)
   dendrites
   axon (nerve fiber)

   Compare the slide to the neuron model and to figure 25.3. You also may note small, darkly stained nuclei of neuroglia around the motor neuron.
6. Sketch and label a motor (efferent) neuron in the space provided in Part B of the laboratory report.
7. Obtain a prepared microscope slide of a dorsal root ganglion. Search the slide and locate a cluster of sensory neuron cell bodies. You also may note bundles of nerve fibers passing among groups of neuron cell bodies (fig. 25.4).
8. Sketch and label a sensory (afferent) neuron cell body in the space provided in Part B of the laboratory report.

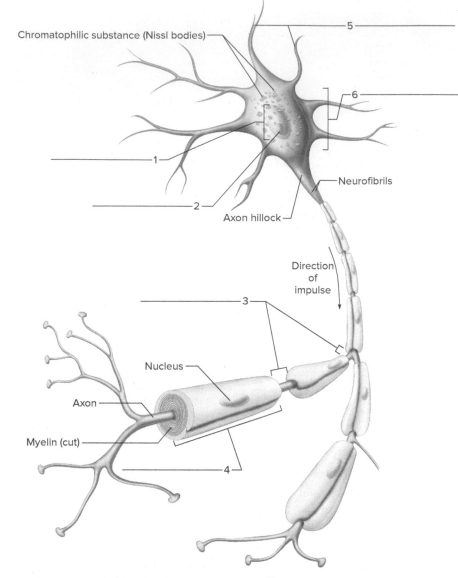

Chromatophilic substance (Nissl bodies)

5

6

Neurofibrils

1

2

Axon hillock

Direction of impulse

3

Nucleus

Axon

Myelin (cut)

4

FIGURE 25.1 Label this diagram of a multipolar motor neuron.

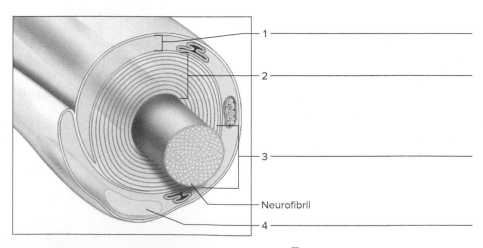

1

2

3

Neurofibril

4

FIGURE 25.2 Label the features of a myelinated axon (nerve fiber).

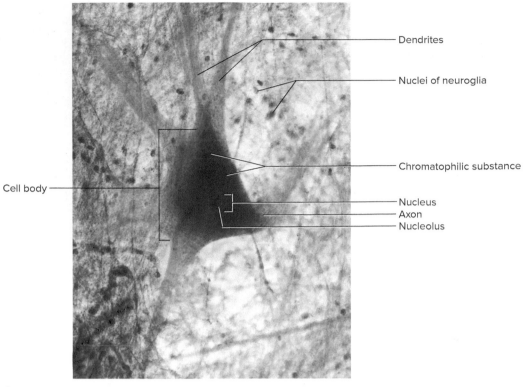

FIGURE 25.3   Micrograph of a multipolar neuron and neuroglia from a spinal cord smear (600×).   Alvin Telser/McGraw-Hill Education

Labels for Figure 25.3:
- Dendrites
- Nuclei of neuroglia
- Chromatophilic substance
- Nucleus
- Axon
- Nucleolus
- Cell body

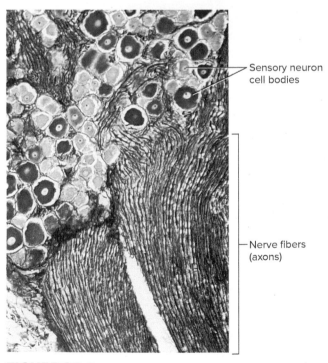

FIGURE 25.4   Micrograph of a dorsal root ganglion containing cell bodies of sensory neurons (100×).   APR
©Ed Reschke

Labels for Figure 25.4:
- Sensory neuron cell bodies
- Nerve fibers (axons)

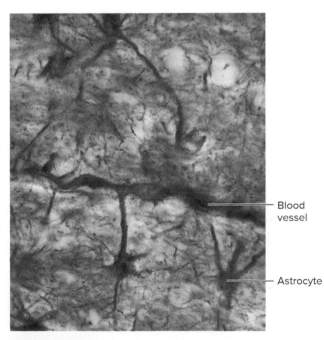

FIGURE 25.5   Micrograph of astrocytes (1,000×).
APR   ©Ed Reschke

Labels for Figure 25.5:
- Blood vessel
- Astrocyte

9. Obtain a prepared microscope slide of neuroglia. Search the slide and locate some darkly stained astrocytes with numerous long, slender processes (fig. 25.5).

10. Sketch neuroglia in the space provided in Part B of the laboratory report.

11. Obtain a prepared microscope slide of a nerve. Locate the cross section of the nerve and note the many round nerve fibers inside. Also note the dense layer of connective tissue (perineurium) that encircles a fascicle of nerve fibers and holds them together in a bundle. The individual nerve fibers are surrounded by a layer of more delicate connective tissue (endoneurium) (fig. 25.6).

12. Using high-power magnification, observe a single nerve fiber. Note the following features:

axon

myelin sheath around the axon of Schwann cell
(most of the myelin may have been dissolved and lost during the slide preparation)

neurilemma of Schwann cell

13. Sketch and label a nerve fiber with Schwann cell (cross section) in the space provided in Part C of the laboratory report.

14. Locate the longitudinal section of the nerve on the slide (fig. 25.7). Note the following:

axon

myelin sheath of Schwann cell

neurilemma of Schwann cell

node of Ranvier

15. Sketch and label a nerve fiber with Schwann cell (longitudinal section) in the space provided in Part C of the laboratory report.

LEARNING EXTENSION

Obtain a prepared microscope slide of Purkinje cells. To locate these neurons, search the slide for large, flask-shaped cell bodies. Each cell body has one or two large, thick dendrites that give rise to extensive branching networks of dendrites. These large cells are located in a particular region of the brain (cerebellar cortex).

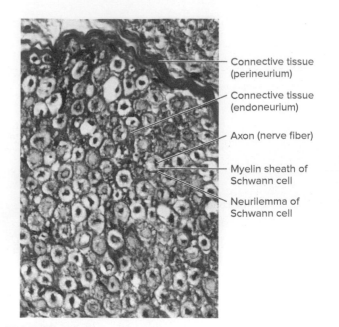

FIGURE 25.6 Cross section of a bundle of axons within a nerve (400×). ©Ed Reschke

Connective tissue (perineurium)
Connective tissue (endoneurium)
Axon (nerve fiber)
Myelin sheath of Schwann cell
Neurilemma of Schwann cell

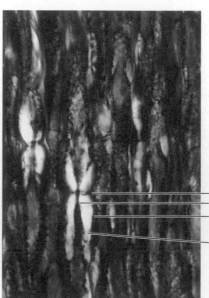

FIGURE 25.7 Longitudinal section of a nerve (650×).
©Ed Reschke

Node of Ranvier
Axon (nerve fiber)
Neurilemma of Schwann cell
Myelin sheath of Schwann cell

The A corresponds to the indicated Learning Outcome(s) found at the beginning of the laboratory exercise.

Nervous Tissue and Nerves

 PART A ASSESSMENTS

Match the terms in column A with the descriptions in column B. Place the letter of your choice in the space provided. Ⓐ Ⓑ

Column A	Column B
a. Astrocyte	_____ **1.** Sheath of Schwann cell containing cytoplasm and nucleus that encloses myelin
b. Axon	
c. Dendrite	_____ **2.** Network of fine threads extending into nerve fiber
d. Interneurons	
e. Microglia	_____ **3.** Substance of Schwann cell composed of lipoprotein that insulates axons and increases impulse speed
f. Motor (efferent) neuron	
g. Myelin sheath	_____ **4.** Neuron process with many branches that conducts an impulse toward the cell body
h. Neurilemma	
i. Neurofibrils	_____ **5.** Star-shaped neuroglia that prevent capillary leakage
j. Oligodendrocyte	
k. Sensory (afferent) neuron	_____ **6.** Nerve fiber arising from a slight elevation of the cell body that conducts an impulse (action potential) away from the cell body
	_____ **7.** Transmits impulse from sensory to motor neuron within central nervous system
	_____ **8.** Transmits impulse out of brain or spinal cord to effectors (muscles and glands)
	_____ **9.** Transmits impulse into brain or spinal cord from receptors
	_____ **10.** Myelin-forming neuroglia in brain and spinal cord
	_____ **11.** Spider-shaped cells that are phagocytic in function and are derived from white blood cells called monocytes

PART B ASSESSMENTS

In the space that follows, sketch the indicated cells. Label any of the cellular structures observed, and indicate the magnification of each sketch. ⓶ ⓷

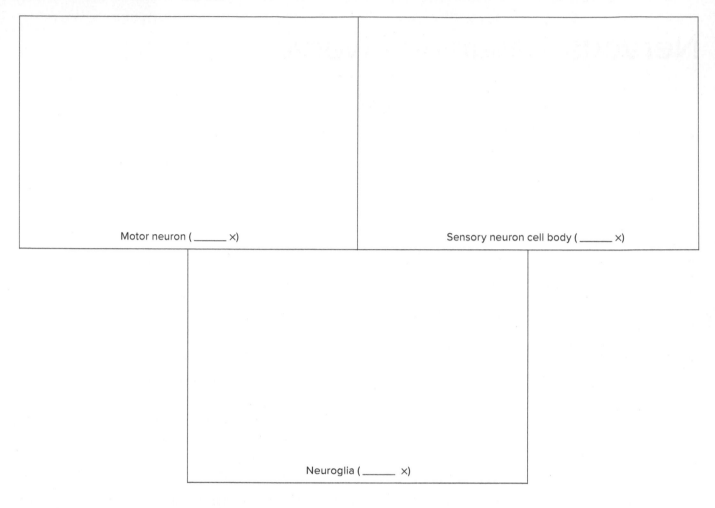

Motor neuron (_____ ×)

Sensory neuron cell body (_____ ×)

Neuroglia (_____ ×)

PART C ASSESSMENTS

In the space that follows, sketch the indicated view of a nerve fiber (axon). Label any structures observed, and indicate the magnification. ⓶ ⓷

Nerve fiber cross section with Schwann cell (_____ ×)

Nerve fiber longitudinal section with Schwann cell (_____ ×)

26

Spinal Cord and Meninges

MATERIALS NEEDED

Textbook
Compound light microscope
Prepared microscope slide of a spinal cord cross
    section with spinal nerve roots
Spinal cord model with meninges

For Demonstration:
Preserved spinal cord with meninges intact

PURPOSE OF THE EXERCISE

To review the characteristics of the spinal cord and meninges and to observe the major features of these structures.

LEARNING OUTCOMES APR

After completing this exercise, you should be able to

1. Identify the major features of the spinal cord.
2. Arrange the layers of the meninges and describe the structure of each.

The spinal cord is a column of nerve fibers that extends down through the vertebral canal. Together with the brain, it makes up the central nervous system.

Neurons within the spinal cord provide a two-way communication system between the brain and the body parts outside the central nervous system. The cord also contains the processing centers for the spinal reflexes.

Three membranes, or meninges, surround the entire CNS. The meninges consist of layers of membranes located between the bones of the skull and vertebral column and the soft tissues of the central nervous system. They include the dura mater, the arachnoid mater, and the pia mater.

 **EXPLORE**

PROCEDURE A—Structure of the Spinal Cord

1. Review section 9.13 entitled "Spinal Cord" in chapter 9 of the textbook.
2. As a review activity, label figure 26.1.
3. Complete Part A of Laboratory Report 26.
4. Obtain a prepared microscope slide of a spinal cord cross section. Use the low power of the microscope to locate the following features:

    posterior median sulcus
    anterior median fissure
    central canal
    gray matter
        gray commissure
        posterior (dorsal) horn
        lateral horn
        anterior (ventral) horn
    white matter
        posterior (dorsal) funiculus (column)
        lateral funiculus (column)
        anterior (ventral) funiculus (column)
    roots of spinal nerve
        dorsal roots
        dorsal root ganglia
        ventral roots

5. Observe the spinal cord model and locate the features listed in step 4.
6. Complete Part B of the laboratory report.

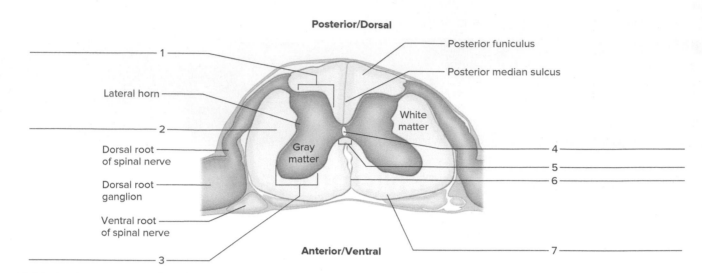

Posterior/Dorsal

Posterior funiculus

Posterior median sulcus

1

Lateral horn

2

White matter

Gray matter

4

Dorsal root of spinal nerve

5

Dorsal root ganglion

6

Ventral root of spinal nerve

Anterior/Ventral

7

3

FIGURE 26.1 Label this cross section of the spinal cord, including the features of the white and gray matter. **⚠️1** **APR**

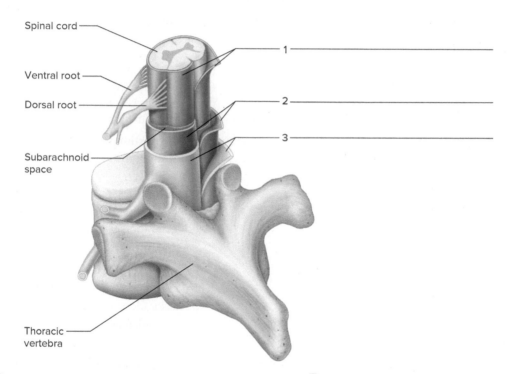

Spinal cord

1

Ventral root

Dorsal root

2

3

Subarachnoid space

Thoracic vertebra

FIGURE 26.2 Label the meninges associated with the spinal cord. **⚠️2** **APR**

EXPLORE

PROCEDURE B—Meninges

1. Review section 9.12 entitled "Meninges" in chapter 9 of the textbook.
2. Observe the spinal cord model with meninges and locate the following features:
   dura mater
   arachnoid mater
   subarachnoid space
   pia mater
3. As a review activity, label figure 26.2.
4. Complete Part C of the laboratory report.

DEMONSTRATION

Observe the preserved section of a spinal cord with the meninges intact. Note the appearance of the central gray matter and the surrounding white matter. Note the heavy covering of dura mater, firmly attached to the cord on each side by ligaments (denticulate ligaments) originating in the pia mater. The intermediate layer of meninges, the arachnoid mater, is devoid of blood vessels, but in a live human being, the space beneath this layer contains cerebrospinal fluid. The pia mater, closely attached to the surface of the spinal cord, contains many blood vessels.

Name _____

Date _____

Section _____

The corresponds to the indicated Learning Outcome(s) found at the beginning of the laboratory exercise.

LABORATORY
REPORT

Spinal Cord and Meninges

PART A ASSESSMENTS

Complete the following statements:

1. The spinal cord gives rise to 31 pairs of _____. A

2. The bulge in the spinal cord that gives off nerves to the upper limbs is called the _____. A

3. The bulge in the spinal cord that gives off nerves to the lower limbs is called the _____. A

4. The _____ is a groove that extends the length of the spinal cord posteriorly. A

5. In a spinal cord cross section, the posterior _____ of the gray matter resemble the upper wings of a butterfly. A

6. The cell bodies of motor neurons are found in the _____ horns of the spinal cord. A

7. The _____ connects the gray matter on the left and right sides of the spinal cord. A

8. The _____ in the gray commissure of the spinal cord contains cerebrospinal fluid and is continuous with the ventricles of the brain. A

9. The white matter of the spinal cord is divided into anterior, lateral, and posterior _____. A

10. The longitudinal bundles of nerve fibers within the spinal cord constitute major nerve pathways called _____. A

PART B ASSESSMENTS

Match the terms in column A with the descriptions in column B. Place the letter of your choice in the space provided. A

Column A	Column B
a. Arachnoid mater	_____ **1.** Outermost layer of meninges
b. Dura mater	_____ **2.** Follows irregular contours of spinal cord surface
c. Epidural space	_____ **3.** Contains cerebrospinal fluid
d. Pia mater	_____ **4.** Thin, weblike middle membrane
e. Subarachnoid space	_____ **5.** Separates dura mater from bone of vertebra

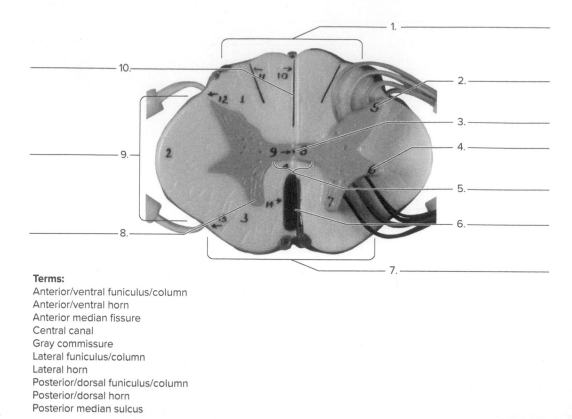

Terms:
Anterior/ventral funiculus/column
Anterior/ventral horn
Anterior median fissure
Central canal
Gray commissure
Lateral funiculus/column
Lateral horn
Posterior/dorsal funiculus/column
Posterior/dorsal horn
Posterior median sulcus

FIGURE 26.3 Using the terms provided identify the structures indicated on the cross section model of the spinal cord image. ©2017 Denoyer-Geppert Science Company denoyer.com

PART C ASSESSMENTS

Identify the features indicated in the spinal cord section of figure 26.4. **A**

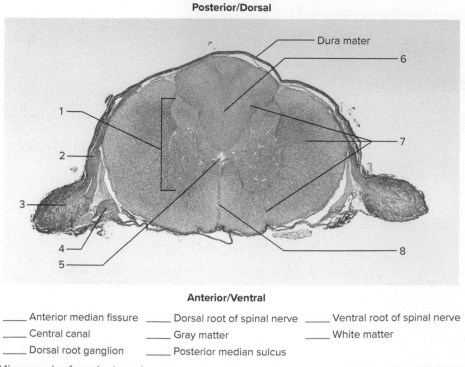

Posterior/Dorsal

Dura mater

Anterior/Ventral

____ Anterior median fissure	____ Dorsal root of spinal nerve	____ Ventral root of spinal nerve
____ Central canal	____ Gray matter	____ White matter
____ Dorsal root ganglion	____ Posterior median sulcus	

FIGURE 26.4 Micrograph of a spinal cord cross section with spinal nerve roots (7.5×). Label the features by placing the correct numbers in the spaces provided. ©Ed Reschke

27

Reflex Arc and Reflexes

MATERIALS NEEDED

Textbook
Rubber percussion hammer

PURPOSE OF THE EXERCISE

To review the characteristics of reflex arcs and reflex behavior and to demonstrate some of the reflexes that occur in the human body.

LEARNING OUTCOMES APR

After completing this exercise, you should be able to

1. Describe the components of a reflex arc.
2. Demonstrate and record stretch reflexes that occur in humans.
3. Analyze the components and patterns of stretch reflexes.

A reflex arc represents the simplest type of neural pathway found in the nervous system. This pathway begins with a receptor at the dendrite end of a sensory (afferent) neuron. The sensory neuron leads into the central nervous system and may communicate with one or more interneurons. Some of these interneurons, in turn, communicate with motor (efferent) neurons, whose axons (nerve fibers) lead outward to effectors.

Thus, when a sensory receptor is stimulated by a change occurring inside or outside the body, impulses may pass through a reflex arc, and, as a result, effectors may respond. Such an automatic, subconscious response is called a *reflex*.

A *stretch reflex* involves a single synapse (monosynaptic) between a sensory and a motor neuron within the gray matter of the spinal cord. Examples of stretch reflexes include the patellar, calcaneal, biceps, triceps, and plantar reflexes. Other, more complex *withdrawal reflexes* involve interneurons in combination with sensory and motor neurons; thus, polysynaptic. Examples of withdrawal reflexes include responses to touching hot objects or stepping on sharp objects.

Reflexes demonstrated in this laboratory exercise are stretch reflexes. When a muscle is stretched by a tap over its tendon, stretch receptors (proprioceptors) called *muscle spindles* are stretched within the muscle, which initiates an impulse over a reflex arc. A sensory neuron conducts an impulse from the muscle spindle into the gray matter of the spinal cord, where it synapses with a motor neuron, which conducts the impulse to the effector muscle. The stretched muscle responds by contracting to resist or reverse further stretching. These stretch reflexes are important to maintaining proper posture, balance, and movements. Observation of many of these reflexes in clinical tests on patients may indicate damage to a level of the spinal cord or peripheral nerves of the particular reflex arc.

EXPLORE

PROCEDURE—Reflex Arc and Reflexes

1. Review sections 9.10 and 9.11 entitled "Types of Nerves" and "Neural Pathways" in chapter 9 of the textbook.
2. Study figure 27.1.
3. Complete Part A of Laboratory Report 27.
4. Work with a laboratory partner to demonstrate each of the reflexes listed. (See fig. 27.2 also.) *It is important that muscles involved in the reflexes be totally relaxed to observe proper responses.* If a person is trying too hard to experience the reflex or is trying to suppress the reflex, assign a multitasking activity while the stimulus with the rubber hammer occurs. For example, assign a physical task with the upper limbs along with a complex mental activity during the patellar (knee-jerk) demonstration. After each demonstration, record your observations in the table provided in Part B of the laboratory report.
   a. *Patellar (knee-jerk) reflex.* Have your laboratory partner sit on a table (or sturdy chair) with legs relaxed and hanging freely over the edge without touching the floor. Gently strike your partner's patellar ligament (just below the patella) with the blunt side of a rubber percussion hammer (fig. 27.2a). The reflex involves the femoral nerve and the spinal cord. The normal response is a moderate extension of the leg at the knee joint.

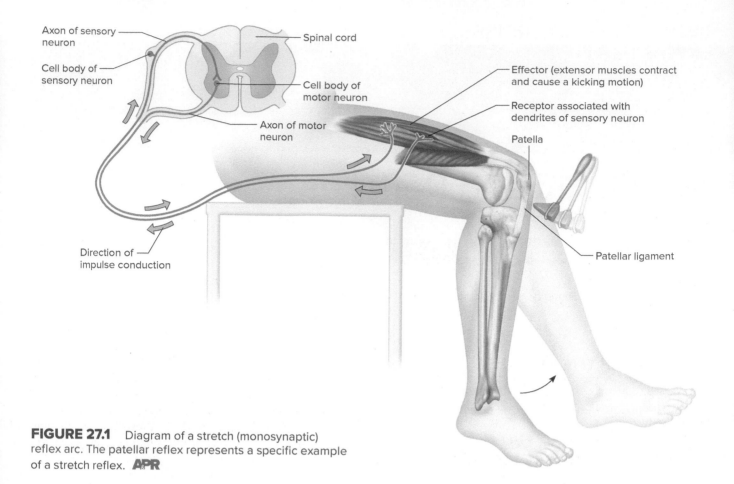

FIGURE 27.1 Diagram of a stretch (monosynaptic) reflex arc. The patellar reflex represents a specific example of a stretch reflex. **APR**

Labels on figure:
- Axon of sensory neuron
- Cell body of sensory neuron
- Spinal cord
- Cell body of motor neuron
- Axon of motor neuron
- Direction of impulse conduction
- Effector (extensor muscles contract and cause a kicking motion)
- Receptor associated with dendrites of sensory neuron
- Patella
- Patellar ligament

b. *Calcaneal (ankle-jerk) reflex.* Have your partner kneel on a chair with their back toward you and feet slightly dorsiflexed over the edge and relaxed. Gently strike the calcaneal tendon (just above its insertion on the calcaneus) with the blunt side of the rubber hammer (fig. 27.2*b*). The reflex involves the tibial nerve and the spinal cord. The normal response is plantar flexion of the foot.

c. *Biceps (biceps-jerk) reflex.* Have your partner place a bare arm bent about 90° at the elbow on the table. Press your thumb on the inside of the elbow over the tendon of the biceps brachii, and gently strike your thumb with the rubber hammer (fig. 27.2*c*). The reflex involves the musculocutaneous nerve and the spinal cord. Watch the biceps brachii for a response. The response might be a slight twitch of the muscle or flexion of the forearm at the elbow joint.

d. *Triceps (triceps-jerk) reflex.* Have your partner lie supine with an upper limb bent about 90° across the abdomen. Gently strike the tendon of the triceps brachii near its insertion just proximal to the olecranon process at the tip of the elbow (fig. 27.2*d*). The reflex involves the radial nerve and the spinal cord. Watch the triceps brachii for a response. The response might be a slight twitch of the muscle or extension of the forearm at the elbow joint.

e. *Plantar reflex.* Have your partner remove a shoe and sock and lie supine with the lateral surface of the foot resting on the table. Draw the metal tip of the rubber hammer, applying firm pressure, over the sole from the heel to the base of the large toe (fig. 27.2*e*). The normal response is flexion (curling) of the toes and plantar flexion of the foot. If the toes spread apart and dorsiflexion of the great toe occurs, the reflex is the abnormal *Babinski reflex* response (normal in infants until the nerve fibers have completed myelinization). If the Babinski reflex occurs later in life, it may indicate damage to the corticospinal tract of the CNS.

5. Complete Part B of the laboratory report.

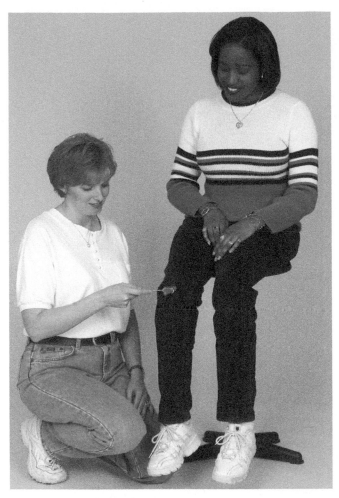

(a) Patellar (knee-jerk) reflex

(b) Calcaneal (ankle-jerk) reflex

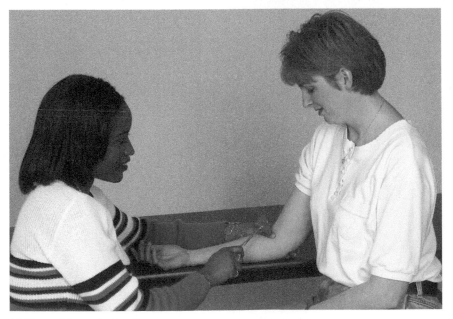

(c) Biceps (biceps-jerk) reflex

FIGURE 27.2 Demonstrate each of the following reflexes, (*a*) patellar reflex, (*b*) calcaneal reflex, (*c*) biceps reflex, (*d*) triceps reflex, and (*e*) plantar reflex. **(a–e):** ©J and J Photography

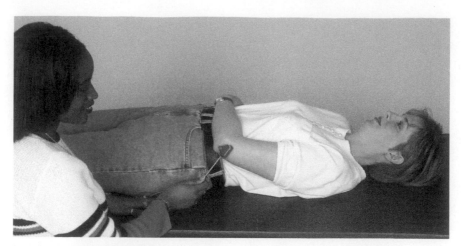

(d) Triceps (triceps-jerk) reflex

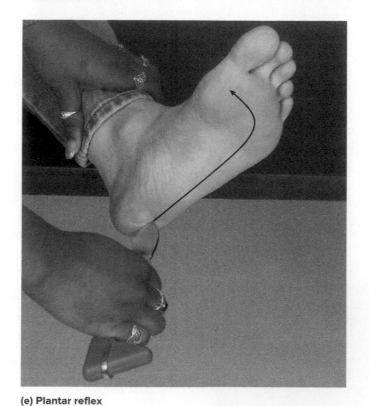

(e) Plantar reflex

FIGURE 27.2 *Continued.*

Name _____

Date _____

Section _____

The corresponds to the indicated Learning Outcome(s) found at the beginning of the laboratory exercise.

Reflex Arc and Reflexes

PART A ASSESSMENTS

Complete the following statements:

1. Neural _____ are routes followed by nerve impulses as they pass through the nervous system. A

2. Interneurons in a withdrawal reflex are located in the _____ . A

3. _____ are automatic subconscious responses to external or internal stimuli. A

4. Effectors of a reflex arc are glands and _____ . A

5. A patellar muscle reflex employs only _____ and motor neurons. A

6. The effector muscle of the patellar reflex is the _____ . A

7. The sensory stretch receptors (muscle spindles) of the patellar reflex are located in the _____ muscle. A

8. The patellar reflex helps the body to maintain upright _____ . A

9. The sensory receptors of a withdrawal reflex are located in the _____ . A

10. _____ muscles in the limbs are the effectors of a withdrawal reflex. A

PART B ASSESSMENTS

1. Complete the following table.

Reflex Tested	Response Observed	Degree of Response (hypoactive, normal, or hyperactive)	Effector Muscle Involved
Patellar (knee-jerk)			
Calcaneal (ankle-jerk)			
Biceps (biceps-jerk)			
Triceps (triceps-jerk)			
Plantar			Gastrocnemius, soleus, and flexor digitorum longus

2. List the major events that occur in the patellar (knee-jerk) reflex, from the striking of the patellar ligament to the resulting response. ⚠ _____

_____

_____

_____

_____

Critical Thinking Application

What characteristics do the reflexes you demonstrated have in common? ⚠

_____

_____

_____

_____

28

Brain and Cranial Nerves

MATERIALS NEEDED

Textbook
Dissectible model of the human brain
Preserved human brain
Anatomical charts of the human brain

PURPOSE OF THE EXERCISE

To review the structural and functional characteristics of the human brain and cranial nerves.

LEARNING OUTCOMES APR

After completing this exercise, you should be able to

1. Identify the major external and internal structures in the human brain.
2. Locate the major functional regions of the brain.
3. Identify the twelve pairs of cranial nerves.
4. Differentiate the functions of the cranial nerves.

The brain, the largest and most complex part of the nervous system, contains nerve centers associated with sensory functions and is responsible for sensations and perceptions. It issues motor commands to skeletal muscles and carries on higher mental activities. It also functions to coordinate muscular movements, and it contains centers and neural pathways necessary for the regulation of internal organs.

Twelve pairs of cranial nerves arise from the ventral surface of the brain and are designated by number and name. The cranial nerves are part of the PNS; most arise from the brainstem region of the brain. Although most of these nerves conduct both sensory and motor impulses, some contain only sensory fibers associated with special sense organs. Others are primarily composed of motor fibers and are involved with the activities of muscles and glands.

EXPLORE

PROCEDURE A—Human Brain

1. Review section 9.14 entitled "Brain" in chapter 9 of the textbook.

2. As a review activity, label figures 28.1 and 28.2.
3. Complete Part A of Laboratory Report 28.
4. Observe the anatomical charts, dissectible model, and preserved specimen of the human brain. Locate each of the following features:

cerebrum
cerebral hemispheres
corpus callosum
gyri (convolutions)
sulci
  central sulcus
  lateral sulcus
fissures
  longitudinal fissure
  transverse fissure
lobes
  frontal lobe
  parietal lobe
  temporal lobe
  occipital lobe
  insula (insular lobe)
cerebral cortex
ventricles
  lateral ventricles
  third ventricle
  fourth ventricle
  choroid plexuses
  cerebral aqueduct
diencephalon
  thalamus
  hypothalamus
  optic chiasma
  mammillary bodies
  pineal gland
cerebellum
  lateral (right and left) hemispheres
  vermis
  cerebellar cortex
brainstem
  midbrain
  pons
  medulla oblongata

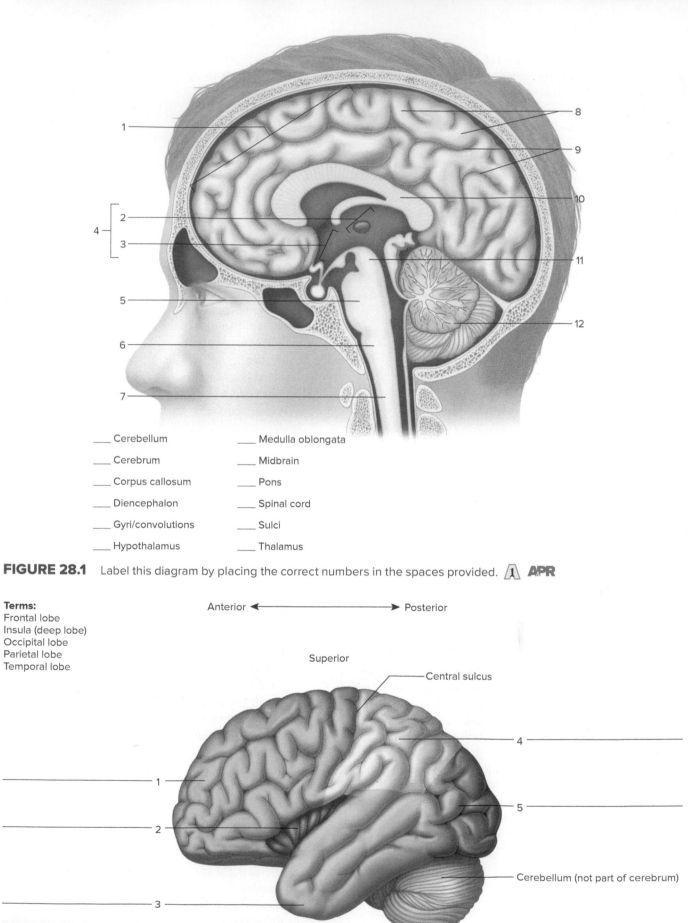

_____ Cerebellum   _____ Medulla oblongata

_____ Cerebrum   _____ Midbrain

_____ Corpus callosum   _____ Pons

_____ Diencephalon   _____ Spinal cord

_____ Gyri/convolutions   _____ Sulci

_____ Hypothalamus   _____ Thalamus

FIGURE 28.1   Label this diagram by placing the correct numbers in the spaces provided. ⚠ **APR**

Terms:
Frontal lobe
Insula (deep lobe)
Occipital lobe
Parietal lobe
Temporal lobe

Anterior ⟷ Posterior

Superior

Central sulcus

Cerebellum (not part of cerebrum)

Inferior

FIGURE 28.2   Label the five lobes of the left cerebral hemisphere, using the terms provided. Lobe 3 is retracted to expose the deep lobe 2 in this figure. ⚠ **APR**

5. Label the areas in figure 28.3 that represent the following functional regions of the cerebrum:

   motor area for voluntary muscle control

   motor speech area (Broca's area)

   cutaneous sensory area

   auditory area

   visual area

   sensory speech area (Wernicke's area)

6. Complete Part B of the laboratory report.

EXPLORE

PROCEDURE B—Cranial Nerves

1. Review the concept heading entitled "Cranial Nerves" in section 9.15 of chapter 9 of the textbook.
2. As a review activity, label figure 28.4.
3. Observe the model and preserved specimen of the human brain, and locate as many of the following cranial nerves as possible as you differentiate some of their associated functions:

   olfactory nerves (I)—smell

   optic nerves (II)—vision

   oculomotor nerves (III)—pupil constriction and opening of eyelid

   trochlear nerves (IV)—stimulation of superior oblique eye muscles

   trigeminal nerves (V)—sensation from face and teeth and mastication movements

   abducens nerves (VI)—lateral eye movements

   facial nerves (VII)—salivation, tear secretions, and taste

   vestibulocochlear nerves (VIII)—hearing and balance (equilibrium)

   glossopharyngeal nerves (IX)—control of swallowing movements

   vagus nerves (X)—regulation of many visceral organs, including the heart rate

   accessory nerves (XI)—control of neck and shoulder muscles

   hypoglossal nerves (XII)—control of tongue movements

The following mnemonic device will help you learn the twelve pairs of cranial nerves in the proper order:

Old **Op**ie **oc**casionally **tri**es **trig**onometry, **a**nd **f**eels **v**ery **glo**omy, **vagu**e, **a**nd **hypo**active.[1]

4. Complete Part C of the laboratory report.

1. From *HAPS-EDucator*, Winter 2002. An official publication of the Human Anatomy & Physiology Society (HAPS).

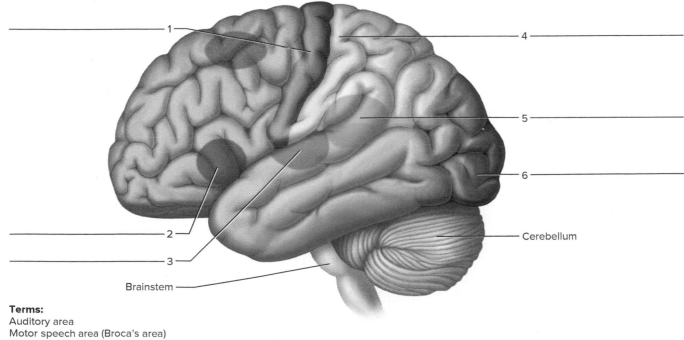

Terms:
Auditory area
Motor speech area (Broca's area)
Cutaneous sensory area
Motor area for voluntary muscle control
Visual area
Sensory speech area (Wernicke's area)

FIGURE 28.3 Label the functional areas of the cerebrum, using the terms provided. (*Note:* These areas are not visible as distinct parts of the brain.) APR

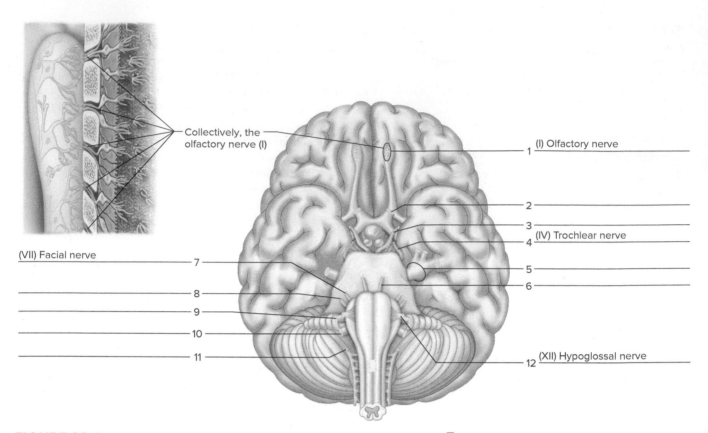

Collectively, the olfactory nerve (I)

(VII) Facial nerve

(I) Olfactory nerve

(IV) Trochlear nerve

(XII) Hypoglossal nerve

1
2
3
4
5
6
7
8
9
10
11
12

FIGURE 28.4   Provide the names of the cranial nerves in this ventral view.

Name _____

Date _____

Section _____

The ⚠ corresponds to the indicated Learning Outcome(s) found at the beginning of the laboratory exercise.

Brain and Cranial Nerves

PART A ASSESSMENTS

Match the terms in column A with the descriptions in column B. Place the letter of your choice in the space provided. ⚠

Column A	Column B
a. Central sulcus	_____ **1.** Cone-shaped gland attached to upper posterior portion of diencephalon
b. Cerebral cortex	_____ **2.** Connects cerebral hemispheres
c. Corpus callosum	_____ **3.** Ridge on surface of cerebrum
d. Diencephalon	_____ **4.** Separates frontal and parietal lobes
e. Gyrus (convolution)	_____ **5.** Part of brainstem between diencephalon and pons
f. Medulla oblongata	_____ **6.** Rounded bulge on underside of brainstem
g. Midbrain	_____ **7.** Part of brainstem continuous with the spinal cord
h. Pineal gland	_____ **8.** Internal brain chamber filled with cerebrospinal fluid
i. Pons	_____ **9.** Contains thalamus and hypothalamus
j. Ventricle	_____ **10.** Thin layer of gray matter on surface of cerebrum

PART B ASSESSMENTS

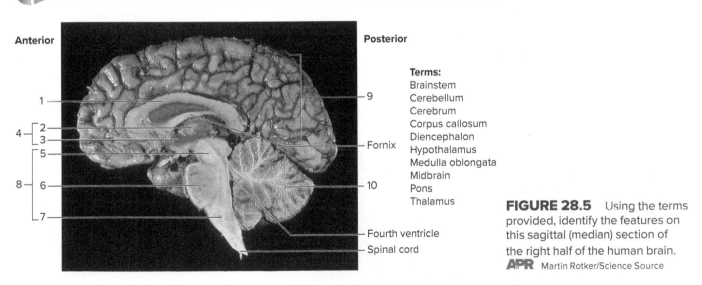

Anterior   Posterior

Terms:
Brainstem
Cerebellum
Cerebrum
Corpus callosum
Diencephalon
Hypothalamus
Medulla oblongata
Midbrain
Pons
Thalamus

Fornix

Fourth ventricle
Spinal cord

FIGURE 28.5 Using the terms provided, identify the features on this sagittal (median) section of the right half of the human brain.
APR Martin Rotker/Science Source

Identify the features indicated in the sagittal (median) section of the right half of the human brain in figure 28.5. ⚠

1. _____   6. _____

2. _____   7. _____

3. _____   8. _____

4. _____   9. _____

5. _____   10. _____

PART C ASSESSMENTS

Match the cranial nerves in column A with the associated functions in column B. Place the letter of your choice in the space provided. ④

Column A	Column B
a. Abducens	_____ **1.** Regulates heart rate
b. Accessory	_____ **2.** Equilibrium and hearing
c. Facial	_____ **3.** Stimulates superior oblique muscle of eye
d. Glossopharyngeal	_____ **4.** Sensory impulses from teeth and face
e. Hypoglossal	_____ **5.** Pupil constriction and eyelid opening
f. Oculomotor	_____ **6.** Smell
g. Olfactory	_____ **7.** Controls neck and shoulder movements
h. Optic	_____ **8.** Controls tongue movements
i. Trigeminal	_____ **9.** Vision
j. Trochlear	_____ **10.** Stimulates lateral rectus muscle of eye
k. Vagus	_____ **11.** Taste, salivation, and secretion of tears
l. Vestibulocochlear	_____ **12.** Swallowing

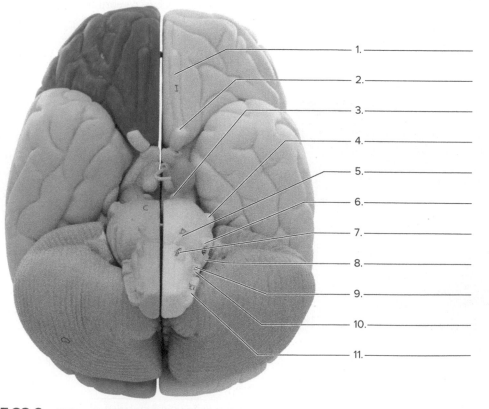

Terms:
Abducens
Accessory
Facial
Glossopharyngeal
Hypoglossal
Oculomotor
Olfactory
Optic
Trigeminal
Vagus
Vestibulocochlear

1.
2.
3.
4.
5.
6.
7.
8.
9.
10.
11.

FIGURE 28.6 Using the terms provided identify the visible cranial nerves on the model image. (Trochlear nerve not visible) C22 Neuro-Anatomical Brain © 3B Scientific GmbH, Germany, 2017 www.3bscientific.com

29

Dissection of the Sheep Brain

MATERIALS NEEDED

Dissectible model of human brain
Preserved sheep brain
Dissecting tray
Dissection instruments
Long knife
Disposable gloves

⚠ SAFETY

- Wear disposable gloves and safety goggles when handling the sheep brains.
- Save or dispose of the brains as instructed.
- Wash your hands before leaving the laboratory.

PURPOSE OF THE EXERCISE

To observe the major features of the sheep brain and to compare these features with those of the human brain.

LEARNING OUTCOMES

After completing this exercise, you should be able to

1. Examine the major structures of the sheep brain.
2. Summarize several differences and similarities between the sheep brain and the human brain.
3. Locate the most developed cranial nerves of the sheep brain.

Mammalian brains have many features in common. Human brains may not be available, so sheep brains often are dissected as an aid to understanding mammalian brain structure. However, the adaptations of the sheep differ from the adaptations of the human, so comparisons of their structural features may not be precise. The sheep is a quadruped; therefore, the spinal cord is horizontal, unlike the vertical orientation in a bipedal human. Preserved sheep brains have a different appearance and are firmer than those removed from the cranial cavity; this is caused by the preservatives used.

⟳ EXPLORE

PROCEDURE—Dissection of the Sheep Brain

1. Obtain a preserved sheep brain and rinse it thoroughly in water to remove as much of the preserving fluid as possible.
2. Examine the surface of the brain for the presence of meninges. (The outermost layers of these membranes may have been mostly lost during removal of the brain from the cranial cavity.) Locate the following:

   dura mater—the thick, opaque outer layer

   arachnoid mater—the weblike, semitransparent middle layer spanning the fissures of the brain

   pia mater—the thin, delicate, innermost layer that adheres to the gyri and sulci of the surface of the brain

3. Remove any remaining dura mater by pulling it gently from the surface of the brain.
4. Position the brain with its ventral surface down in the dissecting tray. Study figure 29.1, and locate the following structures on the specimen:

   cerebral hemispheres

   gyri (convolutions)

   sulci

   longitudinal fissure

   frontal lobe

   parietal lobe

   temporal lobe

   occipital lobe

   cerebellum

   medulla oblongata

   spinal cord

5. Gently separate the cerebral hemispheres along the longitudinal fissure and expose the transverse band of white fibers within the fissure that connects the hemispheres. This band is the *corpus callosum*.

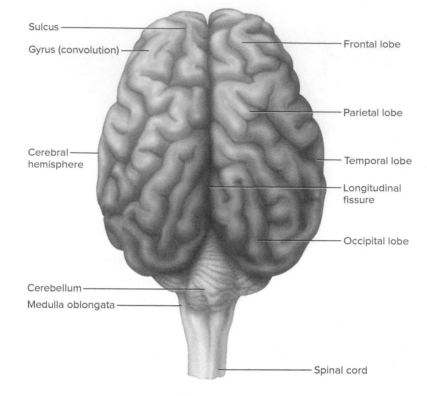

FIGURE 29.1 Dorsal surface of the sheep brain. **APR**

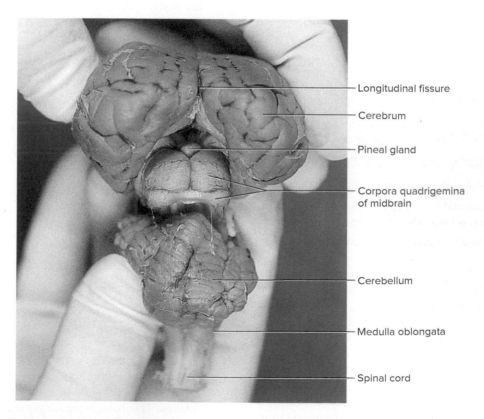

FIGURE 29.2 Gently bend the cerebellum and medulla oblongata away from the cerebrum to expose the pineal gland of the diencephalon and the corpora quadrigemina of the midbrain. ©J and J Photography

6. Bend the cerebellum and medulla oblongata slightly downward and away from the cerebrum (figure 29.2). This will expose the *pineal gland* in the upper midline and the *corpora quadrigemina,* which consists of four rounded structures, called *colliculi,* associated with the midbrain.

7. Position the brain with its ventral surface upward. Study figures 29.3 and 29.4, and locate the following structures on the specimen:

longitudinal fissure
olfactory bulbs
optic nerves
optic chiasma
optic tract
mammillary body (single in sheep)
infundibulum (pituitary stalk)
midbrain
pons

8. Although some of the cranial nerves may be missing or are quite small and difficult to find, locate as many of the following as possible, using figures 29.3 and 29.4 as references:

oculomotor nerves
trochlear nerves
trigeminal nerves
abducens nerves
facial nerves
vestibulocochlear nerves
glossopharyngeal nerves
vagus nerves
accessory nerves
hypoglossal nerves

9. Using a long, sharp knife, cut the sheep brain along the midline to produce a median section. Study figures 29.2 and 29.5, and locate the following structures on the specimen:

cerebrum
olfactory bulb
corpus callosum
cerebellum
    white matter
    gray matter
lateral ventricle (one in each cerebral hemisphere)
third ventricle (within diencephalon)
fourth ventricle (between brainstem and cerebellum)
diencephalon
    optic chiasma
    infundibulum
    pituitary gland (this structure may be missing)
    mammillary body (single in sheep)
    thalamus
    hypothalamus
    pineal gland
midbrain
pons
medulla oblongata

10. Dispose of the sheep brain as directed by the laboratory instructor.

11. Complete Parts A, B, and C of Laboratory Report 29.

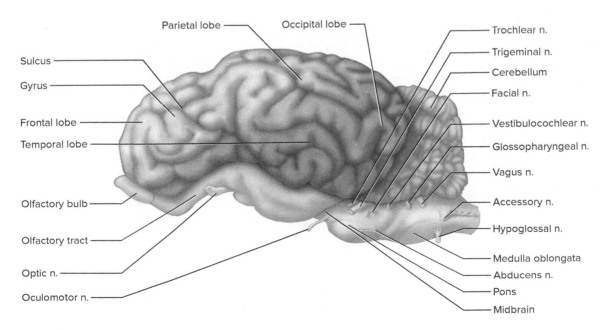

FIGURE 29.3 Lateral surface of the sheep brain. (The "n." stands for nerve.)

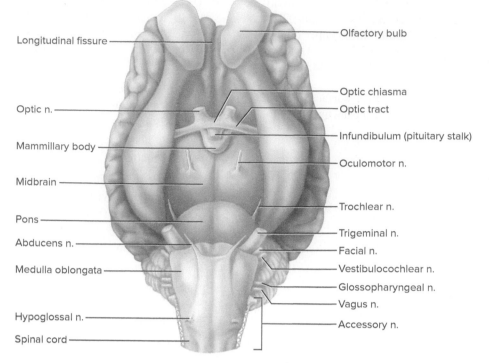

FIGURE 29.4 Ventral surface of the sheep brain. **APR**

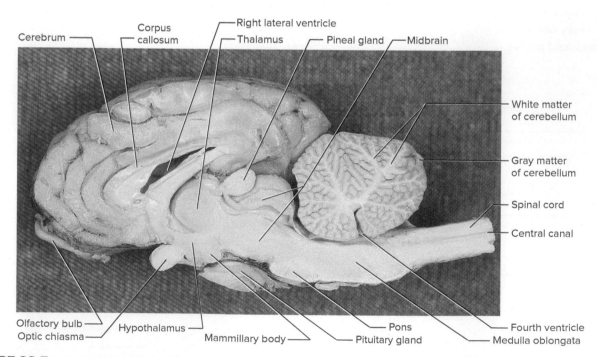

FIGURE 29.5 Sagittal (median) section of the right half of the sheep brain dissection. **APR** ©J and J Photography

Name _____

Date _____

Section _____

The corresponds to the indicated Learning Outcome(s) found at the beginning of the laboratory exercise.

Dissection of the Sheep Brain

PART A ASSESSMENTS

Answer the following questions:

1. Describe the location of any meninges observed to be associated with the sheep brain. 🔲1 _____

_____

_____

2. Compare the relative sizes of the sheep and human cerebral hemispheres. 🔲2 _____

_____

_____

3. How do the gyri and sulci of the sheep cerebrum compare with the human cerebrum in numbers? 🔲2

_____

_____

4. What is the significance of the differences you noted in your answers for questions 2 and 3? 🔲2

_____

_____

5. What difference did you note in the structures of the sheep cerebellum and the human cerebellum? 🔲2

_____

_____

6. How do the sizes of the olfactory bulbs of the sheep brain compare with those of the human brain? 🔲2

_____

_____

7. Based on their relative sizes, which of the cranial nerves seem to be most highly developed in the sheep brain? 🔲3

_____

_____

8. What is the significance of the observations you noted in your answers for questions 6 and 7? 🔲2

_____

_____

Critical Thinking Application

Prepare a list of at least six features to illustrate ways in which the brains of sheep and humans are similar. ⚠2

1. _____

2. _____

3. _____

4. _____

5. _____

6. _____

Interpret the significance of these similarities. ⚠2 _____

_____

_____

_____

PART C ASSESSMENTS

Identify the features indicated in the sagittal (median) section of the sheep brain in figure 29.6. ⚠1

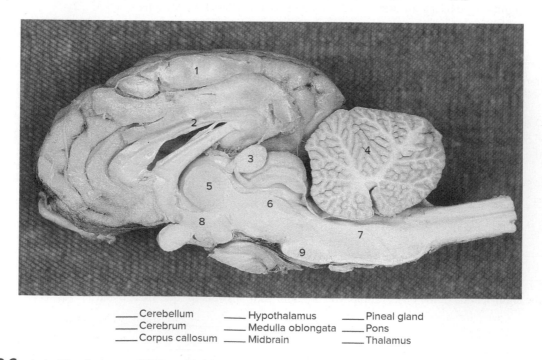

_____ Cerebellum   _____ Hypothalamus   _____ Pineal gland
_____ Cerebrum   _____ Medulla oblongata   _____ Pons
_____ Corpus callosum   _____ Midbrain   _____ Thalamus

FIGURE 29.6   Label the features of this sagittal (median) section of the sheep brain by placing the correct numbers in the spaces provided. ©J and J Photography

30

Ear and Hearing

Textbook
Dissectible ear model
Watch that ticks
Tuning fork (128 or 256 cps)
Rubber hammer
Cotton
Meterstick

For Demonstrations:
Compound light microscope
Prepared microscope slide of cochlea (section)
Audiometer

PURPOSE OF THE EXERCISE

To review the structural and functional characteristics of the ear and to conduct some ordinary hearing tests.

LEARNING OUTCOMES A&PR

After completing this exercise, you should be able to

1. Locate the major structures of the ear.
2. Describe the functions of the structures of the ear.
3. Trace the pathway of sound vibrations from the eardrum to the hearing receptors.
4. Conduct four ordinary hearing tests and summarize the results.

The ear is composed of outer (external), middle, and inner (internal) parts. The outer structures gather sound waves and direct them inward to the eardrum (tympanic membrane). The parts of the middle ear, in turn, transmit vibrations from the eardrum to the inner ear, where the hearing receptors are located. As they are stimulated, these receptors initiate impulses that are conducted over the vestibulocochlear nerve into the auditory cortex of the brain, where the impulses are interpreted and the sensations of hearing are created.

EXPLORE

PROCEDURE A—Structure and Function of the Ear A&PR

1. Review section 10.7 entitled "Sense of Hearing" in chapter 10 of the textbook.
2. As a review activity, label figures 30.1, 30.2, and 30.3.
3. Examine the dissectible model of the ear and locate the following features:

   outer (external) ear
       auricle (pinna)
       external acoustic meatus (external auditory canal)
       eardrum (tympanic membrane)

   middle ear
       tympanic cavity
       auditory ossicles (fig. 30.4)
           malleus
           incus
           stapes
       oval window
   auditory tube (pharyngotympanic tube; eustachian tube)
   inner (internal) ear
       bony (osseous) labyrinth
       membranous labyrinth
       cochlea
       semicircular canals and ducts
       vestibule
           utricle
           saccule
       vestibulocochlear nerve
       vestibular nerve (balance branch)
       cochlear nerve (hearing branch)

4. Complete Parts A and B of Laboratory Report 30.

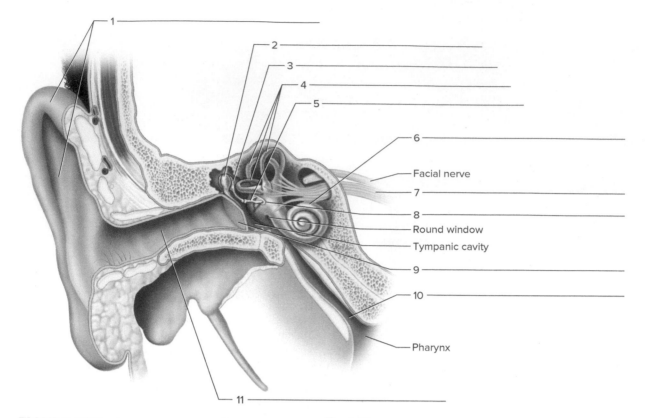

FIGURE 30.1 Label the major structures of the ear. ⚠️ **APR**

Facial nerve
7
8
Round window
Tympanic cavity
9
10
Pharynx

DEMONSTRATION

Observe the section of the cochlea in the demonstration microscope. Locate one of the turns of the cochlea and, using figures 30.3 and 30.5 as a guide, identify the scala vestibuli, cochlear duct, scala tympani, vestibular membrane, basilar membrane, and the spiral organ.

EXPLORE

PROCEDURE B—Hearing Tests

Perform the following tests in a quiet room, using your laboratory partner as the test subject.

1. *Auditory acuity test.* To conduct this test, follow these steps:
   a. Have the test subject sit with eyes closed.
   b. Pack one of the subject's ears with cotton.
   c. Hold a ticking watch close to the open ear, and slowly move it straight out and away from the ear.
   d. Have the subject indicate when the sound of the ticking can no longer be heard.
   e. Use a meterstick to measure the distance in centimeters from the ear to the position of the watch.
   f. Repeat this procedure to test the acuity of the other ear.
   g. Record the test results in Part C of the laboratory report.

2. *Sound localization test.* To conduct this test, follow these steps:
   a. Have the subject sit with eyes closed.
   b. Hold the ticking watch somewhere within the audible range of the subject's ears, and ask the subject to point to the watch.
   c. Move the watch to another position and repeat the request. In this manner, determine how accurately the subject can locate the watch when it is in each of the following positions: in front of the head, behind the head, above the head, on the right side of the head, and on the left side of the head.
   d. Record the test results in Part C of the laboratory report.

3. *Rinne test.* This test is done to assess possible conduction deafness by comparing bone and air conduction. To conduct this test, follow these steps:
   a. Obtain a tuning fork and strike it with a rubber hammer or on the heel of your hand, causing it to vibrate.
   b. Place the end of the fork's handle against the subject's mastoid process behind one ear. Have the prongs of the fork pointed downward and away from the ear, and be sure nothing is touching them (fig. 30.6 *a*). The sound sensation is that of bone conduction. If no sound is experienced, nerve deafness exists.
   c. Ask the subject to indicate when the sound is no longer heard.

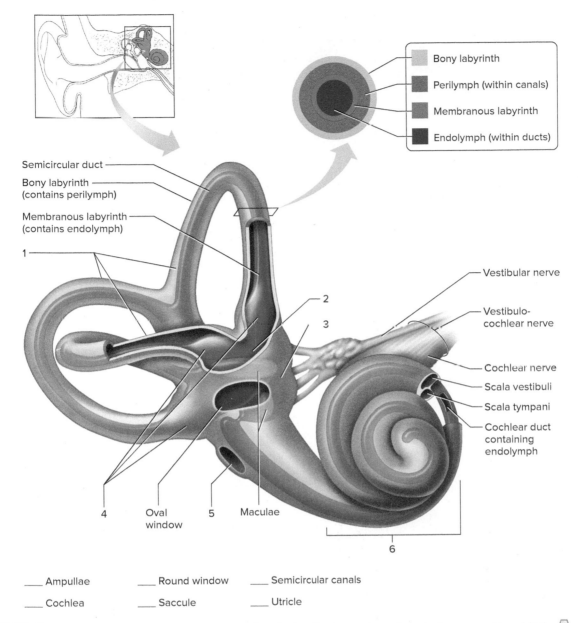

Semicircular duct
Bony labyrinth (contains perilymph)
Membranous labyrinth (contains endolymph)

Bony labyrinth
Perilymph (within canals)
Membranous labyrinth
Endolymph (within ducts)

1

Vestibular nerve
Vestibulo-cochlear nerve
Cochlear nerve
Scala vestibuli
Scala tympani
Cochlear duct containing endolymph

2
3

4   Oval window   5   Maculae

6

___ Ampullae   ___ Round window   ___ Semicircular canals

___ Cochlea   ___ Saccule   ___ Utricle

FIGURE 30.2   Label the structures of the inner ear by placing the correct numbers in the spaces provided.

---

d. Then quickly remove the fork from the mastoid process and position it in the air close to the opening of the nearby external acoustic meatus (fig. 30.6 *b*).

If hearing is normal, the sound (from air conduction) will be heard again; if there is conductive impairment, the sound will not be heard. Conductive impairment involves outer or middle ear defects. Hearing aids can improve hearing for conductive deafness because bone conduction transmits the sound into the inner ear. Surgery could possibly correct this type of defect.

e. Record the test results in Part C of the laboratory report.

4. *Weber test.* This test is used to distinguish possible conduction or sensory deafness. To conduct this test, follow these steps:

a. Strike the tuning fork with the rubber hammer.

b. Place the handle of the fork against the subject's forehead in the midline (fig. 30.7).

c. Ask the subject to indicate if the sound is louder in one ear than in the other or if it is equally loud in both ears.

If hearing is normal, the sound will be equally loud in both ears. If there is conductive impairment, the sound will appear louder in the affected ear. If some degree of sensory (nerve) deafness exists, the sound will be louder in the normal ear. The impairment involves the spiral organ (organ of Corti) or the cochlear nerve. Hearing aids will not improve sensory deafness.

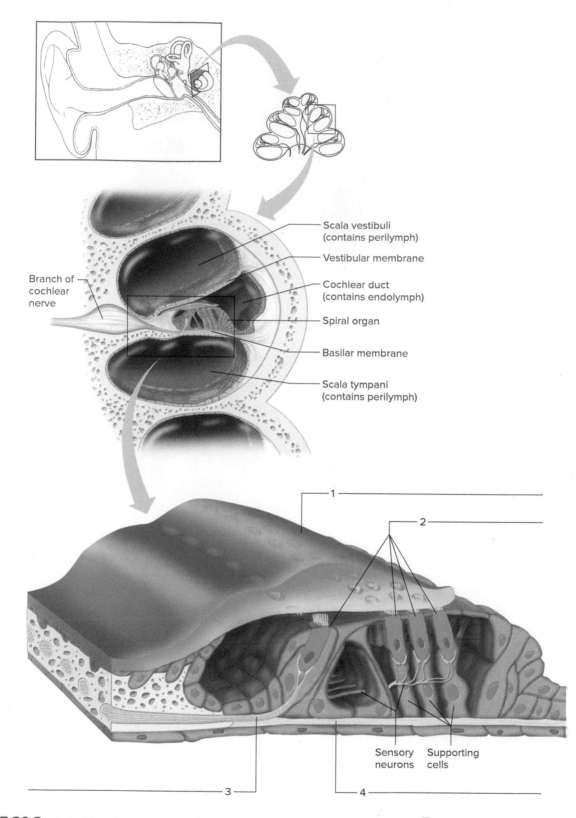

Scala vestibuli
(contains perilymph)

Vestibular membrane

Cochlear duct
(contains endolymph)

Spiral organ

Basilar membrane

Scala tympani
(contains perilymph)

Branch of
cochlear
nerve

Sensory
neurons

Supporting
cells

1

2

3

4

FIGURE 30.3 Label the structures associated with the spiral organ of a cochlea.

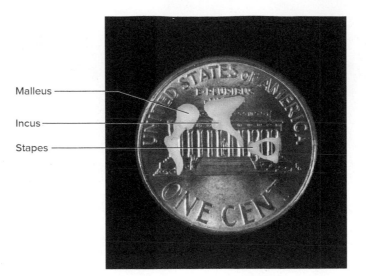

FIGURE 30.4 Middle ear bones (auditory ossicles) superimposed on a penny for a size relationship. The arrangement of the malleus, incus, and stapes in the enlarged photograph resembles the articulations in the middle ear. ©J and J Photography

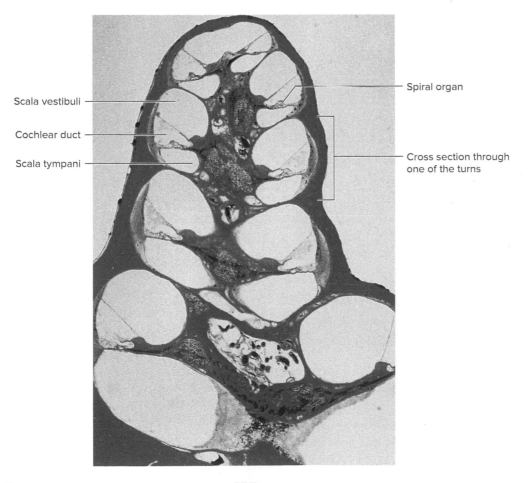

FIGURE 30.5 A section through the cochlea (22×). **APR** Biophoto Associates/Science Source

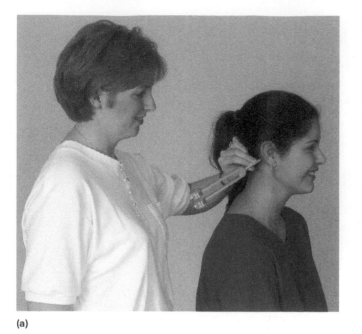

(a)

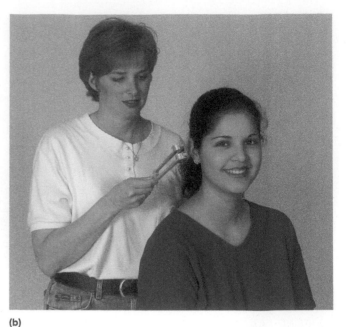

(b)

FIGURE 30.6 Rinne test: (*a*) first placement of vibrating tuning fork until sound is no longer heard; (*b*) second placement of tuning fork to assess air conduction. **(a, b):** ©J and J Photography

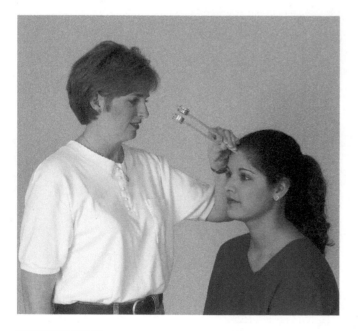

FIGURE 30.7 Weber test. ©J and J Photography

d. Have the subject experience the effects of conductive impairment by packing one ear with cotton and repeating the Weber test. Usually the sound appears louder in the plugged (impaired) ear because extraneous sounds from the room are blocked out.

e. Record the test results in Part C of the laboratory report.

5. Complete Part C of the laboratory report.

 **Critical Thinking Application**

Ear structures from the outer ear into the inner ear are progressively smaller. Using results obtained from the hearing tests, explain this advantage.

_____

_____

_____

_____

_____

DEMONSTRATION

Ask the laboratory instructor to demonstrate the use of the audiometer. This instrument produces sound vibrations of known frequencies transmitted to one or both ears of a test subject through earphones. The audiometer can be used to determine the threshold of hearing for different sound frequencies, and, in the case of a hearing impairment, it can be used to determine the percentage of hearing loss for each frequency.

Name _____

Date _____

Section _____

The  corresponds to the indicated Learning Outcome(s) found at the beginning of the laboratory exercise.

Ear and Hearing

PART A ASSESSMENTS

Match the terms in column A with the descriptions in column B. Place the letter of your choice in the space provided. Ⓐ Ⓐ

Column A	Column B
a. Auditory tube	_____ **1.** Auditory ossicle attached to eardrum
b. Bony labyrinth	_____ **2.** Air-filled space containing auditory ossicles within middle ear
c. Eardrum	_____ **3.** Contacts hairs of hearing receptors
d. External acoustic meatus	_____ **4.** Leads from oval window to apex of cochlea and contains perilymph
e. Malleus	_____ **5.** S-shaped tube leading to eardrum
f. Membranous labyrinth	_____ **6.** Cone-shaped, semitransparent membrane attached to malleus
g. Scala tympani	_____ **7.** Auditory ossicle attached to oval window
h. Scala vestibuli	_____ **8.** Tube containing endolymph within bony labyrinth
i. Stapes	_____ **9.** Bony canal of inner ear in temporal bone
j. Tectorial membrane	_____ **10.** Connects middle ear and pharynx
k. Tympanic cavity	_____ **11.** Extends from apex of cochlea to round window and contains perilymph

PART B ASSESSMENTS

Label the structures indicated in the micrograph of the spiral organ in figure 30.8. Ⓐ

PART C ASSESSMENTS

1. Results of auditory acuity test: Ⓐ

Ear Tested	Audible Distance (cm)
Right	
Left	

Terms:
Basilar membrane
Cochlear duct
Hair cells
Scala tympani
Tectorial membrane

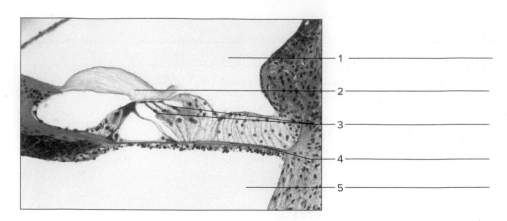

FIGURE 30.8 Using the terms provided, label the structures associated with this spiral organ region of a cochlea (400×). **APR** Biophoto Associates/Science Source

2. Results of sound localization test: 🅰

Actual Location	Reported Location
Front of the head	
Behind the head	
Above the head	
Right side of the head	
Left side of the head	

3. Results of experiments using tuning forks: 🅰

Test	Left Ear (normal or impaired)	Right Ear (normal or impaired)
Rinne		
Weber		

4. Summarize the results of the hearing tests you conducted on your laboratory partner. 🅰

_____

_____

_____

_____

31

Eye Structure

PURPOSE OF THE EXERCISE

To review the structure and function of the eye and to dissect a mammalian eye.

🔄 LEARNING OUTCOMES APR

After completing this exercise, you should be able to

1. Locate the major structures of an eye.
2. Describe the functions of the structures of an eye.
3. Trace the structures through which light passes as it travels from the cornea to the retina.
4. Dissect a mammalian eye and locate its major features.
5. Recognize the anatomical terms pertaining to the eye structure.

The eye contains photoreceptors, modified neurons located on its inner wall. Other parts of the eye provide protective functions or make it possible to move the eyeball. Still other structures focus light entering the eye so that a sharp image is projected onto the receptor cells. Impulses generated when the receptors are stimulated are conducted along the optic nerves to the brain, which interprets the impulses and creates the sensation of sight.

🔄 EXPLORE

PROCEDURE A—Structure and Function of the Eye APR

1. Review the concept headings entitled "Visual Accessory Organs," "Structure of the Eye," and "Photoreceptors" in section 10.9 of chapter 10 of the textbook.
2. As a review activity, label figures 31.1, 31.2, and 31.3.
3. Complete Part A of Laboratory Report 31.
4. Examine the dissectible model of the eye and locate the following features:

eyelid

conjunctiva

orbicularis oculi

levator palpebrae superioris

lacrimal apparatus
    lacrimal gland
    canaliculi
    lacrimal sac
    nasolacrimal duct

extrinsic muscles
    superior rectus
    inferior rectus
    medial rectus
    lateral rectus
    superior oblique
    inferior oblique

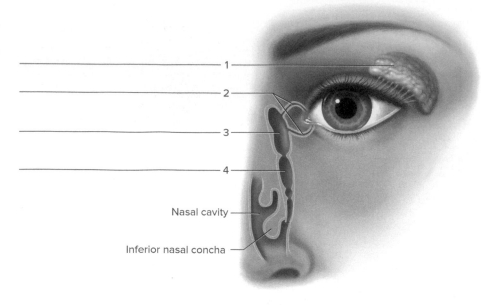

1 _____

2 _____

3 _____

4 _____

Nasal cavity _____

Inferior nasal concha _____

FIGURE 31.1   Label the structures of the lacrimal apparatus. 🄰

Levator palpebrae superioris _____
Pulley (trochlea) _____

1 _____

2 _____

3 _____

4 _____

5 _____

6 _____

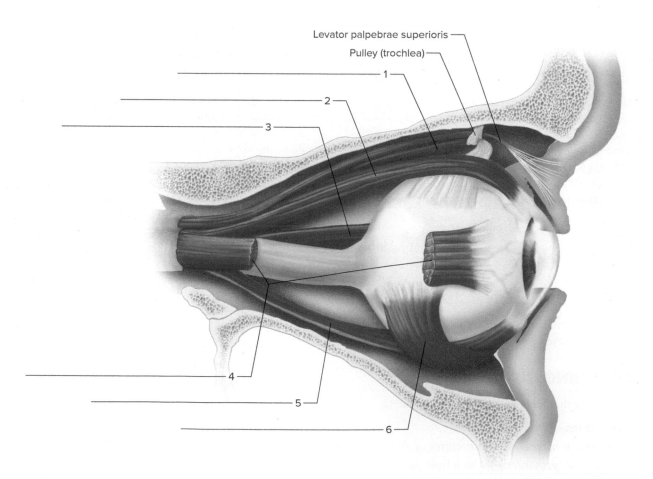

FIGURE 31.2   Label the extrinsic muscles of the right eye (lateral view). 🄰 **APR**

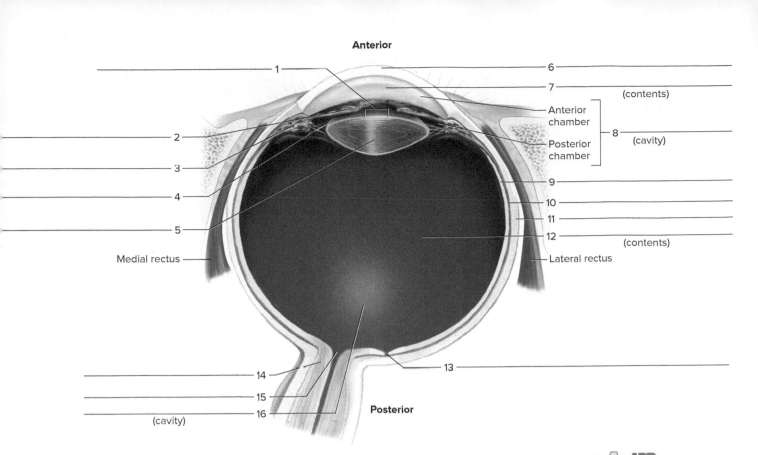

Anterior

1 ——
6 ——
7 —— (contents)
Anterior chamber
Posterior chamber
8 —— (cavity)
2 ——
3 ——
4 ——
9 ——
10 ——
11 ——
12 —— (contents)
5 ——
Medial rectus ——
Lateral rectus ——
14 ——
15 ——
16 ——
(cavity)
13 ——

Posterior

FIGURE 31.3   Label the structures indicated in this transverse section of the right eye (superior view). Ⓐ APR

outer (fibrous) layer
  sclera
  cornea
middle (vascular) layer
  choroid coat
  ciliary body
    ciliary processes
    ciliary muscles
    suspensory ligaments
  iris
  pupil (opening in center of iris)
inner (nervous) layer
  retina
    macula lutea
    fovea centralis
    optic disc
  optic nerve
lens
anterior cavity
  anterior chamber
  posterior chamber
  aqueous humor
posterior cavity
  vitreous humor

LEARNING EXTENSION

Use an ophthalmoscope to examine the interior of your laboratory partner's eye. This instrument consists of a set of lenses held in a rotating disc, a light source, and some mirrors that reflect the light into the test subject's eye (fig. 31.4).

The examination should be conducted in a dimly lit room. Have your partner seated and staring straight ahead at eye level. Move the rotating disc of the ophthalmoscope so that the *O* appears in the lens selection window. Hold the instrument in your right hand with the end of your index finger on the rotating disc. Direct the light at a slight angle from a distance of about 15 cm into the subject's right eye (fig.31.5). The light beam should pass along the inner edge of the pupil. Look through the instrument, and you should see a reddish, circular area—the interior of the eye. Rotate the disc of lenses to higher values until sharp focus is achieved.

Move the ophthalmoscope to within about 5 cm of the eye being examined, *being very careful that the instrument does not touch the eye,* and again rotate the lenses to sharpen the focus (fig. 31.5). Locate the optic disc and the blood vessels that pass through it. Also locate the macula lutea by having your partner stare directly into the light of the instrument (fig. 31.6). *Do not allow light to reach the macula lutea region during the eye exam for longer than 1 second at a time.*

Examine the subject's iris by viewing it from the side and by using a lens with a +15 or +20 value.

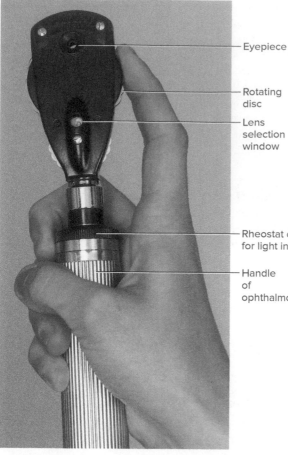

FIGURE 31.4   An ophthalmoscope is used to examine the interior of the eye. ©J and J Photography

- Eyepiece
- Rotating disc
- Lens selection window
- Rheostat control for light intensity
- Handle of ophthalmoscope

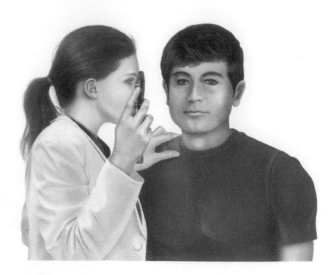

FIGURE 31.5   Rotate the disc of lenses until sharp focus is achieved and then move the ophthalmoscope to within 5 cm of the eye to examine the optic disc of the right eye.

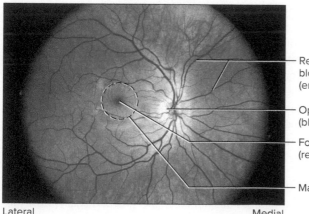

- Retinal blood vessels (emerge from optic disc)
- Optic disc (blind spot)
- Fovea centralis (region of best visual acuity)
- Macula lutea region

Lateral   Medial

FIGURE 31.6   Right retina as viewed through the pupil using an ophthalmoscope. Only small portions of the retina can be viewed at a time. APR Science Source

PROCEDURE B—Eye Dissection

1. Obtain a mammalian eye, place it in a dissecting tray, and dissect it as follows:
   a. Trim away the fat and other connective tissues but leave the stubs of the *extrinsic muscles* and the *optic nerve*. This nerve projects outward from the posterior region of the eyeball.
   b. The *conjunctiva* lines the inside of the eyelid and is reflected over the anterior surface of the eye, except for the cornea. Lift some of this thin membrane away from the eye with forceps and examine it.
   c. Locate and observe the *cornea, sclera,* and *iris.* Also note the *pupil* and its shape. The cornea from a fresh eye is transparent; when preserved, it becomes opaque.
   d. Use sharp scissors to make a frontal section of the eye. To do this, cut through the wall about 1 cm from the margin of the cornea and continue all the way around the eyeball. Try not to damage the internal structures of the eye (fig. 31.7).
   e. Gently separate the eyeball into anterior and posterior portions. Usually, the jellylike vitreous humor will remain in the posterior portion, and the lens may adhere to it. Place the parts in the dissecting tray with their contents facing upward.
   f. Examine the anterior portion of the eye, and locate the *ciliary body,* which appears as a dark, circular structure. Also note the *iris* and the *lens,* if it remained in the anterior portion. The lens is normally attached to the ciliary body by many *suspensory ligaments,* which appear as delicate, transparent threads that detach as a lens is removed.
   g. Use a dissecting needle to gently remove the lens and examine it. If the lens is still transparent, hold it up and look through it at something in the distance and note that the lens inverts the image. The lens of a preserved eye is usually too opaque for this experience. If the lens of the human eye becomes opaque, the defect is called a cataract (fig. 31.8).
   h. Examine the posterior portion of the eye. Note the *vitreous humor.* This jellylike mass helps to hold the lens in place anteriorly and to hold the *retina* against the choroid coat.
   i. Carefully remove the vitreous humor and examine the retina. This layer will appear as a thin, nearly colorless to cream-colored membrane that detaches easily from the choroid coat. Compare the structures identified to figure 31.9.
   j. Locate the *optic disc*—the point where the retina is attached to the posterior wall of the eyeball and where the optic nerve originates. There are no receptor cells in the optic disc, so this region is also called the "blind spot."

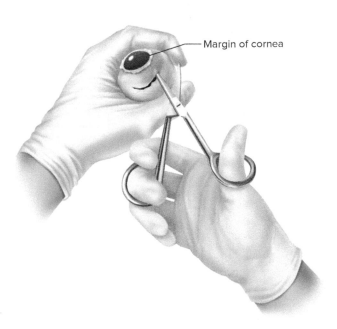

Margin of cornea

FIGURE 31.7 Prepare a frontal section of the eye.

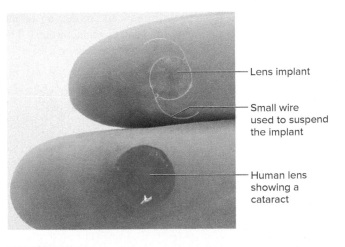

Lens implant

Small wire used to suspend the implant

Human lens showing a cataract

FIGURE 31.8 Human lens showing a cataract and one type of lens implant (intraocular lens replacement) on tips of fingers to show their relative size. A normal lens is transparent. ©J and J Photography

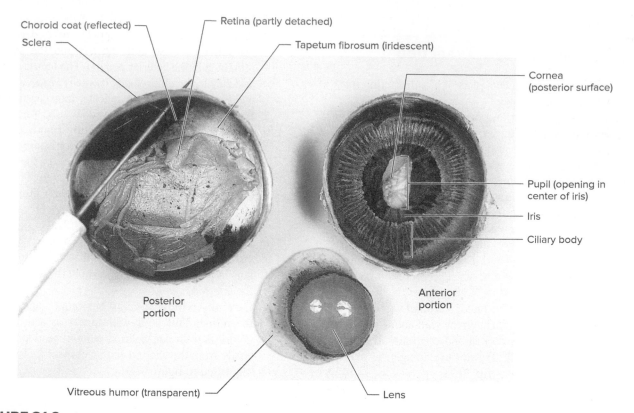

Choroid coat (reflected)
Sclera
Retina (partly detached)
Tapetum fibrosum (iridescent)
Cornea (posterior surface)
Pupil (opening in center of iris)
Iris
Ciliary body
Posterior portion
Anterior portion
Vitreous humor (transparent)
Lens

FIGURE 31.9 Internal structures of the preserved beef eye after dissection. ©J and J Photography

k. Note the iridescent area of the choroid coat beneath the retina. This colored surface in ungulates (mammals having hoofs) is called the *tapetum fibrosum*. It reflects light back through the retina, an action thought to aid the night vision of some animals. The tapetum fibrosum is lacking in the human eye.

2. Discard the tissues of the eye as directed by the laboratory instructor.

3. Complete Parts B and C of the laboratory report.

 **Critical Thinking Application**

A strong blow to the head might cause the retina to detach. From observations made during the eye dissection, explain why this can happen.

_____

_____

_____

_____

Name _____

Date _____

Section _____

The A corresponds to the indicated Learning Outcome(s) found at the beginning of the laboratory exercise.

Eye Structure

PART A ASSESSMENTS

Match the terms in column A with the descriptions in column B. Place the letter of your choice in the space provided. A₁ A₂

Column A	Column B
a. Aqueous humor	_____ **1.** Posterior five-sixths of middle (vascular) layer
b. Choroid coat	_____ **2.** White part of outer (fibrous) layer
c. Ciliary muscles	_____ **3.** Transparent anterior portion of outer layer
d. Conjunctiva	_____ **4.** Inner lining of eyelid
e. Cornea	_____ **5.** Secretes tears
f. Iris	_____ **6.** Fills posterior cavity of eye
g. Lacrimal gland	_____ **7.** Area where optic nerve exits the eye
h. Optic disc	_____ **8.** Smooth muscle that controls the pupil size and light entering the eye
i. Retina	
j. Sclera	_____ **9.** Fills anterior and posterior chambers of the anterior cavity of the eye
k. Suspensory ligament	_____ **10.** Contains photoreceptor cells called rods and cones
l. Vitreous humor	_____ **11.** Connects lens to ciliary body
	_____ **12.** Cause lens to change shape

Complete the following:

13. List the structures and fluids through which light passes as it travels from the cornea to the retina. A₃

_____

_____

_____

14. List three ways in which rods and cones differ in structure or function. A₂ _____

_____

_____

_____

_____

PART B ASSESSMENTS

Complete the following:

1. Which layer of the eye was the most difficult to cut? 4 _____

2. What kind of tissue do you think is responsible for this quality of toughness? 4 _____

3. How do you compare the shape of the pupil in the dissected eye with your pupil? 4 _____

4. Where was aqueous humor in the dissected eye? 4 _____

5. What is the function of the dark pigment in the choroid coat? 2 _____

6. Describe the lens of the dissected eye. 4 _____

7. Describe the vitreous humor of the dissected eye. 4 _____

PART C ASSESSMENTS

Locate the 23 atatomical terms pertaining to the "Eye Structure." 5

W	C	O	N	E	S	E	O	F	P	R	K
R	O	M	U	H	S	G	L	U	O	V	M
I	N	N	E	R	N	X	P	D	T	E	Z
L	J	F	Q	U	E	I	S	P	D	E	K
A	U	D	G	W	L	B	U	I	P	I	R
T	N	I	F	O	V	E	A	R	A	N	M
E	C	O	R	W	V	L	K	I	N	F	S
R	T	R	E	V	R	E	N	S	I	E	G
A	I	O	B	L	I	Q	U	E	T	R	D
L	V	H	W	G	L	A	N	D	E	I	G
M	A	C	U	L	A	A	E	N	R	O	C
D	I	L	E	Y	E	S	C	L	E	R	A

Find and circle words pertaining to Laboratory Exercise 31. The words may be horizontal, vertical, diagonal, or backward.

8 terms (conjunctiva, eyelid, fovea, gland, humor, macula, nerve, oblique)

3 main eye layers; 3 rectus eye muscles; 9 eyeball structures

32

Visual Tests and Demonstrations

MATERIALS NEEDED

Snellen eye chart
3″ × 5″ card (plain)
3″ × 5″ with word typed in center
Astigmatism chart
Meterstick
Metric ruler
Pen flashlight
Ichikawa's or Ishihara's color plates for
  colorblindness test

PURPOSE OF THE EXERCISE

To conduct and interpret the tests for visual acuity, astigmatism, accommodation, color vision, the blind spot, and reflexes of the eye.

LEARNING OUTCOMES

After completing this exercise, you should be able to

(1) Conduct and interpret the tests used to evaluate visual acuity, astigmatism, the ability to accommodate for close vision, and color vision.

(2) Describe four conditions that can lead to defective vision.

(3) Demonstrate and describe the blind spot, photopupillary reflex, accommodation pupillary reflex, and convergence reflex.

Normal vision (emmetropia) results when light rays from objects in the external environment are refracted by the cornea and lens of the eye and focused onto the photoreceptors of the retina. If a person has *nearsightedness (myopia)*, the image focuses in front of the retina as the eyeball is too long; nearby objects are clear, but distant objects are blurred. If a person has *farsightedness (hyperopia)*, the image focuses behind the retina as the eyeball is too short; distant objects are clear, but nearby objects are blurred. A concave lens is required to correct vision for nearsightedness; a convex lens is required to correct vision for farsightedness.

If a defect occurs from unequal curvatures of the cornea or lens, a condition called *astigmatism* results. Focusing in different planes cannot occur simultaneously. Corrective cylindrical lenses allow proper refraction of light onto the retina to compensate for the unequal curvatures.

As part of a natural aging process, the lens's elasticity decreases, which degrades our ability to focus on nearby objects. The near point of accommodation test is used to determine the ability to accommodate. Often a person will use reading glasses to compensate for this condition. Hereditary defects in color vision result from lack of certain cones necessary to absorb certain wavelengths of light. Special color test plates are used to diagnose colorblindness.

As you conduct this laboratory exercise, it does not matter in what order the visual tests or visual demonstrations are performed. The tests and demonstrations do not depend upon the results of any of the others. If you wear glasses, perform the tests with and without the corrective lenses. If you wear contact lenses, it is not necessary to run the tests under both conditions; however, indicate in the laboratory report that all tests were performed with contact lenses in place.

EXPLORE

PROCEDURE A—Visual Tests

Perform the following visual tests using your laboratory partner as a test subject. If your partner usually wears glasses, test each eye with and without the glasses.

1. *Visual acuity test.* Visual acuity (sharpness of vision) can be measured by using a Snellen eye chart (fig. 32.1). This chart consists of several sets of letters in different sizes printed on a white card. The letters near the top of the chart are relatively large, and those in each lower set become smaller. At one end of each set of letters is an acuity value in the form of a fraction. One of the sets near the bottom of the chart, for example, is marked 20/20. The normal eye can clearly see these letters from the standard distance of 20 feet and thus is said to have 20/20 vision. The letter at the top of the chart is marked 20/200. The normal eye can read letters of this size from a distance of 200 feet. Thus, an eye able to read

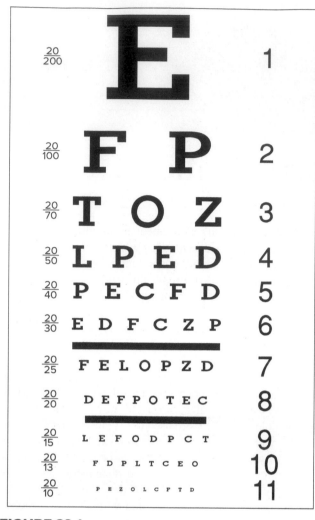

FIGURE 32.1 The Snellen eye chart looks similar to this but is somewhat larger.

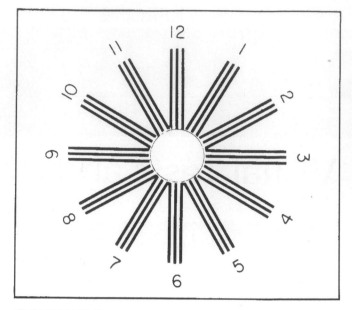

FIGURE 32.2 Astigmatism is evaluated using a chart such as this one.

blurred. Astigmatism can be evaluated by using an astigmatism chart (fig. 32.2). This chart consists of sets of black lines radiating from a central spot like the spokes of a wheel. To a normal eye, these lines appear sharply focused and equally dark; however, if the eye has an astigmatism, some sets of lines appear sharply focused and dark, whereas others are blurred and less dark.

To conduct the astigmatism test, follow these steps:

a. Hang the astigmatism chart on a well-illuminated wall at eye level.

b. Have your partner stand 20 feet in front of the chart, and gently cover your partner's left eye with a 3″ × 5″ card. Ask your partner to focus on the spot in the center of the radiating lines using the right eye, and report which lines, if any, appear more sharply focused and darker.

c. Repeat the procedure, using the left eye for the test.

d. Record the results in Part A of the laboratory report.

3. *Accommodation test.* Accommodation is the changing of the shape of the lens that occurs when the normal eye is focused for near (close) vision. It involves a reflex in which muscles of the ciliary body are stimulated to contract, releasing tension on the suspensory ligaments fastened to the lens capsule. This allows the capsule to rebound elastically, causing the surface of the lens to become more spherical (convex). The ability to accommodate is likely to decrease with age because the tissues involved tend to lose their elasticity. This common aging condition, called *presbyopia,* can be corrected with reading glasses or bifocal lenses.

To evaluate the ability to accommodate, follow these steps:

a. Hold the end of a meterstick against your partner's chin so that the stick extends outward at a right angle to the plane of the face (fig. 32.3).

only the top letter of the chart from a distance of 20 feet is said to have 20/200 vision. This person has less than normal vision. A line of letters near the bottom of the chart is marked 20/15. The normal eye can read letters of this size from a distance of 15 feet, but a person might be able to read it from 20 feet. This person has better-than-normal vision.

To conduct the visual acuity test, follow these steps:

a. Hang the Snellen eye chart on a well-illuminated wall at eye level.

b. Have your partner stand 20 feet in front of the chart, gently cover your partner's left eye with a 3″ × 5″ card, and ask your partner to read the smallest set of letters possible using the right eye.

c. Record the visual acuity value for that set of letters in Part A of Laboratory Report 32.

d. Repeat the procedure, using the left eye for the test.

2. *Astigmatism test.* Astigmatism is a condition that results from a defect in the curvature of the cornea or lens. As a consequence, some portions of the image projected on the retina are sharply focused, but other portions are

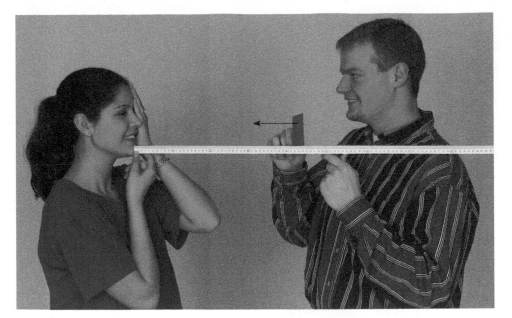

FIGURE 32.3 To determine the near point of accommodation, slide the 3″ × 5″ card along the meterstick toward your partner's open eye until it reaches the closest location where your partner can still see the word sharply focused.
©J and J Photography

b. Have your partner close the left eye.

c. Hold a 3″ × 5″ card with a word typed in the center at the distal end of the meterstick.

d. Slide the card along the stick toward your partner's open right eye, and locate the *point closest to the eye* where your partner can still see the letters of the word sharply focused. This distance is called the *near point of accommodation,* and it tends to increase with age (table 32.1).

e. Repeat the procedure with the right eye closed.

f. Record the results in Part A of the laboratory report.

4. *Color vision test.* Some people exhibit defective color vision because they lack certain cones, usually those sensitive to the reds or greens. This trait is an X-linked (sex-linked) inheritance, so the condition is more prevalent in males (7%) than in females (0.4%). People who lack or possess decreased sensitivity to the red-sensitive cones possess protanopia colorblindness; those who lack or possess decreased sensitivity to green-sensitive cones possess deuteranopia colorblindness. The colorblindness condition is often more of a deficiency or a weakness than a condition of blindness. Laboratory Exercise 48 describes the genetics of colorblindness.

To conduct the color vision test, follow these steps:

a. Examine the color test plates in Ichikawa's or Ishihara's book to test for any red-green color vision deficiency. Also examine figure 32.4.

b. Hold the test plates approximately 30 inches from the subject in bright light. All responses should occur within 3 seconds.

c. Compare your responses with the correct answers in Ichikawa's or Ishihara's book. Determine the percentage of males and females in your class who exhibit any color-deficient vision. If an individual exhibits color-deficient vision, determine if the condition is protanopia or deuteranopia.

d. Record the class results in Part A of the laboratory report.

5. Complete Part A of the laboratory report.

🔄 EXPLORE

PROCEDURE B—Visual Demonstrations

Perform the following demonstrations with the help of your laboratory partner.

1. *Blind-spot demonstration.* There are no photoreceptors in the *optic disc,* located where the nerve fibers of the retina leave the eye and enter the optic nerve. Consequently, this region of the retina is commonly called the *blind spot.*

To demonstrate the blind spot, follow these steps:

a. Close your left eye, hold figure 32.5 about 35 cm away from your face, and stare at the "+" sign in the figure with your right eye.

TABLE 32.1 Near Point of Accommodation

Age (years)	Average Near Point (cm)
10	7
20	10
30	13
40	20
50	45
60	90

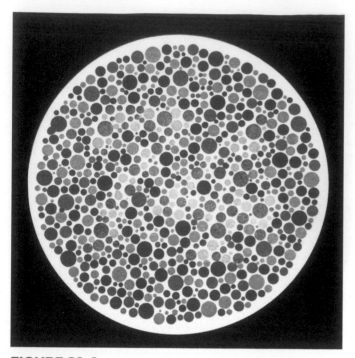

FIGURE 32.4 Sample color test plate to diagnose possible colorblindness. A person with normal color vision perceives the embedded pattern of the number 74; some colorblind individuals only see dots. Tests for colorblindness cannot be conducted with this sample test plate reproduction. For accurate testing, the original plates should be used. Steve Allen/Getty Images

FIGURE 32.5 Blind-spot demonstration.

b. Move the figure closer to your face as you continue to stare at the "+" until the dot on the figure suddenly disappears. This happens when the image of the dot is focused on the optic disc. Measure the right eye distance using a metric ruler or a meterstick.

c. Repeat the procedures with your right eye closed. This time, stare at the dot using your left eye, and the "+" will disappear when the image falls on the optic disc. Measure the distance.

d. Record the results in Part B of the laboratory report.

2. *Photopupillary reflex.* The smooth muscles of the iris control the size of the pupil. For example, when the intensity of light entering the eye increases, a photo-pupillary reflex is triggered, and the circular muscles of the iris are stimulated to contract. As a result, the size of the pupil decreases, and less light enters the eye. This reflex is an autonomic response and protects the photoreceptor cells of the retina from damage.

To demonstrate this photopupillary reflex, follow these steps:

a. Ask your partner to sit with his or her hands thoroughly covering his or her eyes for 2 minutes.

b. Position a pen flashlight close to one eye with the light shining on the hand that covers the eye.

c. Ask your partner to remove the hand quickly.

d. Observe the pupil and note any change in its size.

e. Have your partner remove the other hand, but keep that uncovered eye shielded from extra light.

f. Observe both pupils and note any difference in their sizes.

3. *Accommodation pupillary reflex.* The pupil constricts as a normal accommodation reflex to focusing on close objects. This reduces peripheral light rays and allows the refraction of light to occur closer to the center portion of the lens, resulting in a clearer image.

To demonstrate the accommodation pupillary reflex, follow these steps:

a. Have your partner stare for several seconds at a dimly illuminated object in the room that is more than 20 feet away.

b. Observe the size of the pupil of one eye. Then hold a pencil about 25 cm in front of your partner's face, and have your partner stare at it.

c. Note any change in the size of the pupil.

4. *Convergence reflex.* The eyes converge as a normal convergence reflex to focusing on close objects. The convergence reflex orients the eyeballs toward the object so the image falls on the fovea centralis of both eyes.

To demonstrate the convergence reflex, follow these steps:

a. Repeat the procedure outlined for the accommodation pupillary reflex.

b. Note any change in the position of the eyeballs as your partner changes focus from the distant object to the pencil.

5. Complete Part B of the laboratory report.

The A corresponds to the indicated Learning Outcome(s) found at the beginning of the laboratory exercise.

Visual Tests and Demonstrations

PART A ASSESSMENTS

1. Visual acuity test results: A

Eye Tested	Acuity Values
Right eye	
Right eye with glasses (if applicable)	
Left eye	
Left eye with glasses (if applicable)	

2. Astigmatism test results: A

Eye Tested	Darker Lines
Right eye	
Right eye with glasses (if applicable)	
Left eye	
Left eye with glasses (if applicable)	

3. Accommodation test results: A

Eye Tested	Near Point (cm)
Right eye	
Right eye with glasses (if applicable)	
Left eye	
Left eye with glasses (if applicable)	

4. Color vision test results: ⚠A̲

Condition	Males			Females		
	Class Number	Class Percentage	Expected Percentage	Class Number	Class Percentage	Expected Percentage
Normal color vision			93			99.6
Deficient red-green color vision			7			0.4
Protanopia (lack red-sensitive cones)			less frequent type			less frequent type
Deuteranopia (lack green-sensitive cones)			more frequent type			more frequent type

5. Complete the following: ⚠2̲

  a. What is meant by 20/70 vision? _____

  _____

  b. What is meant by 20/10 vision? _____

  _____

  c. What visual problem is created by astigmatism? _____

  _____

  d. Why does the near point of accommodation often increase with age? _____

  _____

  e. Describe the eye defect that causes color-deficient vision. _____

  _____

♻ PART B ASSESSMENTS

1. Blind-spot results: ⚠3̲
  a. Right eye distance _____
  b. Left eye distance _____
2. Complete the following:

  a. Explain why an eye has a blind spot. ⚠3̲ _____

  _____

  b. Describe the photopupillary reflex. ⚠3̲ _____

  _____

  c. What difference did you note in the size of the pupils when one eye was exposed to bright light and the other eye was shielded from the light? ⚠3̲ _____

  _____

  d. Describe the accommodation pupillary reflex. ⚠3̲ _____

  _____

  e. Describe the convergence reflex. ⚠3̲ _____

  _____

33

Endocrine Histology and Diabetic Physiology

MATERIALS NEEDED

For Procedure A—Endocrine Histology:
Textbook
Human torso model
Compound light microscope
Prepared microscope slides of the following:
  Pituitary gland
  Thyroid gland
  Parathyroid gland
  Adrenal gland
  Pancreas

For Procedure B—Insulin Shock:
500 mL beaker
Live small fish (1″ to 1.5″ goldfish, guppy, rosy red feeder, or other)
Small fish net
Insulin (regular U-100) (HumulinR in a 10 mL vial has 100 units/mL; store in refrigerator—do not freeze)
Syringes for U-100 insulin
10% glucose solution
Clock with second hand or timer

For Learning Extension:
Compound light microscope
Prepared microscope slides of the following:
  Normal human pancreas (stained for alpha and beta cells)
  Human pancreas of diabetic

⚠ SAFETY

- Review all the Laboratory Safety Guidelines in Appendix 1 of your laboratory manual.
- Wear disposable gloves when handling the fish and syringes.
- Dispose of the used syringe and needle in the puncture-resistant container.
- Wash your hands before leaving the laboratory.

PURPOSE OF THE EXERCISE

To examine microscopically the tissues of endocrine glands and to observe behavioral changes that occur during insulin shock.

⟳ LEARNING OUTCOMES APR

After completing this exercise, you should be able to

1. Name and locate the major endocrine glands.
2. Associate the principal hormones and functions of each of the major endocrine glands.
3. Sketch and label tissue sections from the pituitary gland, thyroid gland, parathyroid glands, adrenal glands, and pancreas.
4. Compare the causes, symptoms, and treatments for type 1 and type 2 diabetes mellitus.
5. Observe and record the behavioral changes caused by insulin shock.
6. Observe and record the recovery from insulin shock when sugar is provided to cells.

The endocrine system consists of ductless glands that act together with parts of the nervous system to help control body activities. The endocrine system is unique in that the organs are not anatomically adjacent to each other. Scattered from the head to the genital region, functionally they are glands. These endocrine glands secrete hormones transported in body fluids and affect cells possessing appropriate receptor molecules. In this way, hormones influence the rate of metabolic reactions, the transport of substances through cell membranes, the regulation of water and electrolyte balances, and many other functions. By controlling cellular activities, endocrine glands play important roles in the maintenance of homeostasis.

The primary "fuel" for cells is sugar (glucose). Insulin produced by the beta cells of the pancreatic islets has a primary regulation role in the transport of blood sugar across the cell membranes of most cells of the body. Some individuals do not produce enough insulin; others might not have adequate transport of glucose into the body cells when these target cells lose insulin receptors. This rather common disease is called *diabetes mellitus*. Diabetes mellitus is a functional disease that can have its onset either during childhood or later in life.

The protein hormone insulin has several functions: it stimulates the liver and skeletal muscles to synthesize glycogen from glucose; it inhibits the conversion of noncarbohydrates into glucose; it promotes the facilitated diffusion of glucose across cell membranes of cells possessing insulin receptors (adipose tissue, liver, and skeletal muscle); it decreases blood sugar (glucose); it increases protein synthesis; and it promotes fat storage in adipose cells. In a normal person, as blood sugar increases during nutrient absorption after a meal, the rising glucose levels directly stimulate beta cells of the pancreas to secrete insulin. This increase in insulin prevents blood sugar from sudden surges (hyperglycemia) by promoting glycogen production in the liver and an increase in the entry of glucose into muscle and adipose cells.

Between meals and during sleep, blood glucose levels decrease and insulin decreases. When insulin concentration decreases, more glucose is available to enter cells that lack insulin receptors. Neurons, liver cells, kidney cells, and red blood cells either lack insulin receptors or express limited insulin receptors, and they absorb and utilize glucose without the need for insulin. These cells depend upon blood glucose concentrations to enable cellular respiration and ATP production. When insulin is given as a medication for diabetes mellitus, blood glucose levels may drop drastically (hypoglycemia) if the individual does not eat. This can have significant negative effects on the nervous system, resulting in insulin shock.

Type 1 diabetes mellitus usually has a rapid onset relatively early in life, but it can occur in adults. It occurs when insulin is no longer produced; thus sometimes it is referred to as *insulin-dependent diabetes mellitus (IDDM)*. This autoimmune disorder occurs when the immune system destroys beta cells, resulting in a decrease in insulin production. This disorder affects about 10% to 15% of diabetics. The treatment for type 1 diabetes mellitus is to administer insulin, usually by injection.

Type 2 diabetes mellitus has a gradual onset, usually diagnosed in people over age 40. It is sometimes called *non-insulin-dependent diabetes mellitus (NIDDM)*. This disease results as body cells become less sensitive to insulin or lose insulin receptors and thus cannot respond to insulin even if insulin levels remain normal. This situation is called *insulin resistance*. Risk factors for the onset of type 2 diabetes mellitus, which afflicts about 85% to 90% of diabetics, include heredity, obesity, and lack of exercise. Obesity is a major indicator for predicting type 2 diabetes; a weight loss of a mere 10 pounds will increase the body's sensitivity to insulin as a means of prevention. The treatments for type 2 diabetes mellitus include a diet that avoids foods that stimulate insulin production, weight control, exercise, and medications. (The causes and treatments for type 2 diabetes closely resemble those for coronary artery disease.)

EXPLORE

PROCEDURE A—Endocrine Histology

1. Review the sections entitled "Pituitary Gland," "Thyroid Gland," "Parathyroid Glands," "Adrenal Glands," "Pancreas," and "Pineal, Thymus, and Other Glands." in chapter 11 of the textbook.
2. As a review activity, label figure 33.1.
3. Complete Part A of Laboratory Report 33.
4. Examine the human torso model and locate the following:

hypothalamus
pituitary stalk (infundibulum)
pituitary gland (hypophysis)
    anterior lobe (adenohypophysis)
    posterior lobe (neurohypophysis)

thyroid gland
parathyroid glands
adrenal glands (suprarenal glands)
    adrenal medulla
    adrenal cortex

pancreas
pineal gland
thymus
ovaries
testes

5. Examine the microscopic tissue sections of the following glands, and identify the features described:
Pituitary gland. The pituitary gland is located in the sella turcica of the sphenoid bone. To examine the pituitary tissue, follow these steps:
   a. Observe the tissues using low-power magnification (fig. 33.2).
   b. Locate the *pituitary stalk (infundibulum)*, the *anterior lobe* (the largest part of the gland), and the *posterior lobe*. The anterior lobe makes hormones and appears glandular; the posterior lobe stores and releases hormones made by the hypothalamus and is nervous in appearance.
   c. Observe an area of the anterior lobe with high-power magnification. Locate a cluster of relatively large cells and identify some *acidophil cells*, which contain pink-stained granules, and some *basophil cells*, which contain blue-stained granules. These acidophil and basophil cells are hormone-secreting cells. The hormones secreted include TSH, ACTH, prolactin, growth hormone, FSH, and LH.
   d. Observe an area of the posterior lobe with high-power magnification. Note the numerous unmyelinated nerve fibers present in this lobe. Also locate some *pituicytes*, a type of neuroglia, scattered among the nerve fibers. The posterior lobe stores and releases ADH and oxytocin, which are synthesized in the hypothalamus.

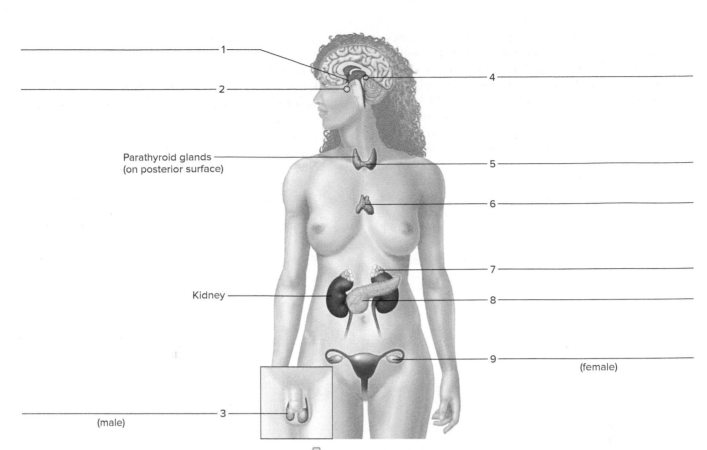

1 ————

2 ————

Parathyroid glands
(on posterior surface) ————

Kidney ————

3 ————
(male)

4 ————

5 ————

6 ————

7 ————

8 ————

9 ————
(female)

FIGURE 33.1 Label the major endocrine glands. **A**

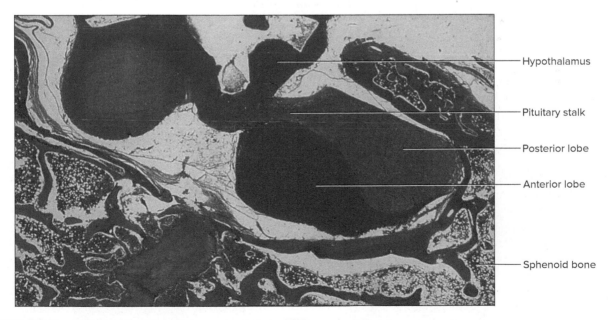

Hypothalamus

Pituitary stalk

Posterior lobe

Anterior lobe

Sphenoid bone

FIGURE 33.2 Micrograph of the pituitary gland (6×). **APR** Biophoto Associates/Science Source

e. Prepare labeled sketches of representative portions of the anterior and posterior lobes of the pituitary gland in Part B of the laboratory report.

Thyroid gland. To examine the thyroid tissue, follow these steps:

a. Use low-power magnification to observe the tissue (fig. 33.3). Note the numerous *follicles,* each of which consists of a layer of cells surrounding a colloid-filled cavity.

b. Observe the tissue using high-power magnification. The cells forming the wall of a follicle are simple cuboidal epithelial cells. These cells secrete thyroid hormone (T_3 and T_4). The extrafollicular (parafollicular) cells secrete calcitonin.

c. Prepare a labeled sketch of a representative portion of the thyroid gland in Part B of the laboratory report.

Parathyroid gland. Typically, four parathyroid glands are attached to the posterior surface of the thyroid gland. To examine the parathyroid tissue, follow these steps:

a. Use low-power magnification to observe the tissue (fig. 33.4). The gland consists of numerous tightly packed secretory cells.

b. Switch to high-power magnification and locate two types of cells—a smaller form (chief cells) arranged in cordlike patterns and a larger form (oxyphil cells) that have distinct cell boundaries and are present in clusters. *Chief cells* secrete parathyroid hormone, whereas the function of *oxyphil cells* is not clearly understood.

c. Prepare a labeled sketch of a representative portion of the parathyroid gland in Part B of the laboratory report.

Adrenal gland. To examine the adrenal tissue, follow these steps:

a. Use low-power magnification to observe the tissue (fig. 33.5). Note the thin capsule of connective tissue that covers the gland. Just beneath the capsule, there is a relatively thick *adrenal cortex.* The central portion of the gland is the *adrenal medulla.*

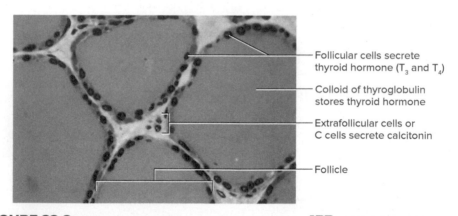

Follicular cells secrete thyroid hormone (T_3 and T_4)

Colloid of thyroglobulin stores thyroid hormone

Extrafollicular cells or C cells secrete calcitonin

Follicle

FIGURE 33.3 Micrograph of the thyroid gland (300×). **APR** Biophoto Associates/Science Source

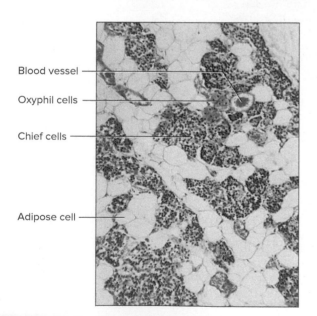

Blood vessel

Oxyphil cells

Chief cells

Adipose cell

FIGURE 33.4 Micrograph of the parathyroid gland (65×). **APR** Biophoto Associates/Science Source

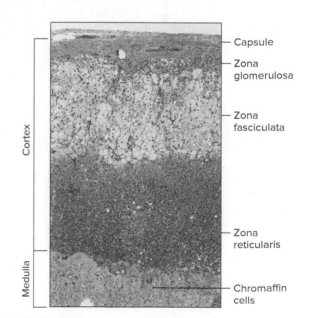

Cortex

Medulla

Capsule

Zona glomerulosa

Zona fasciculata

Zona reticularis

Chromaffin cells

FIGURE 33.5 Micrograph of the adrenal cortex and the adrenal medulla (75×). ©Ed Reschke

The cells of the cortex are in three poorly defined layers. Those of the outer layer (zona glomerulosa) are arranged irregularly; those of the middle layer (zona fasciculata) are in long cords; and those of the inner layer (zona reticularis) are arranged in an interconnected network of cords. The most noted hormones secreted from the cortex include aldosterone from the zona glomerularis, cortisol from the zona fasciculata, and estrogens and androgens from the zona reticularis. The cells of the medulla are relatively large and irregularly shaped, and they often occur in clusters. The medulla cells secrete epinephrine (adrenaline) and norepinephrine (noradrenaline).

b. Using high-power magnification, observe each of the layers of the cortex and the cells of the medulla.

c. Prepare labeled sketches of representative portions of the adrenal cortex and medulla in Part B of the laboratory report.

Pancreas. To examine the pancreas tissue, follow these steps:

a. Use low-power magnification to observe the tissue (fig. 33.6). The gland largely consists of deeply stained exocrine cells arranged in clusters around secretory ducts. These exocrine cells (acinar cells) secrete pancreatic juice rich in digestive enzymes. There are circular masses of differently stained cells scattered throughout the gland. These clumps of cells constitute the *pancreatic islets (islets of Langerhans),* and they represent the endocrine portion of the pancreas. The beta cells secrete insulin and the alpha cells secrete glucagon.

b. Examine an islet, using high-power magnification. Unless special stains are used, it is not possible to distinguish alpha and beta cells.

c. Prepare a labeled sketch of a representative portion of the pancreas in Part B of the laboratory report.

6. Complete Part C of the laboratory report.

EXPLORE

PROCEDURE B—Insulin Shock

If too much insulin is present relative to the amount of glucose available to the cells, insulin shock can occur. In this procedure, insulin shock will be created in a fish by introducing insulin into the water with a fish. Insulin will be absorbed into the blood of the gills of the fish. When insulin in the blood reaches levels above normal, rapid hypoglycemia occurs. As hypoglycemia happens from the increased insulin absorption, the brain cells obtain less of the much-needed glucose, and epinephrine is secreted. Consequently, a rapid heart rate, anxiety, sweating, mental disorientation, impaired vision, dizziness, convulsions, and possible unconsciousness are complications during insulin shock. Although these complications take place in humans, some of them are visually noticeable in a fish that is in insulin shock. The most observable components of insulin shock in a fish are related to behavioral changes, including rapid and irregular movements of the entire fish and increased gill cover and mouth movements.

1. Review the sections entitled "Pancreas" and "Diabetes Mellitus—Clinical Application" in chapter 11 of the

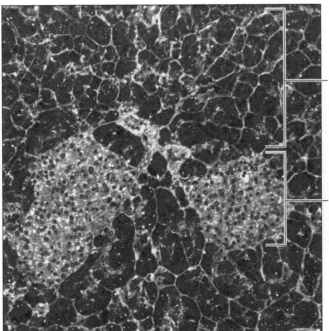

Exocrine (acinar) cells that secrete digestive enzymes

Pancreatic islet (islet of Langerhans) composed of endocrine cells that secrete hormones

FIGURE 33.6   Micrograph of the pancreas (200×).    APR ©From Kent M. Van De Graaff and Stuart Ira Fox, Concepts of Human Anatomy and Physiology, 2/e. ©1989 Wm. C. Brown Publisher, Dubuque, Iowa. All Rights Reserved. Reprinted with permission.

textbook, as well as the introduction to this laboratory exercise.

2. Complete the table in Part D of the laboratory report.

3. A fish can be used to observe normal behavior, induced insulin shock, and recovery from insulin shock. The pancreatic islets are located in the pyloric ceca or other scattered regions in fish. The behavioral changes of fish when they experience insulin shock and recovery mimic those of humans. Observe a fish in 200 mL of aquarium water in a 500 mL beaker. Make your observations of swimming, gill cover (operculum), and mouth movements for 5 minutes. Record your observations in Part E of the laboratory report.

4. Add 200 units of room-temperature insulin slowly, using the syringe, into the water with the fish to induce insulin shock. Record the time or start the timer when the insulin is added. Watch for changes of the swimming, gill cover, and mouth movements as insulin diffuses into the blood at the gills. Signs of insulin shock might include rapid and irregular movements. Record your observations of any behavioral changes and the time of the onset in Part E of the laboratory report.

5. After definite behavioral changes are observed, use the small fish net to move the fish into another 500 mL beaker containing 200 mL of a 10% room-temperature glucose solution. Record the time when the fish is transferred into this container. Make observations of any recovery of normal behaviors, including the amount of time involved. Record your observations in Part E of the laboratory report.

6. When the fish appears to be fully recovered, return it to its normal container. Dispose of the syringe according to directions from your laboratory instructor.

7. Complete Part E of the laboratory report.

LEARNING EXTENSION

Examine a normal stained pancreas under high-power magnification (fig. 33.6). Locate the clumps of cells that constitute the *pancreatic islets (islets of Langerhans)* that represent the endocrine portions of the organ. A pancreatic islet contains four distinct cells that secrete different hormones: *alpha cells,* which secrete glucagon; *beta cells,* which secrete insulin; *delta cells,* which secrete somatostatin; and *F cells,* which secrete pancreatic polypeptide. Some of the specially stained pancreatic islets allow for distinguishing the different types of cells (figs. 33.7 and 33.8). The normal pancreatic islet has about 80% beta cells concentrated in the central region of an islet, about 15% alpha cells near the periphery of an islet, about 5% delta cells, and a small number of F cells. There are complex interactions among the four hormones affecting blood sugar, but in general, glucagon will increase blood sugar, whereas insulin will lower blood sugar; thus, these two hormones are considered antagonistic in their function. Somatostatin acts as a paracrine secretion to inhibit alpha and beta cell secretions, and pancreatic polypeptide inhibits secretions from delta cells and digestive enzyme secretions from the pancreatic acini.

Examine the pancreas of a diabetic under high-power magnification (fig. 33.9). Locate a pancreatic islet and note the number of beta cells as compared to a normal number of beta cells in the normal pancreatic islet. Prepare a labeled sketch of a pancreatic islet showing indications of diabetes mellitus in Part B of the laboratory report.

Describe the differences that you observed between a normal pancreatic islet and a pancreatic islet of a person with diabetes mellitus.

_____
_____
_____
_____
_____
_____

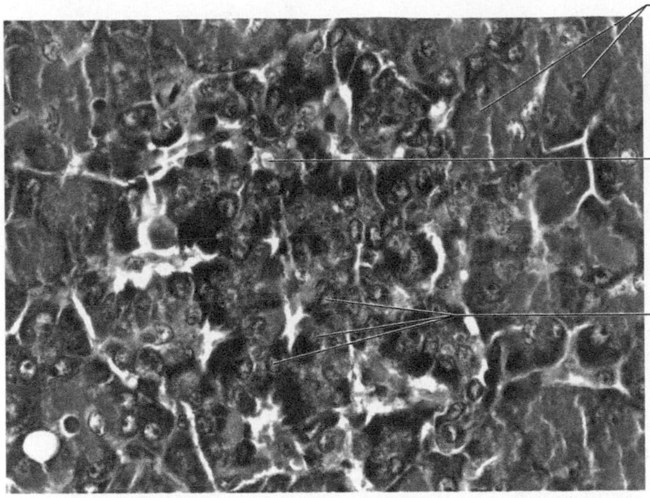

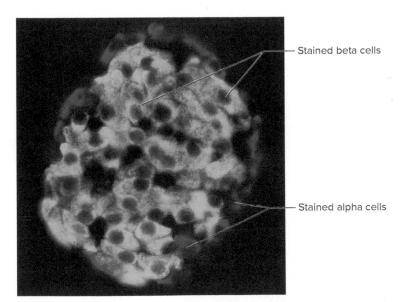

Secretory granules of
exocrine (acinar) cells

Capillary

Stained beta cells
within pancreatic islet

FIGURE 33.7 Micrograph of a specially stained pancreatic islet surrounded by exocrine (acinar) cells (500×). The insulin-containing beta cells are stained dark purple. **APR** ©Alvin Telser, Ph.D.

Stained beta cells

Stained alpha cells

FIGURE 33.8 Micrograph of a pancreatic islet using immuno fluorescence stains to visualize alpha and beta cells. Only the areas of interest are visible in the fluorescence microscope. **APR** ©Alvin Telser, Ph.D.

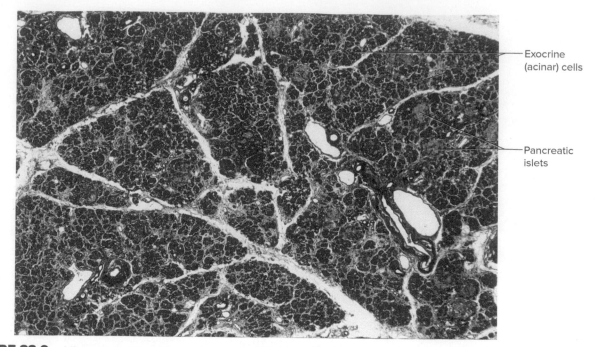

Exocrine (acinar) cells

Pancreatic islets

FIGURE 33.9   Micrograph of pancreas showing indications of changes from diabetes mellitus (100×). The changes from diabetes mellitus include areas of fibrosis and amyloid replacement of islet cells. Michael Abbey/Science Source

Name _____

Date _____

Section _____

The 🄰 corresponds to the indicated Learning Outcome(s) found at the beginning of the laboratory exercise.

Endocrine Histology and Diabetic Physiology

🔄 PART A ASSESSMENTS

Complete the following:

1. Name six hormones secreted by the anterior lobe of the pituitary gland. 🄰 _____

_____

2. Name two hormones secreted by the posterior lobe of the pituitary gland. 🄰 _____

3. Name the pituitary hormone responsible for the following actions: 🄰

    a. Causes kidneys to conserve water _____

    b. Stimulates cells to increase in size and divide more rapidly _____

    c. Stimulates secretion from thyroid gland _____

    d. Causes contraction of uterine wall muscles _____

    e. Stimulates secretion from adrenal cortex _____

    f. Stimulates milk production _____

4. Name two thyroid hormones that affect metabolic rate. 🄰 _____

5. Name a hormone secreted by the thyroid gland that acts to lower blood calcium. 🄰 _____

6. Name a hormone that acts to raise blood calcium. 🄰 _____

7. Name three target organs of parathyroid hormone. 🄰 _____

8. Name two hormones secreted by the adrenal medulla. 🄰 _____

9. List five different effects produced by these medullary hormones. 🄰 _____

_____

10. Name the most important mineralocorticoid secreted by the adrenal cortex. 🄰 _____

11. List two actions of this mineralocorticoid. 🄰 _____

_____

12. Name the most important glucocorticoid secreted by the adrenal cortex. 🄰 _____

13. List three actions of this glucocorticoid. 🄰 _____

_____

14. Distinguish the two hormones secreted by the alpha and beta cells of the pancreatic islets. 🄰 _____

_____

Critical Thinking Application

Briefly explain how the actions of pancreatic hormones complement one another. 🔼2

_____

_____

PART B ASSESSMENTS

Sketch and label representative portions of the following endocrine glands: 🔼3

Pituitary gland (_____×) (anterior and posterior lobes)	Thyroid gland (_____×)
Parathyroid gland (_____×)	Adrenal gland (_____×) (cortex and medulla)
Pancreatic islet (_____×) (normal)	Pancreatic islet (_____×) (showing changes from diabetes mellitus)

PART C ASSESSMENTS

Match the endocrine gland in column A with a characteristic of the gland in column B. Place the letter of your choice in the space provided. 🔺1 🔺2

Column A	Column B
a. Adrenal cortex	_____ **1.** Located in sella turcica of sphenoid bone
b. Adrenal medulla	_____ **2.** Contains alpha and beta cells
c. Hypothalamus	_____ **3.** Contains colloid-filled cavities
d. Pancreatic islets	_____ **4.** Attached to posterior surface of thyroid gland
e. Parathyroid gland	_____ **5.** Secretes cortisol and aldosterone
f. Pituitary gland	_____ **6.** Attached to pituitary gland by a stalk
g. Thymus	_____ **7.** Gland inside another gland near kidneys
h. Thyroid gland	_____ **8.** Located in mediastinum

PART D ASSESSMENTS

Complete the missing parts of the table on diabetes mellitus: 🔺4

Characteristic	Type 1 Diabetes	Type 2 Diabetes
Onset age	Early age or adult	
Onset of symptoms		Slow
Percentage of diabetics		85%–90%
Natural insulin levels	Below normal	
Beta cells of pancreatic islets		Not destroyed
Pancreatic islet cell antibodies	Present	
Risk factors of having the disease	Heredity	
Typical treatments	Insulin administration	
Untreated blood sugar levels		Hyperglycemia

1. Record your observations of normal behavior of the fish.

    a. Swimming movements: _____

    _____

    b. Gill cover movements: _____

    _____

    c. Mouth movements: _____

2. Record your observations of the fish after insulin was added to the water. /5\

    a. Recorded time that insulin was added: _____

    b. Swimming movements: _____

    _____

    c. Gill cover movements: _____

    _____

    d. Mouth movements: _____

    _____

    e. Elapsed time until the insulin shock symptoms occurred: _____

3. Record your observations after the fish was transferred into a glucose solution. /6\

    a. Recorded time that fish was transferred into the glucose solution: _____

    b. Describe any changes in the behavior of the fish that indicate a recovery from insulin shock. _____

    _____

    _____

    c. Elapsed time until indications of a recovery from insulin shock occurred: _____

 Critical Thinking Application

Describe the importance for a type 1 diabetic to regulate insulin administration, meals, and exercise. /4\

_____

_____

_____

_____

_____

_____

_____

_____

34

Blood Cells and Blood Typing

MATERIALS NEEDED

Textbook
Compound light microscope
Prepared microscope slides of human blood
    (Wright's stain)
Colored pencils
ABO blood-typing kit
Simulated blood-typing kits are suggested as a
    substitute for collected blood.

For Demonstration:
Microscope slide
Alcohol swabs
Sterile blood lancet
Toothpicks
Anti-D serum
Slide warming table (Rh blood-typing box)
Disposable gloves

For Learning Extension:
Prepared slides of pathological blood, such as
    eosinophilia, leukocytosis, leukopenia, and
    lymphocytosis

⚠ SAFETY

- It is important that students learn and practice correct procedures for handling body fluids. Consider using simulated blood-typing kits, mammal blood other than human, or contaminant-free blood that has been tested and is available from various laboratory supply houses. Some of the procedures might be accomplished as demonstrations only. If student blood is used, it is important that students handle only their blood.
- Use an appropriate disinfectant to wash the laboratory tables before and after the procedures.
- Wear disposable gloves and safety goggles when handling blood samples.
- Clean end of a finger with an alcohol swab before the puncture is performed.
- Use the sterile blood lancet only once.
- Dispose of used lancets and blood-contaminated items in an appropriate container (never use the wastebasket).
- Wash your hands before leaving the laboratory.

PURPOSE OF THE EXERCISE

To review the characteristics of blood cells, to examine them microscopically, to perform a differential white blood cell count, to determine the ABO blood type of a blood sample, and to observe an Rh blood-typing test.

🔄 LEARNING OUTCOMES APR

After completing this exercise, you should be able to

1. Describe the structure and function of red blood cells, white blood cells, and platelets.
2. Identify and sketch red blood cells, five types of white blood cells, and platelets.
3. Perform and interpret the results of a differential white blood cell count.
4. Analyze the basis of ABO blood typing.
5. Match the ABO type of a blood sample.
6. Explain the basis of Rh blood typing.
7. Interpret how the Rh type of a blood sample is determined.

Blood is the only type of connective tissue with a fluid extracellular matrix, called plasma. Plasma is a mixture of substances dissolved or suspended in water. The formed elements, or blood cells, are produced in red bone marrow and include red blood cells, white blood cells, and cell fragments called platelets (fig. 34.1).

Red blood cells contain hemoglobin and transport gases between the body cells and the lungs, white blood cells defend the body against infections, and platelets play an important role in stoppage of bleeding (hemostasis).

Blood typing involves identifying protein substances called *antigens* present on the outer surface of red blood cell membranes. Although many different antigens are associated with human red blood cells, only a few of them are of clinical importance. These include the antigens of the ABO group and those of the Rh group.

To determine which antigens are present, a blood sample is mixed with blood-typing sera that contain known types of antibodies. If a particular antibody contacts a corresponding antigen, a reaction occurs, and the red blood cells clump together (agglutination). Thus, if blood cells are mixed with serum containing antibodies that react with

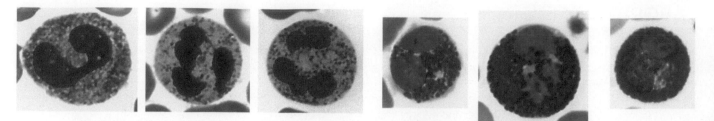

Neutrophils (3 of many variations)
- Fine light-purple granules
- Nucleus single to five lobes (highly variable)
- Immature neutrophils, called bands, have a single C-shaped nucleus
- Mature neutrophils, called segs, have a lobed nucleus
- Often called polymorphonuclear leukocytes when older

Eosinophils (3 of many variations)
- Coarse reddish granules
- Nucleus usually bilobed

Basophils (3 of many variations)
- Coarse deep blue to almost black granules
- Nucleus often almost hidden by granules

Lymphocytes (3 of many variations)
- Slightly larger than RBCs
- Thin rim of nearly clear cytoplasm
- Nearly round nucleus appears to fill most of cell in smaller lymphocytes
- Larger lymphocytes hard to distinguish from monocytes

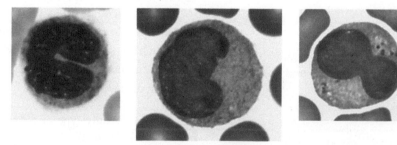

Monocytes (3 of many variations)
- Largest WBC; 2–3x larger than RBCs
- Cytoplasm nearly clear
- Nucleus round, kidney-shaped, oval, or lobed

Platelets (several variations)
- Cell fragments
- Single to small clusters

Erythrocytes (several variations)
- Lack nucleus (mature cell)
- Biconcave discs
- Thin centers appear almost hollow

FIGURE 34.1 Micrographs of blood cells illustrating some of the numerous variations of each type. Appearance characteristics for each cell pertain to a thin blood film using Wright's stain (1,000×). **APR** ©Alvin Telser, Ph.D.

antigen A and the cells clump together, antigen A must be present in those cells.

> **WARNING:** *Because of the possibility of blood infections being transmitted from one student to another if blood testing is performed in the classroom, it is suggested that commercially prepared blood slides and blood-typing kits containing virus-free human blood be used for ABO blood typing and for Rh blood typing. The instructor may wish to demonstrate Rh blood typing. Observe all of the safety procedures listed for this lab.*

EXPLORE

PROCEDURE A—Types of Blood Cells

1. Review the concept headings entitled "Red Blood Cells," "White Blood Cells," and "Blood Platelets" in section 12.2 of chapter 12 of the textbook.
2. Complete Part A of Laboratory Report 34.
3. Refer to figure 34.1 as an aid in identifying the various types of blood cells. Study the functions of the blood cells listed in table 34.1. Use the prepared slide of blood and locate each of the following:

   red blood cell (erythrocyte)
   white blood cell (leukocyte)
   granulocytes
   neutrophil
   eosinophil
   basophil
   agranulocytes
   lymphocyte
   monocyte
   platelet (thrombocyte)

TABLE 34.1 Cellular Components of Blood

Component	Function
Red blood cell (erythrocyte)	Transports oxygen and carbon dioxide
White blood cell (leukocyte)	Destroys pathogenic microorganisms and parasites, removes worn cells, provides immunity
Granulocytes—Have Granular Cytoplasm	
1. Neutrophil	Phagocytizes small particles
2. Eosinophil	Destroys parasites, helps control inflammation and allergic reactions
3. Basophil	Releases heparin and histamine
Agranulocytes—Lack Granular Cytoplasm	
1. Lymphocyte	Provides immunity
2. Monocyte	Phagocytizes large particles
Platelet (thrombocyte)	Helps control blood loss from broken vessels

4. In Part B of the laboratory report, prepare sketches of single blood cells to illustrate each type. Pay particular attention to the relative size, nuclear shape, and color of granules in the cytoplasm (if present). The sketches should be accomplished using either the high-power objective or the oil immersion objective of the compound light microscope.

EXPLORE

PROCEDURE B—Differential White Blood Cell Count

A differential white blood cell count is performed to determine the percentage of each of the various types of white blood cells present in a blood sample. The test is useful because the relative proportions of white blood cells may change in particular diseases, as indicated in table 34.2. Neutrophils, for example, usually increase during bacterial infections, whereas eosinophils may increase during certain parasitic infections and allergic reactions.

1. To make a differential white blood cell count, follow these steps:
   a. Using high-power magnification or an oil immersion objective, focus on the cells at one end of a prepared blood slide where the cells are well distributed.

LEARNING EXTENSION

> Obtain a prepared slide of pathological blood that has been stained with Wright's stain. Perform a differential white blood cell count using this slide, and compare the results with the values for normal blood listed in table 34.2. What differences do you note?
>
> _____
> _____
> _____
> _____

TABLE 34.2 Differential White Blood Cell Count

Cell Type	Normal Value (percent)	Elevated Levels May Indicate
Neutrophil	50–70	Bacterial infections, stress
Lymphocyte	25–33	Mononucleosis, whooping cough, viral infections
Monocyte	3–9	Malaria, tuberculosis, fungal infections
Eosinophil	1–3	Allergic reactions, autoimmune diseases, parasitic worms
Basophil	<1	Cancers, chicken pox, hypothyroidism

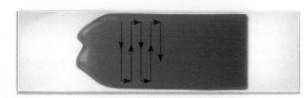

FIGURE 34.2 Move the blood slide back and forth to avoid passing the same cells twice.

b. Slowly move the blood slide back and forth, following a path that avoids passing over the same cells twice (fig. 34.2).

c. Each time you encounter a white blood cell, identify its type and record it in Part C of the laboratory report.

d. Continue searching for and identifying white blood cells until you have recorded 100 cells in the data table. *Percent* means "parts of 100" for each type of white blood cell, so the total number observed is equal to its percentage in the blood sample.

2. Complete Part C of the laboratory report.

EXPLORE

PROCEDURE C—ABO Blood Typing

1. Review the concept headings entitled "Antigens and Antibodies" and "ABO Blood Group" in section 12.5 of chapter 12 of the textbook.

2. Complete Part D of the laboratory report.

3. Perform the ABO blood type test using the blood-typing kit. To do this, follow these steps:

   a. Obtain a clean microscope slide and mark across its center with a wax pencil to divide it into right and left halves. Also write "Anti-A" near the edge of the left half and "Anti-B" near the edge of the right half (fig. 34.3).

   b. Place a small drop of blood on each half of the microscope slide. Work quickly so that the blood will not have time to clot.

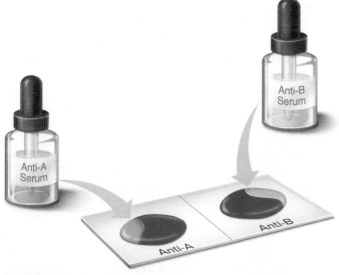

FIGURE 34.3 Slide prepared for ABO blood typing.

c. Add a drop of anti-A serum to the blood on the left half and a drop of anti-B serum to the blood on the right half. Note the color coding of the anti-A and anti-B typing sera. To avoid contaminating the serum, avoid touching the blood with the serum while it is in the dropper; instead, allow the serum to fall from the dropper onto the blood.

d. Use separate toothpicks to stir each sample of serum and blood together, and spread each over an area about as large as a quarter. Dispose of toothpicks in an appropriate container.

e. Examine the samples for clumping of red blood cells (agglutination) after 2 minutes.

f. See table 34.3 and figure 34.4 for aid in interpreting the test results.

g. Discard contaminated materials as directed by the laboratory instructor.

4. Complete Part E of the laboratory report.

Critical Thinking Application

Judging from the observations of the blood-typing results, suggest which components in the anti-A and anti-B sera caused clumping (agglutination).

_____

_____

_____

_____

DEMONSTRATION—Rh Blood Typing

1. Review the concept heading entitled "Rh Blood Group" in section 12.5 of chapter 12 of the textbook.

2. Complete Part F of the laboratory report.

3. To determine the Rh blood type of a blood sample, follow these steps:

   a. Thoroughly wash hands with soap and water, and dry them with paper towels.

   b. Cleanse the end of the middle finger with an alcohol swab, and let the finger dry in the air.

   c. Lance the tip of a finger using a sterile disposable blood lancet. Place a small drop of blood in the center of a clean microscope slide. Cover the lanced finger with a bandage.

   d. Add a drop of anti-D serum to the blood and mix them together with a clean toothpick.

   e. Place the slide on the plate of a slide warming table (Rh blood-typing box) prewarmed to 45°C (113°F) (fig. 34.5).

   f. Slowly rock the table (box) back and forth to keep the mixture moving, and watch for clumping (agglutination) of the blood cells. When clumping occurs in anti-D serum, the clumps usually are smaller than those that appear in anti-A or anti-B sera, so they may be less obvious. However, if clumping occurs,

TABLE 34.3 Possible Reactions of ABO Blood-Typing Sera and Donor Compatibilities

| Reactions | | Blood Type | Preferred Donor Type | Permissible Donor in Limited Amounts | Incompatible Donor |
Anti-A Serum	Anti-B Serum				
Clumping	No clumping	Type A	A	O	B, AB
No clumping	Clumping	Type B	B	O	A, AB
Clumping	Clumping	Type AB (universal recipient)	AB	A, B, O	None
No clumping	No clumping	Type O (universal donor)	O	No alternative types	A, B, AB

Note: The inheritance of the ABO blood groups is described in Laboratory Exercise 48.

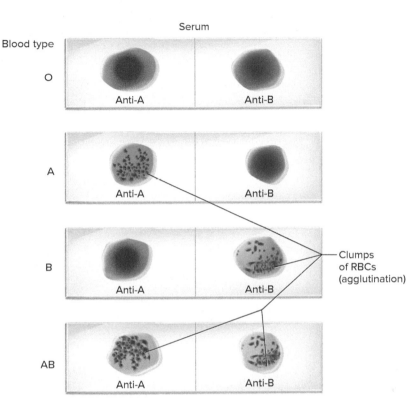

FIGURE 34.4 Four possible results of the ABO test.

the blood is called Rh positive; if no clumping occurs *within 2 minutes,* the blood is called Rh negative.

g. Discard all contaminated materials in appropriate containers.

4. Complete Parts G and H of the laboratory report.

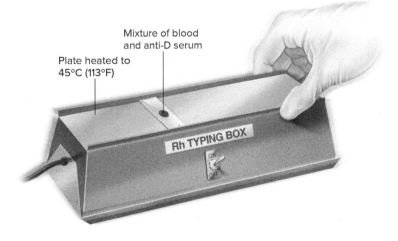

FIGURE 34.5 Slide warming box (table) used for Rh blood typing.

Notes

Name _____

Date _____

Section _____

The  corresponds to the indicated Learning Outcome(s) found at the beginning of the laboratory exercise.

LABORATORY
REPORT

34

Blood Cells and Blood Typing

♻ PART A ASSESSMENTS

Complete the following:

1. Red blood cells (RBCs) are also called _____ . ⒜

2. The functions of red blood cells are _____ . ⒜

3. _____ is the oxygen-carrying substance in a red blood cell. ⒜

4. White blood cells are also called _____ . ⒜

5. White blood cells with granular cytoplasm are called _____ . ⒜

6. White blood cells lacking granular cytoplasm are called _____ . ⒜

7. Normally, the most numerous white blood cells are _____ . ⒜

8. A platelet, a fragment of a cell, lacks a(n) _____ . ⒜

♻ PART B ASSESSMENTS

Sketch a single blood cell of each type in the following spaces. Use colored pencils to represent the stained colors of the cells. Label any features that can be identified. ⒝

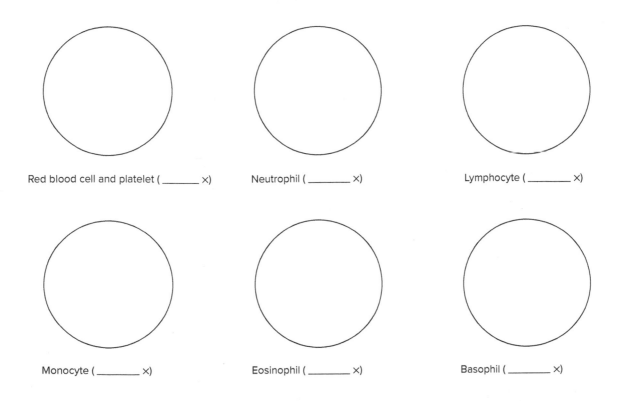

Red blood cell and platelet (_____ ×)   Neutrophil (_____ ×)   Lymphocyte (_____ ×)

Monocyte (_____ ×)   Eosinophil (_____ ×)   Basophil (_____ ×)

1. *Differential White Blood Cell Count Data Table.* As you identify white blood cells, record them in the table by using a tally system, such as IIII/II. Place tally marks in the "Number Observed" column, and total each of the five WBCs when the differential count is completed. Obtain a total of all five WBCs counted to determine the percentage of each WBC type. ⒊

Type of WBC	Number Observed	Total	Percent
Neutrophil			
Lymphocyte			
Monocyte			
Eosinophil			
Basophil			
		Total of column	

2. How do the results of your differential white blood cell count compare with the normal values listed in table 34.2? ⒊

_____

_____

_____

Critical Thinking Application

Explain the difference between a differential white blood cell count and a total white blood cell count. ⒊

_____

_____

_____

_____

PART D ASSESSMENTS

Complete the following statements:

1. The antigens of the ABO blood group are located on the red blood cell _____ . ⓐ

2. The blood of every person contains one of (how many possible?) _____ combinations of antigens. ⓐ

3. Type A blood contains antigen _____ . ⓐ

4. Type B blood contains antigen _____ . ⓐ

5. Type A blood contains _____ antibody in the plasma. ⓐ

6. Type B blood contains _____ antibody in the plasma. ⓐ

7. Persons with ABO blood type _____ are sometimes called universal recipients. ⓐ

8. Persons with ABO blood type _____ are sometimes called universal donors. ⓐ

PART E ASSESSMENTS

Complete the following:

1. What was the ABO type of the blood tested? 5 _____
2. What ABO antigens are present on the red blood cells of this type of blood? 5 _____
3. What ABO antibodies are present in the plasma of this type of blood? 5 _____
4. If a person with this blood type needed a blood transfusion, what ABO type(s) of blood could be received
   safely? 5 _____
5. If a person with this blood type were serving as a blood donor, what ABO blood type(s) could receive the blood safely? 5

_____

PART F ASSESSMENTS

Complete the following:

1. The Rh blood group was named after the _____ . 6
2. Of the antigens in the Rh group, the most important is _____ . 6
3. If red blood cells lack Rh antigens, the blood is called _____ . 6
4. If an Rh-negative person sensitive to Rh-positive blood receives a transfusion of Rh-positive blood, the
   donor's cells are likely to _____ . 6
5. An Rh-negative woman, who might be carrying a(n) _____ fetus, is given an
   injection of RhoGAM to prevent hemolytic disease of the fetus and newborn (erythroblastosis fetalis). 6

PART G ASSESSMENTS

Complete the following:

1. What was the Rh type of the blood tested in the demonstration? 7 _____
2. What Rh antigen is present on the red blood cells of this type of blood? 7 _____
3. If a person with this blood type needed a blood transfusion, what type of blood could be received safely? 7 _____
4. If a person with this blood type were serving as a blood donor, a person with what type of blood could receive the blood
   safely? 7 _____

PART H ASSESSMENTS

Examine figure 34.6. The figure contains test results for both ABO type and Rh type performed in this laboratory exercise. Observe whether clumping (agglutination) occured to answer the questions.

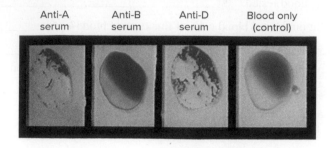

| Anti-A serum | Anti-B serum | Anti-D serum | Blood only (control) |

FIGURE 34.6   Blood test results adding anti-A serum, anti-B serum, and anti-D serum to separate drops of blood. An unaltered drop of blood is shown as a control; it does not indicate any clumping (agglutination) reaction.
Image Source/Getty Images

1. This individual has ABO type _____. ⑤
2. This individual has Rh type _____. ⑦

35

Heart Structure

MATERIALS NEEDED

Textbook
Dissectible human heart model
Preserved sheep or other mammalian heart
Dissecting tray
Dissecting instruments
Disposable gloves

For Learning Extension:
Colored pencils (red and blue)

⚠ SAFETY

- Wear disposable gloves when working on the heart dissection.
- Save or dispose of the dissected heart as instructed.
- Wash the dissecting tray and instruments as instructed.
- Wash your laboratory table.
- Wash your hands before leaving the laboratory.

PURPOSE OF THE EXERCISE

To review the structural characteristics of the human heart and to examine the major features of a mammalian heart.

LEARNING OUTCOMES APR

After completing this exercise, you should be able to

1. Identify and label the major structural features of the human heart and closely associated blood vessels.
2. Match heart structures with appropriate locations and functions.
3. Compare the features of the human heart with those of another mammal.

The heart is a muscular pump located within the mediastinum and resting upon the diaphragm. It is enclosed by the lungs, thoracic vertebrae, and sternum, and attached at its superior end (the base) are several large blood vessels. Its inferior end extends downward to the left and terminates as a bluntly pointed apex.

The heart and the proximal ends of the attached blood vessels are enclosed by a double-layered pericardium. The innermost layer of this membrane (visceral pericardium) consists of a thin covering closely applied to the surface of the heart, whereas the outer layer (parietal pericardium with fibrous pericardium) forms a tough, protective sac surrounding the heart. Between the parietal and visceral layers of the pericardium is a space, the pericardial cavity, that contains a small volume of serous (pericardial) fluid.

EXPLORE

PROCEDURE A—The Human Heart

1. Review section 13.2 entitled "Structure of the Heart" in chapter 13 of the textbook.
2. As a review activity, label figures 35.1, 35.2, and 35.3.
3. Complete Part A of Laboratory Report 35.
4. Examine the human heart model, and locate the following features:

heart
    base (superior region where blood vessels emerge)
    apex (inferior, rounded end)

pericardium (pericardial sac)
    fibrous pericardium (outer layer)
    parietal pericardium (inner lining of fibrous pericardium)
    visceral pericardium (epicardium)

pericardial cavity (between parietal and visceral pericardial membranes)

myocardium (cardiac muscle)

endocardium (lines heart chambers)

atria
    right atrium
    left atrium
    auricles

ventricles
    right ventricle
    left ventricle

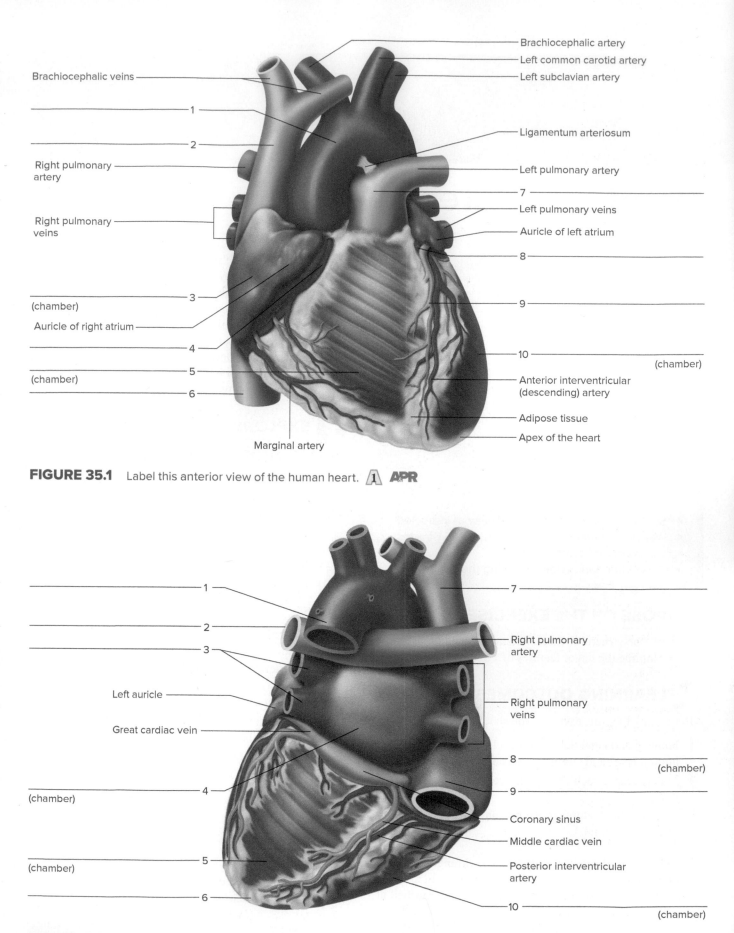

Brachiocephalic veins —

Right pulmonary
artery —

Right pulmonary
veins —

(chamber)

Auricle of right atrium —

(chamber)

——— 1

——— 2

——— 3
(chamber)

——— 4

——— 5

——— 6

Marginal artery

Brachiocephalic artery
Left common carotid artery
Left subclavian artery

Ligamentum arteriosum

Left pulmonary artery

——— 7

Left pulmonary veins

Auricle of left atrium

——— 8

——— 9

——— 10
(chamber)

Anterior interventricular
(descending) artery

Adipose tissue

Apex of the heart

FIGURE 35.1 Label this anterior view of the human heart. 🔺 **APR**

——— 1

——— 2

——— 3

Left auricle —

Great cardiac vein —

(chamber)

(chamber)

——— 4

——— 5

——— 6

——— 7

Right pulmonary
artery

Right pulmonary
veins

——— 8
(chamber)

——— 9

Coronary sinus

Middle cardiac vein

Posterior interventricular
artery

——— 10
(chamber)

FIGURE 35.2 Label this posterior view of the human heart. 🔺

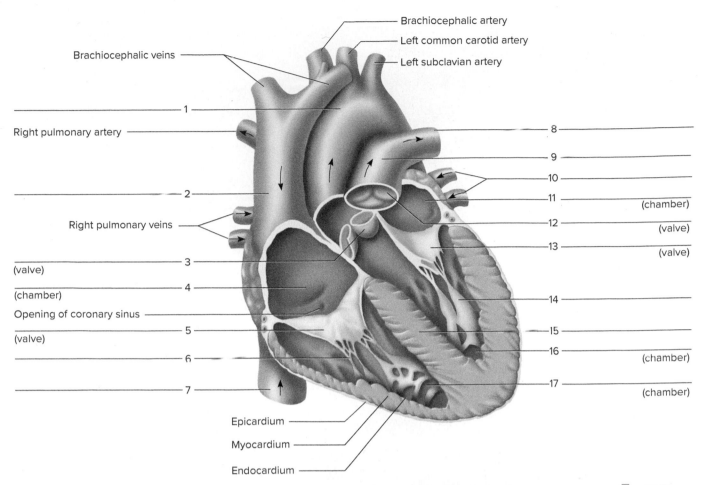

Brachiocephalic veins

Brachiocephalic artery
Left common carotid artery
Left subclavian artery

Right pulmonary artery

Right pulmonary veins

(valve)

(chamber)

Opening of coronary sinus

(valve)

1

2

3

4

5

6

7

8

9

10

11   (chamber)

12   (valve)

13   (valve)

14

15

16   (chamber)

17   (chamber)

Epicardium

Myocardium

Endocardium

FIGURE 35.3   Label this frontal section of the human heart. The arrows indicate the direction of blood flow. 🄰 **APR**

atrioventricular valves (AV valves)
   tricuspid valve (right atrioventricular valve)
   mitral valve (bicuspid valve; left atrioventricular valve)
semilunar valves
   pulmonary valve
   aortic valve
chordae tendineae
papillary muscles
superior vena cava
inferior vena cava
pulmonary trunk
pulmonary arteries
pulmonary veins
aorta
left coronary artery
right coronary artery
cardiac veins
coronary sinus (for return of blood from cardiac veins into right atrium)

5. Label the human heart model in figure 35.4.

LEARNING EXTENSION

Use red and blue colored pencils to color the blood vessels in figure 35.3. Use red to illustrate a blood vessel with oxygen-rich (oxygenated) blood, and use blue to illustrate a blood vessel with oxygen-poor (deoxygenated) blood. You can check your work by referring to the corresponding figures in the textbook, presented in full color.

♻ EXPLORE

PROCEDURE B—Dissection of a Sheep Heart

1. Obtain a preserved sheep heart. Rinse it in water thoroughly to remove as much of the preservative as possible. Also run water into the large blood vessels to force any blood clots out of the heart chambers. The structures of a sheep heart are similar to those of a human heart and will therefore be used for further exploration. Although sheep are quadrupeds, human terminology will be used for the sheep heart.

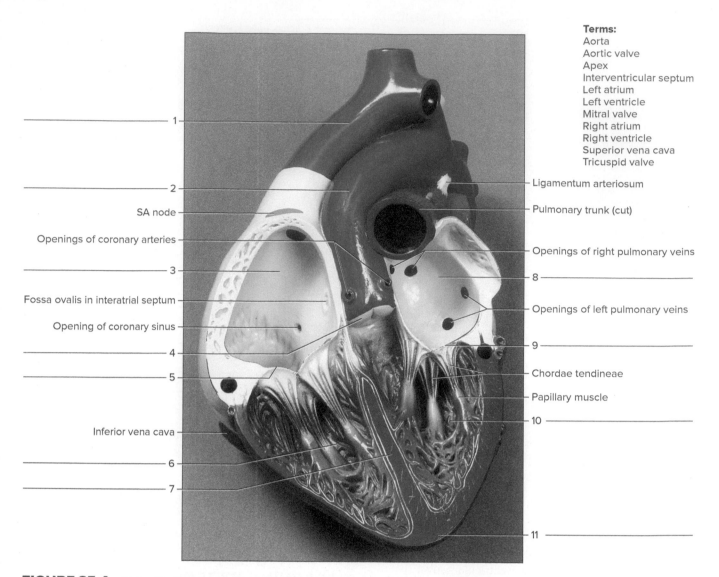

Terms:
Aorta
Aortic valve
Apex
Interventricular septum
Left atrium
Left ventricle
Mitral valve
Right atrium
Right ventricle
Superior vena cava
Tricuspid valve

1

2
SA node
Openings of coronary arteries
3
Fossa ovalis in interatrial septum
Opening of coronary sinus
4
5
Inferior vena cava
6
7

Ligamentum arteriosum
Pulmonary trunk (cut)
Openings of right pulmonary veins
8
Openings of left pulmonary veins
9
Chordae tendineae
Papillary muscle
10
11

FIGURE 35.4 Using the terms provided, identify the features indicated on this anterior view of a frontal section of a human heart model. (*Note: The pulmonary valve is not shown on the portion of the model photographed.*) **1** ©J and J Photography

2. Place the heart in a dissecting tray with its anterior surface up (fig. 35.5), and proceed as follows:
   a. Although the relatively thick *pericardial sac* is probably missing, look for traces of this membrane around the origins of the large blood vessels.
   b. Locate the *visceral pericardium,* which appears as a thin, transparent layer on the surface of the heart. Use a scalpel to remove a portion of this layer and expose the *myocardium* beneath. Also note the abundance of fat along the paths of various blood vessels. This adipose tissue occurs in the loose connective tissue that underlies the visceral pericardium.
   c. Identify the following:

      right atrium
      right ventricle
      left atrium
      left ventricle
      coronary arteries

3. Examine the posterior surface of the heart (fig. 35.6). Locate the stumps of two relatively thin-walled blood vessels that enter the right atrium. Demonstrate this connection by passing a slender probe through them. The upper vessel is the *superior vena cava,* and the lower one is the *inferior vena cava.*
4. Open the right atrium. To do this, follow these steps:
   a. Insert a blade of the scissors into the superior vena cava and cut downward through the atrial wall (fig. 35.6).
   b. Open the chambers, locate the *right atrioventricular valve (tricuspid valve),* and examine its cusps.
   c. Also locate the opening to the *coronary sinus* between the valve and the inferior vena cava.
   d. Run some water through the right atrioventricular valve to fill the chamber of the right ventricle.
   e. Gently squeeze the ventricles and watch the cusps of the valve as the water moves up against them.

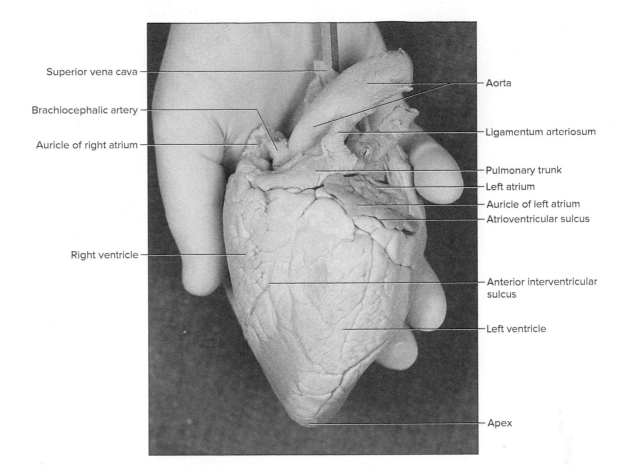

Superior vena cava

Brachiocephalic artery

Auricle of right atrium

Right ventricle

Aorta

Ligamentum arteriosum

Pulmonary trunk

Left atrium

Auricle of left atrium

Atrioventricular sulcus

Anterior interventricular sulcus

Left ventricle

Apex

FIGURE 35.5   Anterior surface of sheep heart. ©J and J Photography

5. Open the right ventricle as follows:
   a. Continue cutting downward through the right atrio-ventricular valve and the right ventricular wall until you reach the apex of the heart.
   b. Locate the *chordae tendineae* and the *papillary muscles.*
   c. Find the opening to the *pulmonary trunk* and use the scissors to cut upward through the wall of the right ventricle. Follow the pulmonary trunk until you have exposed the *pulmonary valve.*
   d. Examine the valve and its cusps.
6. Open the left side of the heart. To do this, follow these steps:
   a. Insert the blade of the scissors through the wall of the left atrium and cut downward on the anterior portion to the apex of the heart.
   b. Open the left atrium and locate the four openings of the *pulmonary veins.* Pass a slender probe through each opening and locate the stump of its vessel.

   c. Examine the *left atrioventricular valve (bicuspid valve)* and its cusps.
   d. Also examine the left ventricle and compare the thickness of its wall with that of the right ventricle.
7. Locate the aorta, which leads away from the left ventricle, and proceed as follows:
   a. Compare the thickness of the aortic wall with that of the pulmonary trunk.
   b. Use scissors to cut along the length of the aorta to expose the *aortic valve* at its base.
   c. Examine the cusps of the valve and locate the openings of the *coronary arteries* just distal to them.
8. As a review, locate and identify the stumps of each of the major blood vessels associated with the heart.
9. Discard or save the specimen as directed by the laboratory instructor.
10. Complete Part B of the laboratory report.

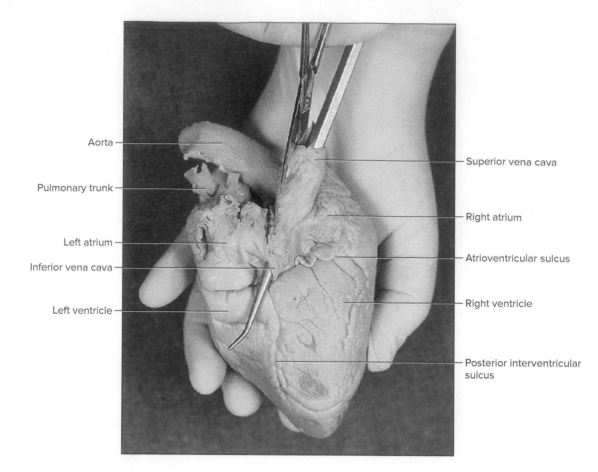

FIGURE 35.6 Posterior surface of sheep heart. To open the right atrium, insert a blade of the scissors into the superior (anterior in sheep) vena cava and cut downward. ©J and J Photography

Labels on figure:
- Aorta
- Pulmonary trunk
- Left atrium
- Inferior vena cava
- Left ventricle
- Superior vena cava
- Right atrium
- Atrioventricular sulcus
- Right ventricle
- Posterior interventricular sulcus

Name _____

Date _____

Section _____

The 🄰 corresponds to the indicated Learning Outcome(s) found at the beginning of the laboratory exercise.

LABORATORY
REPORT

35

Heart Structure

♻ PART A ASSESSMENTS

Match the terms in column A with the descriptions in column B. Place the letter of your choice in the space provided. 🄰

Column A	Column B
a. Aorta	_____ **1.** Upper chamber of heart
b. Atrium	
c. Cardiac vein	_____ **2.** Structure from which chordae tendineae originate
d. Coronary artery	_____ **3.** Prevents blood movement from right ventricle to right atrium
e. Endocardium	
f. Mitral valve	_____ **4.** Membranes around heart
g. Myocardium	_____ **5.** Prevents blood movement from left ventricle to left atrium
h. Papillary muscle	
i. Pericardial cavity	_____ **6.** Gives rise to left and right pulmonary arteries
j. Pericardial sac	_____ **7.** Inner lining of heart chamber
k. Pulmonary trunk	_____ **8.** Layer largely composed of cardiac muscle tissue
l. Tricuspid valve	
	_____ **9.** Space containing serous fluid to reduce friction during heartbeats
	_____ **10.** Drains blood from myocardial capillaries
	_____ **11.** Supplies blood to heart muscle
	_____ **12.** Distributes blood to body organs (systemic circuit) except lungs

♻ PART B ASSESSMENTS

Complete the following:

1. Compare the structure of the right atrioventricular (tricuspid) valve with that of the pulmonary valve. 🄰

_____

_____

_____

_____

2. Describe the action of the right atrioventricular (tricuspid) valve when you squeeze the water-filled right ventricle. 🄰

_____

_____

_____

_____

3. Describe the function of the chordae tendineae and the papillary muscles.  _____

_____

_____

_____

4. What is the significance of the difference in thickness between the wall of the aorta and the wall of the pulmonary trunk? _____

_____

_____

_____

5. List the correct pathway through which blood must flow in relation to the region of the heart. Assume blood is currently in a vena cava and will eventually enter the aorta. Include the major blood vessels (arteries and veins) attached directly to a heart chamber, the four heart chambers, and the four heart valves in your list. A completed figure 35.3 would be a helpful reference to use as you compose the answer.

_____

_____

_____

_____

Critical Thinking Application

What is the significance of the difference in thickness of the ventricular walls?

_____

_____

_____

36

Cardiac Cycle

MATERIALS NEEDED

For Procedure A—Heart Sounds:
Textbook
Stethoscope
Alcohol swabs

For Procedure B—Electrocardiogram:
Electrocardiograph (or other instrument for recording an ECG)
Cot or table
Alcohol swabs
Electrode paste (gel)
Electrodes and cables
Lead selector switch

PURPOSE OF THE EXERCISE

To review the events of a cardiac cycle, to become acquainted with normal heart sounds, and to record an electrocardiogram.

LEARNING OUTCOMES APR

After completing this exercise, you should be able to

① Interpret the major events of a cardiac cycle.

② Associate the sounds produced during a cardiac cycle with the valves closing.

③ Correlate the components of a normal ECG pattern with the phases of a cardiac cycle.

④ Record and interpret an electrocardiogram.

A set of atrial contractions while the ventricular walls relax, followed by ventricular contractions while the atrial walls relax, constitutes a *cardiac cycle.* Such a cycle is accompanied by blood pressure changes within the heart chambers, movement of blood into and out of the chambers, and opening and closing of heart valves. These closing valves produce vibrations in the tissues and thus create the sounds associated with the heartbeat. If backflow of blood occurs when a heart valve is closed, creating a turbulence noise, the condition is known as a murmur.

The regulation and coordination of the cardiac cycle involves the *cardiac conduction system.* The pathway of electrical signals originates from the *SA (sinoatrial) node* located in the right atrium near the entrance of the superior vena cava. The stimulation of the heartbeat and the heart rate originate from the SA node, so it is often called the *pacemaker.* As the signals pass through the atrial walls toward the *AV (atrioventricular) node,* contractions of the atria take place. When the AV node has been signaled, the rapid continuation of the electrical signals occurs through the *AV bundle,* passes through the *right and left bundle branches* within the interventricular septum, and terminates via the *Purkinje fibers* throughout the ventricular walls. Once the myocardium has been stimulated, ventricular contractions happen, completing one cardiac cycle. The sympathetic and parasympathetic subdivisions of the autonomic nervous system influence the activity of the pacemaker under various conditions. An increased rate results from sympathetic responses; a decreased rate results from parasympathetic responses.

A number of electrical changes also occur in the myocardium as it contracts and relaxes. These changes can be detected by using metal electrodes and an instrument called an *electrocardiograph.* The recording produced by the instrument is an *electrocardiogram,* or *ECG (EKG).*

EXPLORE

PROCEDURE A—Heart Sounds

1. Review the concept headings entitled "Pressure and Volume Changes During a Cardiac Cycle" and "Heart Sounds" of section 13.3 in chapter 13 of the textbook.

2. Complete Part A of Laboratory Report 36.

3. Listen to your heart sounds. To do this, follow these steps:

    a. Obtain a stethoscope, and clean its earpieces and the diaphragm by using alcohol swabs.

    b. Fit the earpieces into your ear canals so that the angles are positioned in the forward direction.

    c. Firmly place the diaphragm (bell) of the stethoscope on the chest over the fifth intercostal space near the apex of the heart (fig. 36.1) and listen to

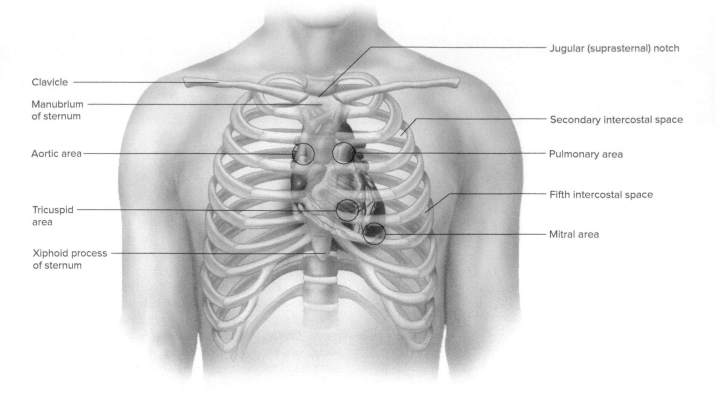

Clavicle

Manubrium of sternum

Aortic area

Tricuspid area

Xiphoid process of sternum

Jugular (suprasternal) notch

Secondary intercostal space

Pulmonary area

Fifth intercostal space

Mitral area

FIGURE 36.1 The first sound (lubb) of a cardiac cycle can be heard by placing the diaphragm of a stethoscope over the fifth intercostal space near the apex of the heart. The second sound (dupp) can be heard over the second intercostal space, just left of the sternum. The thoracic regions circled indicate where the sounds of each heart valve are most easily heard. Note that where the sound is heard may differ slightly from the anatomical location of the structure. **APR**

the sounds. This is a good location to hear the first sound (*lubb*) of a cardiac cycle when the AV valves are closing, which occurs during ventricular *systole* (contraction).

d. Move the diaphragm to the second intercostal space just to the left of the sternum, and listen to the sounds from this region. You should be able to hear the second sound (*dupp*) of the cardiac cycle clearly when the semilunar valves are closing, which occurs during ventricular *diastole* (relaxation).

e. It is possible to hear sounds associated with the aortic and pulmonary valves by listening from the second intercostal space on either side of the sternum. The aortic valve sound comes from the right and the pulmonary valve sound from the left. The sound associated with the mitral valve can be heard from the fifth intercostal space at the nipple line on the left. The sound of the tricuspid valve can be heard at the fourth intercostal space just to the left of the sternum (fig. 36.1). The four points for listening (ausculating) the four heart valve sounds when closing are not directly superficial to the valve locations. This is because the sounds associated with the closing valves travel in a slanting direction to the surface of the chest wall where the stethoscope placements occur.

4. Inhale slowly and deeply, and exhale slowly while you listen to the heart sounds from each of the locations as before. Note any changes that have occurred in the sounds.

5. Exercise moderately outside the laboratory for a few minutes so that other students listening to heart sounds will not be disturbed. After the exercise period, listen to the heart sounds and note any changes that have occurred in them. *This exercise should be avoided by anyone with health risks.*

6. Complete Part B of the laboratory report.

⬙ EXPLORE

PROCEDURE B—Electrocardiogram

1. Review the concept headings entitled "Cardiac Conduction System" and "Electrocardiogram" of section 13.3 in chapter 13 of the textbook.

2. Complete Part C of the laboratory report.

3. The laboratory instructor will demonstrate the proper adjustment and use of the instrument available to record an electrocardiogram.

4. Record your laboratory partner's ECG. To do this, follow these steps:

a. Have your partner lie on a cot or table close to the electrocardiograph, remaining as relaxed and still as possible.

b. Scrub the electrode placement locations with alcohol swabs (fig. 36.2). Apply a small quantity of electrode paste to the skin on the insides of the wrists and ankles. (Any jewelry on the wrists or ankles should be removed.)

c. Spread some electrode paste over the inner surfaces of four electrodes and attach one to each of the prepared skin areas (fig. 36.2). Make sure there is good contact between the skin and the metal of the electrodes. The electrode on the right ankle is the grounding system.

d. Attach the electrodes to the corresponding cables of a lead selector switch. When an ECG recording is made, only two electrodes are used at a time, and the selector switch allows various combinations of electrodes (leads) to be activated. Three standard limb leads placed on the two wrists and the left ankle are used for an ECG. This arrangement has become known as *Einthoven's triangle,*[1] which permits the recording of the potential difference between any two of the electrodes.

The standard leads I, II, and III are called bipolar leads because they are the potential difference between two electrodes (a positive and a negative). Lead I measures the potential difference between the right wrist (negative) and the left wrist (positive). Lead II measures the potential difference between the right wrist and the left ankle, and lead III measures the potential difference between the left wrist and the left ankle. The right ankle is always the ground.

e. Turn on the recording instrument and adjust it as previously demonstrated by the laboratory instructor. The paper speed should be set at 2.5 cm/second. This is the standard speed for ECG recordings.

f. Set the lead selector switch to lead I (right wrist, left wrist electrodes), and record the ECG for 1 minute.

g. Set the lead selector switch to lead II (right wrist, left ankle electrodes), and record the ECG for 1 minute.

h. Set the lead selector switch to lead III (left wrist, left ankle electrodes), and record the ECG for 1 minute.

i. Remove the electrodes and clean the paste from the metal and skin.

j. Use figure 36.3 to label the ECG components of the results from leads I, II, and III. The PQ interval is often called the PR interval because the Q wave is often small or absent. The normal PQ interval is 0.12–0.20 seconds. The normal QRS complex duration is less than 0.10 second.

5. Complete Part D of the laboratory report.

1. Willem Einthoven (1860–1927), a Dutch physiologist, received the Nobel Prize for Physiology or Medicine for his work with electrocardiograms.

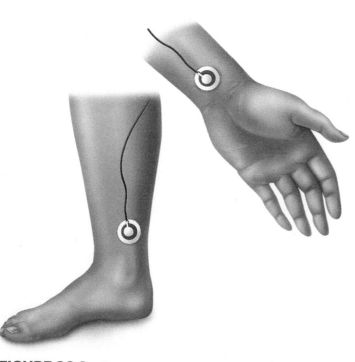

FIGURE 36.2 To record an ECG, attach electrodes to the wrists and ankles.

FIGURE 36.3 Components of a normal ECG pattern with a time scale. **APR**

Name _____

Date _____

Section _____

The ⚠ corresponds to the indicated Learning Outcome(s) found at the beginning of the laboratory exercise.

Cardiac Cycle

♻ PART A ASSESSMENTS

Complete the following statements:

1. The period during which a heart chamber is contracting is called _____ . ⚠

2. The period during which a heart chamber is relaxing is called _____ . ⚠

3. During ventricular contraction, the AV valves (tricuspid and mitral valves) are _____ . ⚠

4. During ventricular relaxation, the AV valves are _____ . ⚠

5. Heart sounds are due to _____ in the heart tissues created by changes in blood flow. ⚠

6. The first sound of the cardiac cycle occurs when the _____ valves are closing. ⚠

7. The second sound of the cardiac cycle occurs when the _____ valves are closing. ⚠

♻ PART B ASSESSMENTS

Complete the following:

1. What changes did you note in the heart sounds when you inhaled deeply? ⚠ _____

_____

_____

2. What changes did you note in the heart sounds following the exercise period? ⚠ _____

_____

_____

267

Complete the following statements:

1. The cells (fibers) of the cardiac conduction system are specialized _____ tissue. ⑶

2. Normally, the _____ node serves as the pacemaker of the heart. ⑶

3. The _____ node is located in the inferior portion of the interatrial septum. ⑶

4. The large fibers on the distal side of the AV node make up the _____. ⑶

5. The fibers that carry cardiac impulses from the interventricular septum into the myocardium are called _____ fibers. ⑶

6. A(n) _____ is a recording of electrical changes occurring in the myocardium during the cardiac cycle. ⑶

7. Between cardiac cycles, cardiac muscle fibers remain _____ with no detectable electrical changes. ⑶

8. The P wave corresponds to the depolarization of the muscle fibers of the _____. ⑶

9. The QRS complex corresponds to the depolarization of the muscle fibers of the _____. ⑶

10. The T wave corresponds to the repolarization of the muscle fibers of the _____. ⑶

![icon] **PART D ASSESSMENTS**

1. Attach a short segment of the ECG recording from each of the three leads you used, and label the waves of each. ⑷

Lead I

Lead II

Lead III

2. What differences do you find in the ECG patterns of these leads? 🔺 _____
_____
_____
_____
_____
_____

3. How much time passed from the beginning of the P wave to the beginning of the QRS complex (PQ interval or PR interval) in the ECG from lead I? A _____

4. What is the significance of this PQ (PR) interval? A _____

_____

_____

_____

5. How can you determine heart rate from an electrocardiogram? A _____

_____

_____

_____

6. What was your heart rate as determined from the ECG? A _____

Critical Thinking Application

If a person's heart rate were 72 beats per minute, determine the number of QRS complexes that would have appeared on an ECG during the first 30 seconds. A

_____

_____

_____

37

Blood Vessel Structure, Arteries, and Veins

MATERIALS NEEDED

Textbook
Dissectible human heart model
Human torso model
Anatomical charts of the cardiovascular system
Compound light microscope
Prepared microscope slides:
 Artery cross section
 Vein cross section
Live frog or goldfish
Frog Ringer's solution
Paper towel
Rubber bands
Frog board or heavy cardboard (with a 1-inch hole
 cut in one corner)
Dissecting pins
Thread
Masking tape

For Learning Extension:
Ice
Hot plate
Thermometer

⚠ SAFETY

- Wear disposable gloves when handling the live
 frogs.
- Return the frogs to the location indicated after
 the experiment.
- Wash your hands before leaving the laboratory.

LEARNING OUTCOMES A&PR

After completing this exercise, you should be able to

1. Distinguish the structures and functions of arteries,
 veins, and capillaries.
2. Interpret the types of blood vessels in the web of a
 frog's foot.
3. Locate the major arteries of the systemic circuit on a
 diagram, chart, or model.
4. Locate the major veins of the systemic circuit on a
 diagram, chart, or model.
5. Recognize the anatomical terms pertaining to the
 blood vessel structure, arteries, and veins.

The blood vessels form a closed system of tubes that carry blood to and from the heart, lungs, and body cells. These tubes include *arteries* and *arterioles* that transport blood away from the heart; *capillaries* in which exchanges of substances occur between the blood and surrounding tissues; and *venules* and *veins* that return blood to the heart.

The blood vessels of the cardiovascular system can be divided into two major pathways—the *pulmonary circuit* and the *systemic circuit*. Within each circuit, arteries transport blood away from the heart. After exchanges of gases, nutrients, and wastes have occurred between the blood and the surrounding tissues, veins return the blood to the heart.

Variations exist in anatomical structures among humans, especially in arteries and veins. The illustrations in this laboratory manual represent normal (meaning the most common variation) anatomy.

PURPOSE OF THE EXERCISE

To review the structure and functions of blood vessels and locate the major arteries and veins.

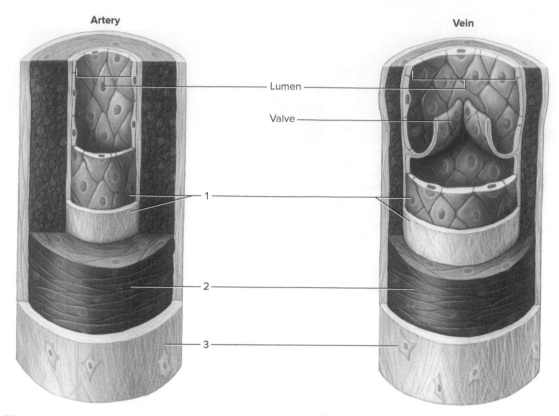

Artery

Vein

Lumen

Valve

1

2

3

FIGURE 37.1 Label the tunics of the wall of this artery and vein.

 **EXPLORE**

PROCEDURE A—Blood Vessel Structure

1. Review section 13.4 entitled "Blood Vessels" in chapter 13 of the textbook.
2. As a review activity, label figure 37.1.
3. Complete Part A of Laboratory Report 37.
4. Obtain a microscope slide of an artery cross section, and examine it using low-power and high-power magnification. Identify the three distinct layers (tunics) of the arterial wall. The inner layer (**tunica interna**) is composed of an endothelium (simple squamous epithelium) and appears as a wavy line due to an abundance of elastic fibers that have recoiled just beneath it. The middle layer (**tunica media**) consists of numerous concentrically arranged smooth muscle cells with elastic fibers scattered among them. The outer layer (**tunica externa**) contains connective tissue rich in collagen fibers (fig. 37.2).
5. Prepare a labeled sketch of the arterial wall in Part B of the laboratory report.
6. Obtain a slide of a vein cross section and examine it as you did the artery cross section. Note the thinner wall and larger lumen relative to an artery of comparable locations. Identify the three layers of the wall, and

prepare a labeled sketch in Part B of the laboratory report.
7. Complete Part B of the laboratory report.
8. Observe the blood vessels in the webbing of a frog's foot. (As a substitute for a frog, a live goldfish could be used to observe circulation in the tail.) To do this, follow these steps:
   a. Obtain a live frog. Wrap its body in a moist paper towel, leaving one foot extending outward. Secure the towel with rubber bands, but be careful not to wrap the animal so tightly that it could be injured. Try to keep the nostrils exposed.
   b. Place the frog on a frog board or on a piece of heavy cardboard with the foot near the hole in one corner.
   c. Fasten the wrapped body to the board with masking tape.
   d. Carefully spread the web of the foot over the hole and secure it to the board with dissecting pins and thread (fig. 37.3). Keep the web moist with frog Ringer's solution.
   e. Secure the board on the stage of a microscope with heavy rubber bands, and position it so that the web is beneath the objective lens.

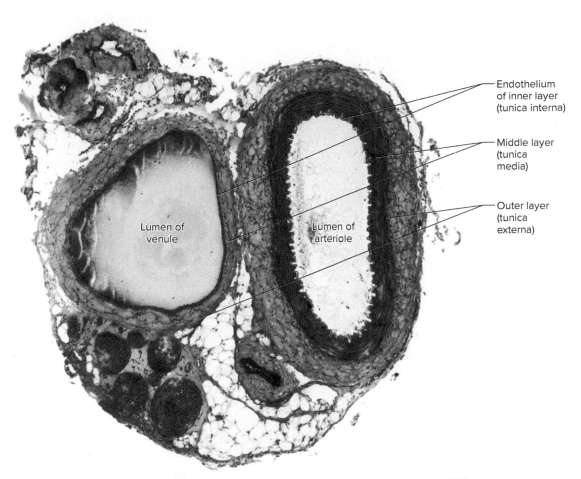

Endothelium
of inner layer
(tunica interna)

Middle layer
(tunica
media)

Outer layer
(tunica
externa)

Lumen of
venule

Lumen of
arteriole

FIGURE 37.2 Cross section of a small artery (arteriole) and a small vein (venule) (200×). **APR** National Geographic Image Collection/Alamy Stock Photo

LEARNING EXTENSION

Investigate the effect of temperature change on the blood vessels of the frog's foot by flooding the web with a small quantity of ice water. Observe the blood vessels with low-power magnification and note any changes in their diameters or the rate of blood flow. Remove the ice water and replace it with water heated to about 35°C (95°F). Repeat your observations. What do you conclude from this experiment?

_____

_____

_____

f. Focus on the web using low-power magnification, and locate some blood vessels. Note the movement of the blood cells and the direction of the blood flow. You might notice that red blood cells of frogs are nucleated. Identify an arteriole, a capillary, and a venule.

g. Examine each of these vessels with high-power magnification.

h. When finished, return the frog to the location indicated by your instructor. The microscope lenses and stage will likely need cleaning after the experiment.

9. Complete Part C of the laboratory report.

↻ EXPLORE

PROCEDURE B—Arterial System

1. Review section 13.6 entitled "Arterial System" in chapter 13 of the textbook.

2. As a review activity, label figures 37.4, 37.5, 37.6, 37.7, and 37.8.

3. Locate the following arteries of the systemic circuit on the charts and human torso model:

aorta

    ascending aorta

    aortic arch (arch of aorta)

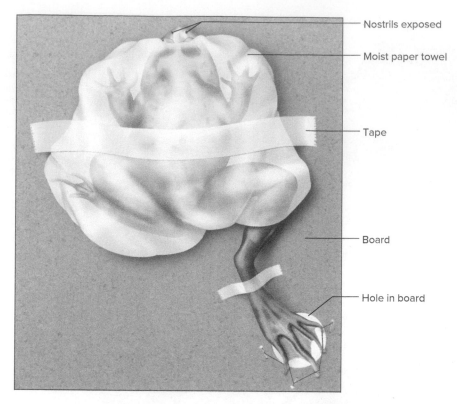

FIGURE 37.3 Spread the web of the foot over the hole and secure it to the board with pins and thread.

- Nostrils exposed
- Moist paper towel
- Tape
- Board
- Hole in board

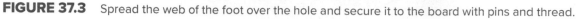

thoracic aorta
abdominal aorta

branches of the aorta
  coronary artery
  brachiocephalic trunk (artery)
  left common carotid artery
  left subclavian artery
  celiac trunk (artery)
  superior mesenteric artery
  renal artery
  inferior mesenteric artery
  common iliac artery

arteries to neck, head, and brain
  vertebral artery
  common carotid artery
  external carotid artery
  internal carotid artery
  facial artery

arteries of base of brain (fig. 37.6)
  vertebral artery

basilar artery
internal carotid artery
cerebral arterial circle (circle of Willis)

arteries to shoulder and upper limb
  subclavian artery
  axillary artery
  brachial artery
  deep brachial artery
  ulnar artery
  radial artery

arteries to pelvis and lower limb
  common iliac artery
  internal iliac artery
  external iliac artery
  femoral artery
  popliteal artery
  anterior tibial artery
  dorsalis pedis artery (dorsal artery of foot)
  posterior tibial artery

4. Complete Part D of the laboratory report.

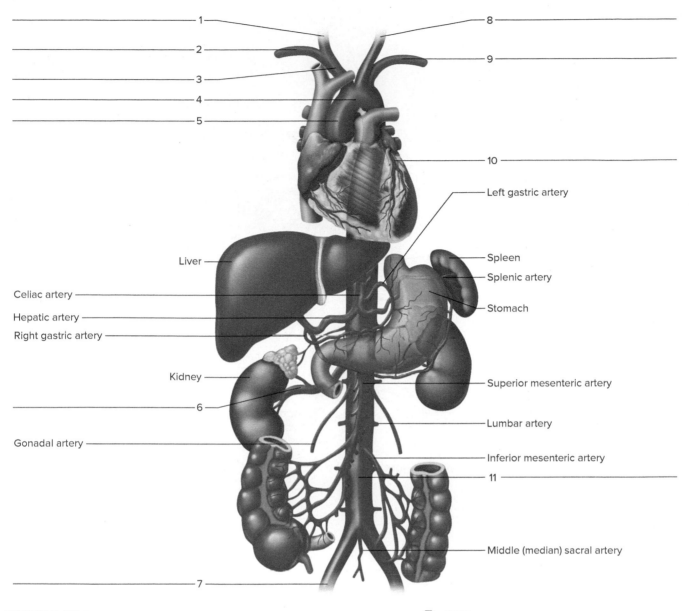

1 _____

2 _____

3 _____

4 _____

5 _____

8 _____

9 _____

10 _____

Left gastric artery

Liver

Celiac artery

Hepatic artery

Right gastric artery

Kidney

6 _____

Gonadal artery

7 _____

Spleen

Splenic artery

Stomach

Superior mesenteric artery

Lumbar artery

Inferior mesenteric artery

11 _____

Middle (median) sacral artery

FIGURE 37.4 Label the portions of the aorta and its principal branches. 🄐 APR

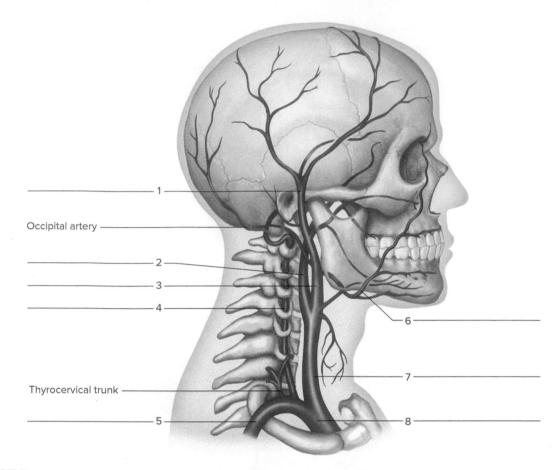

Occipital artery

Thyrocervical trunk

1

2

3

4

5

6

7

8

FIGURE 37.5 Label the arteries supplying the right side of the neck and head. (The clavicle has been removed.) **4** **APR**

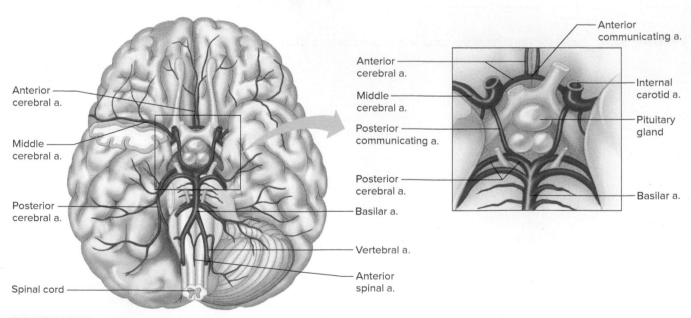

Anterior
cerebral a.

Middle
cerebral a.

Posterior
cerebral a.

Spinal cord

Anterior
cerebral a.

Middle
cerebral a.

Posterior
communicating a.

Posterior
cerebral a.

Basilar a.

Vertebral a.

Anterior
spinal a.

Anterior
communicating a.

Internal
carotid a.

Pituitary
gland

Basilar a.

FIGURE 37.6 View of inferior surface of the brain. The cerebral arterial circle (circle of Willis) is formed by the anterior and posterior cerebral arteries, which join the internal carotid arteries (*a.* strands for artery). **APR**

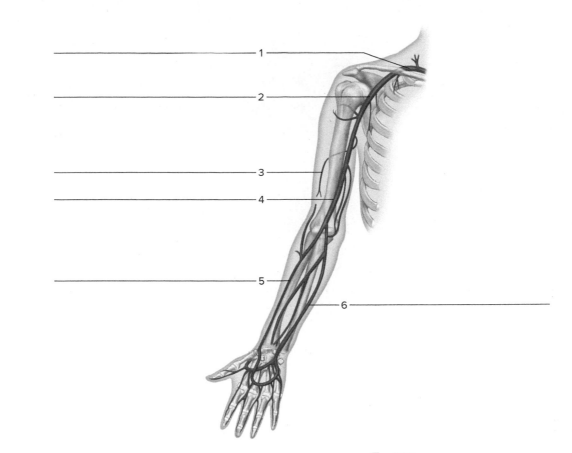

FIGURE 37.7 Label the major arteries of the shoulder and upper limb. 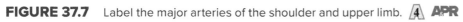 APR

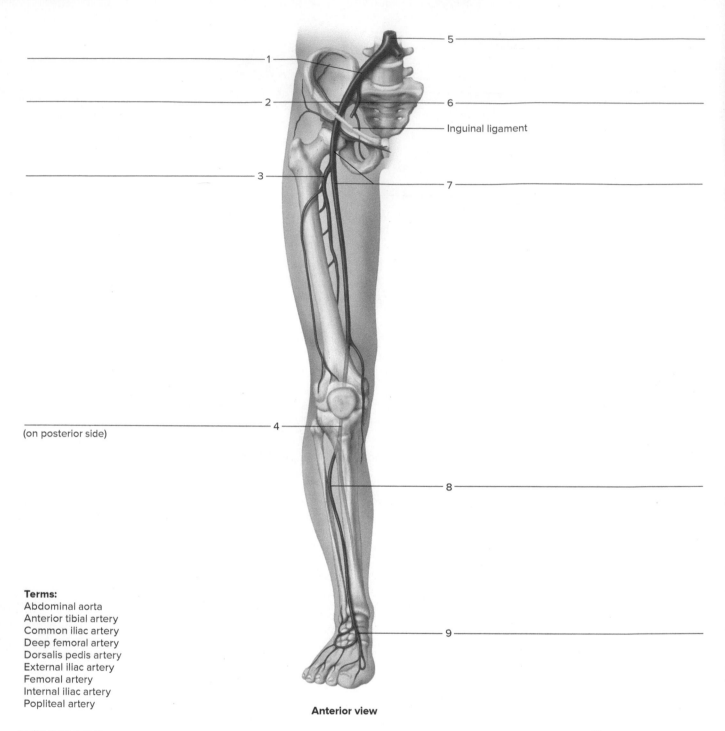

5 —————————

1 ———

2 ———

—— 6

—— Inguinal ligament

3 ———

—— 7

4 ———

(on posterior side)

—— 8

Terms:
Abdominal aorta
Anterior tibial artery
Common iliac artery
Deep femoral artery
Dorsalis pedis artery
External iliac artery
Femoral artery
Internal iliac artery
Popliteal artery

—— 9

Anterior view

FIGURE 37.8 Using the terms provided, label the major arteries supplying the pelvis and lower limb. ⒜ **APR**

PROCEDURE C—Venous System

1. Review section 13.7 entitled "Venous System" in chapter 13 of the textbook.
2. As a review activity, label figures 37.9, 37.10, 37.11, and 37.12.
3. Locate the following veins of the systemic circuit on the charts and human torso model:

   veins from brain, head, and neck
   external jugular vein
   internal jugular vein
   vertebral vein
   subclavian vein
   brachiocephalic vein
   superior vena cava

   veins from upper limb and shoulder
   radial vein
   ulnar vein
   brachial vein
   basilic vein
   cephalic vein
   median cubital vein (antecubital vein)

   axillary vein
   subclavian vein

   veins of abdominal viscera
   hepatic portal vein
   gastric vein
   superior mesenteric vein
   splenic vein
   inferior mesenteric vein
   hepatic vein
   renal vein

   veins from lower limb and pelvis
   anterior tibial vein
   posterior tibial vein
   popliteal vein
   femoral vein
   great (long) saphenous vein
   small (short) saphenous vein
   external iliac vein
   internal iliac vein
   common iliac vein
   inferior vena cava

4. Complete Parts E, F, and G of the laboratory report.

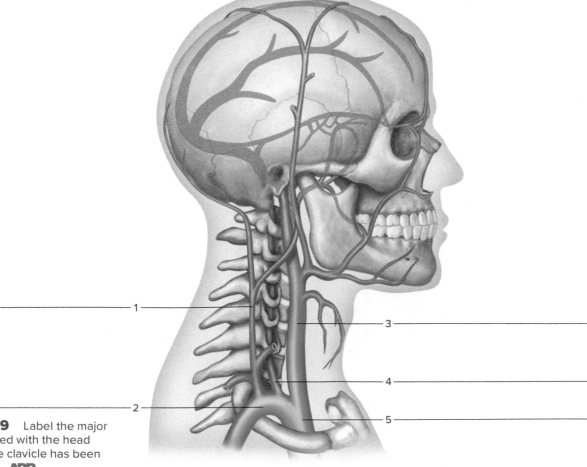

FIGURE 37.9 Label the major veins associated with the head and neck. (The clavicle has been removed.) **5** **APR**

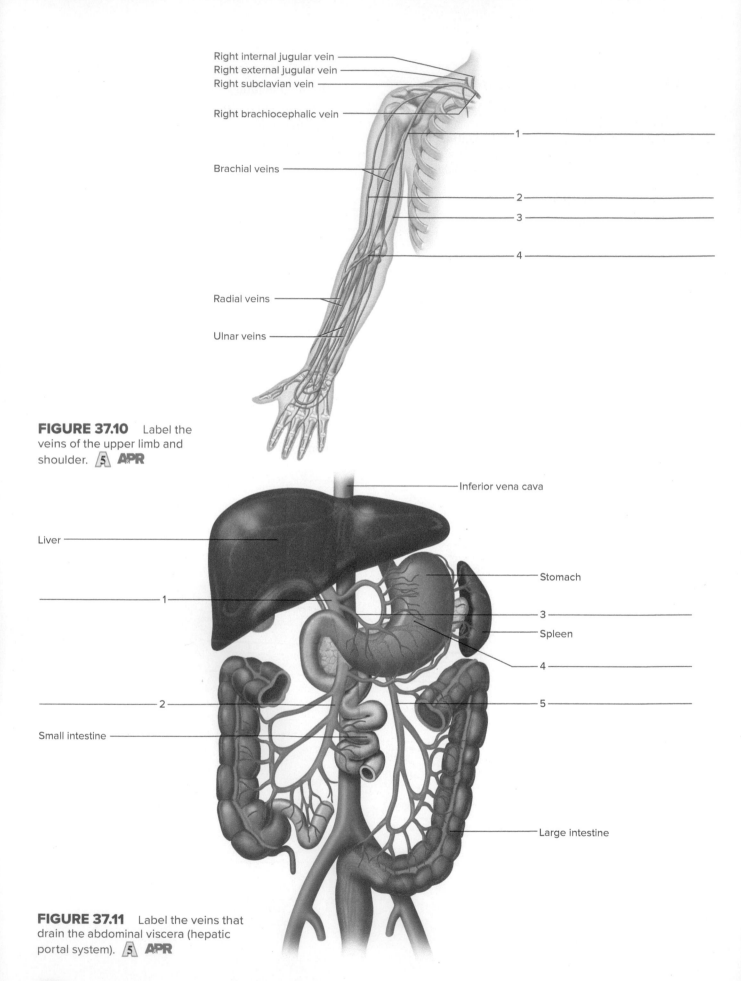

Right internal jugular vein

Right external jugular vein

Right subclavian vein

Right brachiocephalic vein

Brachial veins

Radial veins

Ulnar veins

1

2

3

4

FIGURE 37.10 Label the veins of the upper limb and shoulder. 🖐 **APR**

Inferior vena cava

Liver

Stomach

1

3

Spleen

4

2

5

Small intestine

Large intestine

FIGURE 37.11 Label the veins that drain the abdominal viscera (hepatic portal system). 🖐 **APR**

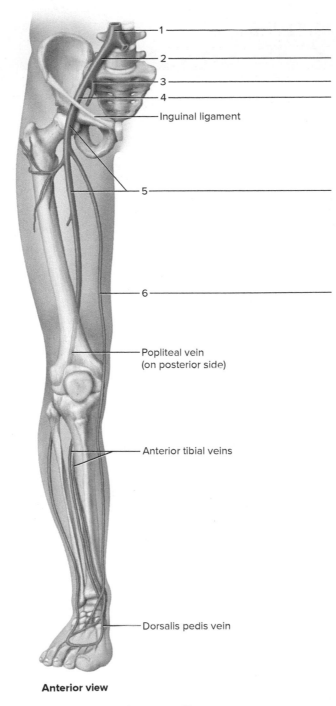

1 ————————————————————
2 ————————————————————
3 ————————————————————
4 ————————————————————
Inguinal ligament

5 ————————————————————

6 ————————————————————

Popliteal vein
(on posterior side)

Anterior tibial veins

Dorsalis pedis vein

Anterior view

FIGURE 37.12   Label the veins of the pelvis and lower limb.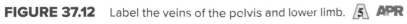

Notes

The A corresponds to the indicated Learning Outcome(s) found at the beginning of the laboratory exercise.

Blood Vessel Structure, Arteries, and Veins

PART A ASSESSMENTS

Complete the following statements:

1. Simple squamous epithelial tissue called _____ forms the inner linings of the tunica interna of blood vessels. A

2. The _____ of an arterial wall contains many smooth muscle cells. A

3. The _____ of an arterial wall is largely composed of connective tissue. A

4. When contraction of the smooth muscle in a blood vessel wall occurs, the vessel is referred to as being in a condition of _____ . A

5. Relaxation of the smooth muscle in a blood vessel wall results in the vessel being in a condition of _____ . A

6. The smallest blood vessels are called _____ . A

7. Filtration results when substances are forced through capillary walls by _____ pressure. A

8. The presence of plasma proteins in the blood increases its _____ pressure as compared to tissue fluids. A

9. _____ in certain veins close if the blood begins to back up in that vein. A

PART B ASSESSMENTS

1. Sketch and label a section of an artery wall and a vein wall. A

Artery wall (_____ ×)	Vein wall (_____ ×)

2. Describe the differences you noted in the structures of the walls of an artery and a vein. Mention each of the three layers of the wall. ⟨1⟩ _____

_____

_____

Critical Thinking Application

Explain the functional significance of the differences you noted in the structures of the arterial and venous walls. ⟨1⟩

_____

_____

PART C ASSESSMENTS

1. How did you distinguish between arterioles and venules when you observed the vessels in the web of the frog's foot? ⟨2⟩ _____

_____

2. How did you recognize capillaries in the web? ⟨2⟩ _____

_____

3. What differences did you note in the rate of blood flow through the arterioles, capillaries, and venules? ⟨2⟩

_____

_____

PART D ASSESSMENTS

Match the arteries in column A with the descriptions in column B. Place the letter of your choice in the space provided. ⟨4⟩

Column A	Column B
a. Anterior tibial	_____ **1.** Jaw, teeth, and face
b. Brachial	_____ **2.** Kidney
c. Celiac	_____ **3.** Upper digestive tract, spleen, and liver
d. External carotid	_____ **4.** Foot and toes
e. Internal carotid	_____ **5.** Gluteal muscles
f. Internal iliac	_____ **6.** Biceps muscle
g. Phrenic	_____ **7.** Knee joint
h. Popliteal	_____ **8.** Adrenal gland
i. Renal	_____ **9.** Diaphragm
j. Suprarenal	_____ **10.** Brain

PART E ASSESSMENTS

Match each vein in column A with the vein it drains into from column B. Place the letter of your choice in the space provided.

Column A	Column B
a. Anterior tibial	_____ **1.** Popliteal
b. Basilic	_____ **2.** Axillary
c. Brachiocephalic	_____ **3.** Inferior vena cava
d. Common iliac	_____ **4.** Subclavian
e. External jugular	_____ **5.** Brachial
f. Femoral	_____ **6.** Superior vena cava
g. Popliteal	_____ **7.** Femoral
h. Radial	_____ **8.** External iliac

PART F ASSESSMENTS

Locate the 20 anatomical terms pertaining to the "Blood Vessel Structure, Arteries, and Veins." 6

B	A	S	I	L	I	C	C	I	N	U	T
W	J	Y	K	S	F	A	C	I	A	L	H
V	U	R	A	U	M	G	L	O	A	N	O
F	G	A	R	O	K	A	R	E	Z	A	R
C	U	N	T	N	N	T	T	B	Z	R	A
E	L	O	E	E	A	I	L	I	A	C	C
P	A	M	R	H	L	A	T	R	O	P	I
H	R	L	I	P	V	C	A	I	L	E	C
A	W	U	O	A	X	I	L	L	A	R	Y
L	U	P	L	S	Y	S	T	E	M	I	C
I	H	V	E	R	T	E	B	R	A	L	G
C	A	P	I	L	L	A	R	Y	T	M	D

Find and circle words pertaining to Laboratory Exercise 37. The words may be horizontal, vertical, diagonal, or backward.

7 terms (arteriole, capillary, portal, pulmonary, systemic, thoracic, tunic)

9 arteries; 4 veins

Label the major arteries and veins indicated in figure 37.13. 🄰 🄰 🄰

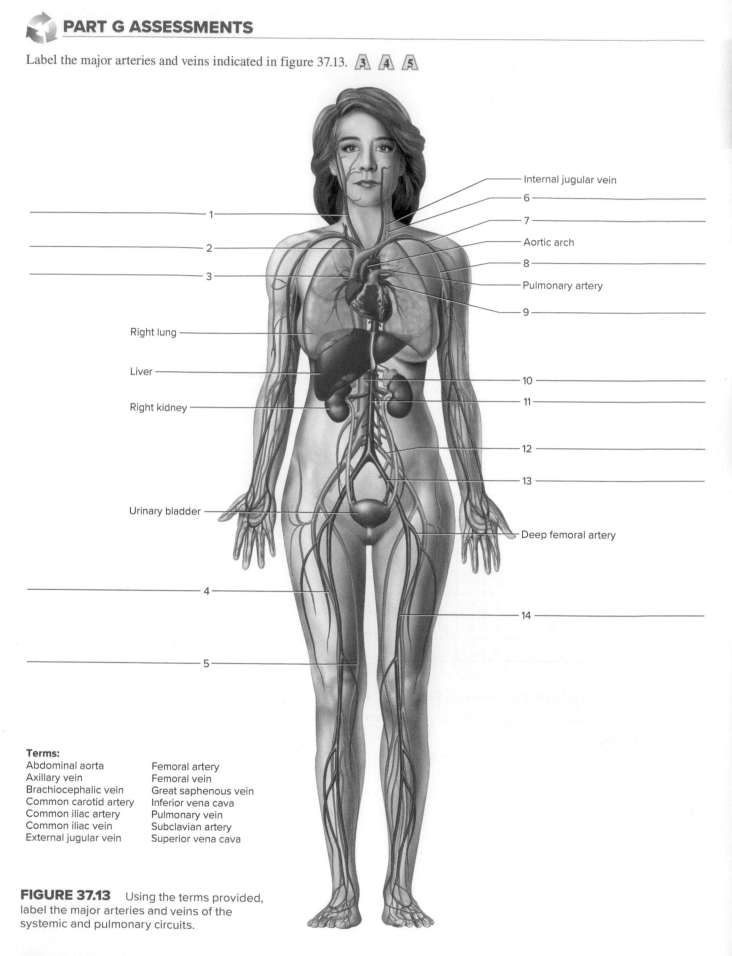

Internal jugular vein

1

2

3

6

7

Aortic arch

8

Pulmonary artery

9

Right lung

Liver

Right kidney

10

11

12

13

Deep femoral artery

Urinary bladder

4

14

5

Terms:

Abdominal aorta
Axillary vein
Brachiocephalic vein
Common carotid artery
Common iliac artery
Common iliac vein
External jugular vein

Femoral artery
Femoral vein
Great saphenous vein
Inferior vena cava
Pulmonary vein
Subclavian artery
Superior vena cava

FIGURE 37.13 Using the terms provided, label the major arteries and veins of the systemic and pulmonary circuits.

38

Pulse Rate and Blood Pressure

MATERIALS NEEDED

Textbook
Clock with second hand
Sphygmomanometer
Stethoscope
Alcohol swabs

PURPOSE OF THE EXERCISE

To examine the pulse, to determine the pulse rate, to measure blood pressure, and to investigate the effects of body position and exercise on pulse rate and blood pressure.

LEARNING OUTCOMES

After completing this exercise, you should be able to

1. Determine pulse rate and correlate pulse characteristics.
2. Test the effects of various factors on pulse rate.
3. Describe and measure blood pressure using a sphygmomanometer.
4. Test the effects of various factors on blood pressure.

The surge of blood that enters the arteries each time the ventricles of the heart contract causes the elastic walls of these vessels to swell. Then, as the ventricles relax, the walls recoil. This alternate expanding and recoiling of an arterial wall can be felt as a pulse in vessels that run close to the surface of the body.

The force exerted by the blood pressing against the inner walls of arteries also creates blood pressure. This pressure reaches a maximum during ventricular contraction and then drops to its lowest level while the ventricles are relaxed.

EXPLORE

PROCEDURE A—Pulse Rate

1. Review section 13.5 entitled "Blood Pressure" in chapter 13 of the textbook.
2. Complete Part A of Laboratory Report 38.

3. Examine your laboratory partner's radial pulse. To do this, follow these steps:
   a. Have your partner sit quietly, remaining as relaxed as possible.
   b. Locate the pulse by placing your index and middle fingers over the radial artery on the anterior surface near the lateral side of the wrist. Do not use your thumb for sensing the pulse because you may feel a pulse coming from an artery in the thumb. The radial artery is most often used because it is easy and convenient to locate.
   c. To determine the pulse rate, count the number of pulses that occur in 1 minute. This can be accomplished by counting pulses in 30 seconds and multiplying that number by 2. (A pulse count for a full minute, although taking a little longer, would give a better opportunity to detect pulse characteristics and irregularities.)
   d. Note the characteristics of the pulse. That is, can it be described as regular or irregular, strong or weak, hard or soft? The pulse should be regular and the amplitude (magnitude) will decrease as the distance from the left ventricle increases. The strength of the pulse reveals some indications of blood pressure. Under high blood pressure, the pulse feels very hard and strong; under low blood pressure, the pulse feels weak and can be easily compressed.
   e. Record the pulse rate and pulse characteristics while sitting, in Part B of the laboratory report.
4. Repeat the procedure and determine the pulse rate and pulse characteristics in each of the following conditions:
   a. sitting;
   b. immediately after lying down;
   c. 3–5 minutes after lying down;
   d. immediately after standing;
   e. 3–5 minutes after standing quietly;
   f. immediately after 3 minutes of moderate exercise (*omit if the person has health problems*);
   g. 3–5 minutes after exercise has ended.
   h. Record the pulse rates and pulse characteristics in Part B of the laboratory report.
5. Complete Part B of the laboratory report.

Determine the pulse rate and pulse characteristics in two additional locations on your laboratory partner. Locate and record the pulse rate from the common carotid artery and the dorsalis pedis artery. ⚠️

Common carotid artery pulse rate _____

Dorsalis pedis artery pulse rate _____

Compare the amplitude of the pulse characteristics between these two pulse locations and interpret any of the variations noted. _____

_____

_____

_____

EXPLORE

PROCEDURE B—Blood Pressure

1. Although a blood pressure is present in all blood vessels, the standard location to record blood pressure is the brachial artery.

2. Measure your laboratory partner's arterial blood pressure. To do this, follow these steps:

   a. Obtain a sphygmomanometer and a stethoscope.

   b. Clean the earpieces and the diaphragm of the stethoscope with alcohol swabs.

   c. Have your partner sit quietly with their upper limb resting on a table at heart level. Have the person remain as relaxed as possible.

   d. Locate the brachial artery at the antecubital space. Wrap the cuff of the sphygmomanometer around the arm so that its lower border is about 2.5 cm above the bend of the elbow. Center the bladder of the cuff in line with the *brachial pulse* (fig. 38.1).

   e. Palpate the *radial pulse.* Close the valve on the neck of the rubber bulb connected to the cuff, and pump air from the bulb into the cuff. Inflate the cuff while watching the sphygmomanometer, and note the pressure when the pulse disappears. (This is a rough estimate of the systolic pressure.) Immediately deflate the cuff. Do not leave the cuff inflated for more than 1 minute.

   f. Position the stethoscope over the brachial artery. Reinflate the cuff to a level 30 mm Hg higher than the point where the pulse disappeared during palpation.

   g. Slowly open the valve of the bulb until the pressure in the cuff drops at a rate of about 2 or 3mm Hg per second.

   h. Listen for sounds (Korotkoff sounds) from the brachial artery. When the first loud tapping sound is heard, record the reading as the systolic pressure. This indicates the pressure exerted against the arterial wall during systole.

   i. Continue to listen to the sounds as the pressure drops, and note the level when the last sound is heard. Record this reading as the diastolic pressure, which measures the constant arterial resistance.

   j. Release all of the pressure from the cuff.

   k. Repeat the procedure until you have two blood pressure measurements from each arm, allowing 2–3 minutes of rest between readings.

   l. Average your readings and enter them in the table in Part C of the laboratory report.

3. Measure and record your partner's blood pressure in each of the following conditions:

   a. 3–5 minutes after lying down;

   b. 3–5 minutes after standing quietly;

   c. immediately after 3 minutes of moderate exercise *(omit if the person has health problems);*

   d. 3–5 minutes after exercise has ended.

   e. Record the blood pressures in Part C of the laboratory report.

4. Complete Part C of the laboratory report.

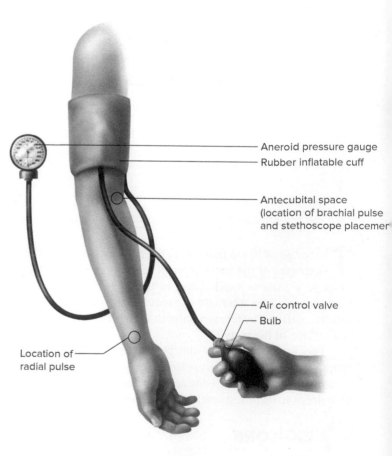

FIGURE 38.1 Blood pressure is commonly measured by using a sphygmomanometer (blood pressure cuff). The use of a column of mercury is the most accurate measurement, but due to environmental concerns, it has been replaced by alternative gauges and digital readouts.

Aneroid pressure gauge
Rubber inflatable cuff
Antecubital space (location of brachial pulse and stethoscope placement)
Air control valve
Bulb
Location of radial pulse

The corresponds to the indicated Learning Outcome(s) found at the beginning of the laboratory exercise.

Pulse Rate and Blood Pressure

♻ PART A ASSESSMENTS

Complete the following statements:

1. Blood pressure is the force exerted against the _____. 🔺3

2. The term *blood pressure* is most commonly used to refer to systemic _____ pressure. 🔺3

3. The maximum pressure achieved during ventricular contraction is called _____ pressure. 🔺3

4. The lowest pressure that remains in the arterial system during ventricular relaxation is called _____ pressure. 🔺3

5. Blood pressure rises and falls in a pattern corresponding to the phases of the _____. 🔺3

6. A pulse is caused by the _____. 🔺1

7. The _____ artery in the arm is the standard systemic artery in which blood pressure is measured. 🔺3

♻ PART B ASSESSMENTS

1. Enter your observations of pulse rates and pulse characteristics in the table. 🔺1 🔺2

Test Subject	Pulse Rate (beats/min)	Pulse Characteristics
Sitting		
Lying down		
3–5 minutes later		
Standing		
3–5 minutes later		
After exercise		
3–5 minutes later		

2. Summarize the effects of body position and exercise on the pulse rates and pulse characteristics. 🔺2

_____

_____

PART C ASSESSMENTS

1. Enter the initial measurements of blood pressure from a relaxed sitting position in the table. 🔺3

Reading	Blood Pressure in Right Arm	Blood Pressure in Left Arm
First		
Second		
Average		

2. Enter your test results in the table. 🔺4

Test Subject	Blood Pressure
3–5 minutes after lying down	
3–5 minutes after standing	
After 3 minutes of moderate exercise	
3–5 minutes later	

3. Summarize the effects of body position and exercise on blood pressure. 🔺4 _____

_____

_____

4. Summarize any correlations between pulse rate and blood pressure from any of the experimental conditions. 🔺2 🔺4 _____

_____

_____

Critical Thinking Application

When a pulse is palpated and counted, which blood pressure (systolic or diastolic) is characteristic at that moment? Explain your answer. 🔺1

_____

_____

_____

39

Lymphatic System

MATERIALS NEEDED

Textbook
Human torso model
Anatomical chart of the lymphatic system
Compound light microscope
Prepared microscope slides:
    Lymph node section
    Human thymus section
    Human spleen section

PURPOSE OF THE EXERCISE

To review the structure of the lymphatic system and to observe the microscopic structure of a lymph node, the thymus, and the spleen.

LEARNING OUTCOMES A&PR

After completing this exercise, you should be able to

1. Trace the major lymphatic pathways in an anatomical chart or model.
2. Locate and identify the major clusters of lymph nodes (lymph glands) in an anatomical chart or a model.
3. Describe the structure and function of a lymphatic vessel, a lymph node, the thymus, and the spleen.
4. Locate and sketch the major microscopic structures of a lymph node, the thymus, and the spleen.

The lymphatic system is a vast collection of cells, fluid, and biochemicals that travel in lymphatic capillaries and vessels and the organs and glands that produce them. The system is closely associated with the cardiovascular system and includes a network of vessels that assist in the circulation of body fluids. These vessels provide pathways through which excess fluid can be transported away from interstitial spaces within most tissues and returned to the bloodstream. Without the lymphatic system, this fluid would accumulate in tissue spaces, producing edema.

The organs of the lymphatic system also help defend the tissues against infections by filtering particles from lymph and by supporting the activities of lymphocytes that furnish immunity against specific disease-causing agents or pathogens.

EXPLORE

PROCEDURE A—Lymphatic Pathways A&PR

1. Review section 14.2 entitled "Lymphatic Pathways" in chapter 14 of the textbook.
2. As a review activity, label figure 39.1.
3. Complete Part A of Laboratory Report 39.
4. Observe the human torso model and the anatomical chart of the lymphatic system, and locate the following features:
   lymphatic vessels
   lymph nodes
   lymphatic trunks
   collecting ducts
       thoracic (left lymphatic) duct
       right lymphatic duct
   internal jugular veins
   subclavian veins

EXPLORE

PROCEDURE B—Lymph Nodes

1. Review the concept heading entitled "Lymph Nodes" in section 14.4 of chapter 14 of the textbook.
2. As a review activity, label figure 39.2.
3. Complete Part B of the laboratory report.
4. Observe the anatomical chart of the lymphatic system and the human torso model and locate the clusters of lymph nodes in the following regions:
   cervical region
   axillary region
   inguinal region
   pelvic cavity
   abdominal cavity
   thoracic cavity

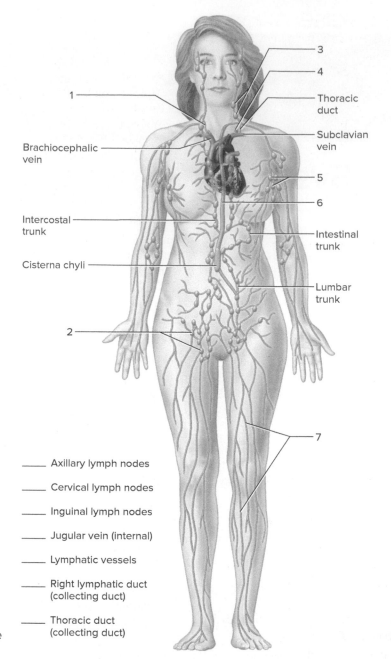

3
4
Thoracic duct
Subclavian vein
1
Brachiocephalic vein
5
6
Intercostal trunk
Intestinal trunk
Cisterna chyli
Lumbar trunk
2
7

_____ Axillary lymph nodes

_____ Cervical lymph nodes

_____ Inguinal lymph nodes

_____ Jugular vein (internal)

_____ Lymphatic vessels

_____ Right lymphatic duct (collecting duct)

_____ Thoracic duct (collecting duct)

FIGURE 39.1 Label the diagram by placing the correct numbers in the spaces provided. ⚠1

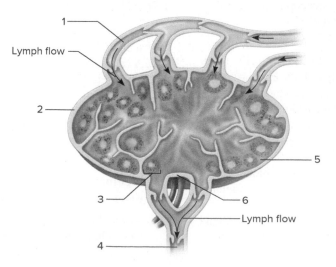

1
Lymph flow
2
3
6
Lymph flow
4
5

_____ Afferent lymphatic vessel

_____ Capsule

_____ Efferent lymphatic vessel

_____ Hilum

_____ Lymphatic nodule

_____ Lymphatic sinus

FIGURE 39.2 Label this diagram of a lymph node by placing the correct numbers in the spaces provided. ⚠3

5. Palpate the lymph nodes in your cervical region. They are located along the lower border of the mandible and between the mandible and the sternocleidomastoid muscle. They feel like small, firm lumps.

6. Study figure 39.3.

7. Obtain a prepared microscope slide of a lymph node and observe it using low-power magnification (fig. 39.3*b*). Identify the *capsule* that surrounds the node and is mainly composed of collagen fibers, the *lymphatic nodules* that appear as dense masses near the surface of the node, and the *lymphatic sinuses* that appear as narrow spaces where lymph circulates between the nodules and the capsule.

8. Using high-power magnification, examine a nodule within the lymph node. The nodule contains densely packed *lymphocytes.*

9. Prepare a labeled sketch of a representative section of a lymph node in Part D of the laboratory report.

EXPLORE

PROCEDURE C—Thymus and Spleen

1. Review the concept headings entitled "Thymus" and "Spleen" in section 14.4 of chapter 14 of the textbook.

2. Locate the thymus and spleen in the anatomical chart of the lymphatic system and on the human torso model.

3. Complete Part C of the laboratory report.

4. Obtain a prepared microscope slide of human thymus and observe it using low-power magnification (fig. 39.4). Note how the thymus is subdivided into *lobules* by *septa* of connective tissue that contain blood vessels. Identify the *capsule* of loose connective tissue that surrounds the thymus; the outer *cortex* of a lobule composed of densely packed cells and deeply stained; and the inner *medulla* of a lobule composed of more loosely packed lymphocytes and epithelial cells and lightly stained.

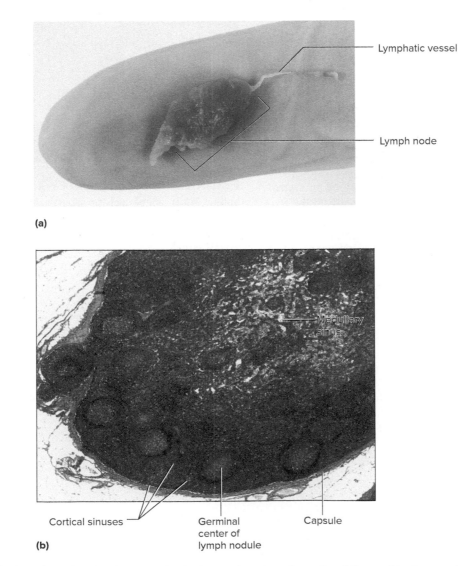

(a)

(b)

FIGURE 39.3 Lymph nodes: (*a*) photograph of a human lymph node on tip of finger; (*b*) micrograph of a lymph node (20×).

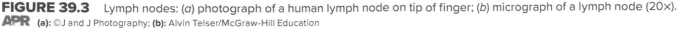

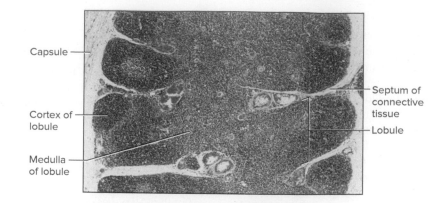

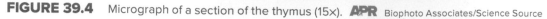

FIGURE 39.4 Micrograph of a section of the thymus (15×). **APR** Biophoto Associates/Science Source

5. Examine the cortex tissue of a lobule using high-power magnification. The cells of the cortex are composed of densely packed *lymphocytes* among some epithelial cells and *macrophages*. Some of these cortical cells may be undergoing mitosis, so their chromosomes may be visible.

6. Prepare a labeled sketch of a representative section of the thymus in Part D of the laboratory report.

7. Obtain a prepared slide of the human spleen and observe it using low-power magnification (fig. 39.5). Identify the *capsule* of dense connective tissue that surrounds the spleen. The tissues of the spleen include circular *nodules* of *white pulp* enclosed in a matrix of *red pulp*.

8. Using high-power magnification, examine a nodule of white pulp and red pulp. The cells of the white pulp are mainly *lymphocytes*. Also, there may be an arteriole centrally located in the nodule. The cells of the red pulp are mostly red blood cells with many lymphocytes and macrophages. The macrophages engulf and destroy cellular debris and bacteria.

9. Prepare a labeled sketch of a representative section of the spleen in Part D of the laboratory report.

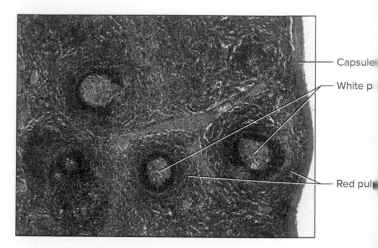

FIGURE 39.5 Micrograph of a section of the spleen (40×). **APR** Biophoto Associates/Science Source

Name _____

Date _____

Section _____

The  corresponds to the indicated Learning Outcome(s) found at the beginning of the laboratory exercise.

Lymphatic System

♻ PART A ASSESSMENTS

Complete the following statements:

1. Lymphatic pathways begin as _____ that merge to form lymphatic vessels. Ⓐ

2. The wall of a lymphatic capillary consists of a single layer of _____ cells. Ⓐ

3. Once tissue (interstitial) fluid is inside a lymphatic capillary, the fluid is called _____. Ⓐ

4. Lymphatic vessels have walls similar to those of _____ but thinner. Ⓐ

5. Lymphatic vessels contain _____ that help prevent the backflow of lymph. Ⓐ

6. Lymphatic vessels usually lead to lymph _____ that filter the fluid being transported. Ⓐ

7. The _____ duct is the larger and longer of the two lymphatic collecting ducts. Ⓐ

♻ PART B ASSESSMENTS

Complete the following:

1. Lymph nodes contain large numbers of white blood cells called _____ and macrophages that fight invading microorganisms. Ⓐ

2. The indented region where blood vessels and nerves join a lymph node is called the _____. Ⓐ

3. Lymphatic _____ that contain germinal centers are the structural units of a lymph node. Ⓐ

4. The spaces within a lymph node are called lymphatic _____ through which lymph circulates. Ⓐ

5. Lymph enters a node through a(n) _____ lymphatic vessel. Ⓐ

6. A connective tissue _____ encloses a lymph node. Ⓐ

Complete the following statements:

1. The _____ gland is located in the mediastinum, anterior to the aortic arch. ⒊

2. The thymus is composed of lymphatic tissue, subdivided into _____. ⒊

3. The lymphocytes that leave the thymus function in providing _____. ⒊

4. The hormones secreted by the thymus are called _____. ⒊

5. The _____ is the largest lymphatic organ. ⒊

6. The sinuses within the spleen contain _____. ⒊

7. The tiny islands (nodules) of tissue within the spleen that contain many lymphocytes constitute the _____ pulp. ⒊

8. The _____ pulp of the spleen contains large numbers of red blood cells, lymphocytes, and macrophages. ⒊

9. _____ within the spleen engulf and destroy foreign particles and cellular debris. ⒊

PART D ASSESSMENTS

Sketch and label structural features of the following lymphatic organs: ④

Lymph node (_____ X)	Thymus (_____ X)

Spleen (_____ X)

40

Digestive Organs

MATERIALS NEEDED

Textbook
Human torso model
Skull with teeth
Teeth, sectioned
Tooth model, sectioned
Paper cup
Compound light microscope
Prepared microscope slides of the following:
  Salivary gland
  Esophagus
  Stomach (fundus)
  Small intestine (jejunum)
  Large intestine

PURPOSE OF THE EXERCISE APR

To review the structure and function of the digestive organs and to examine the tissues of these organs.

LEARNING OUTCOMES APR

After completing this exercise, you should be able to

1. Locate and label the major digestive organs and their major structures.
2. Describe the functions of these organs.
3. Examine and sketch the structures of a section of the small intestine.

The digestive system includes the organs associated with the alimentary canal and several accessory structures. The alimentary canal, a muscular tube, passes through the body from the opening of the mouth to the anus. It includes the mouth, pharynx, esophagus, stomach, small intestine, and large intestine. The function of the canal is to move substances throughout its length. It is specialized in various regions to store, digest, and absorb food materials and to eliminate the residues. The accessory organs, which include the salivary glands, liver, gallbladder, and pancreas, secrete products into the alimentary canal that aid digestive functions.

EXPLORE

PROCEDURE A—Mouth And Salivary Glands

1. Review sections 15.3 and 15.4 entitled "Mouth" and "Salivary Glands" in chapter 15 of the textbook.
2. As a review activity, label figures 40.1, 40.2, and 40.3.
3. Examine the mouth of the human torso model and a skull. Locate the following structures:

mouth
oral cavity
vestibule
tongue
  lingual frenulum
  papillae
palate
  hard palate
  soft palate
  uvula
palatine tonsils
gums (gingivae)
teeth
  incisors
  canines (cuspids)
  premolars (bicuspids)
  molars

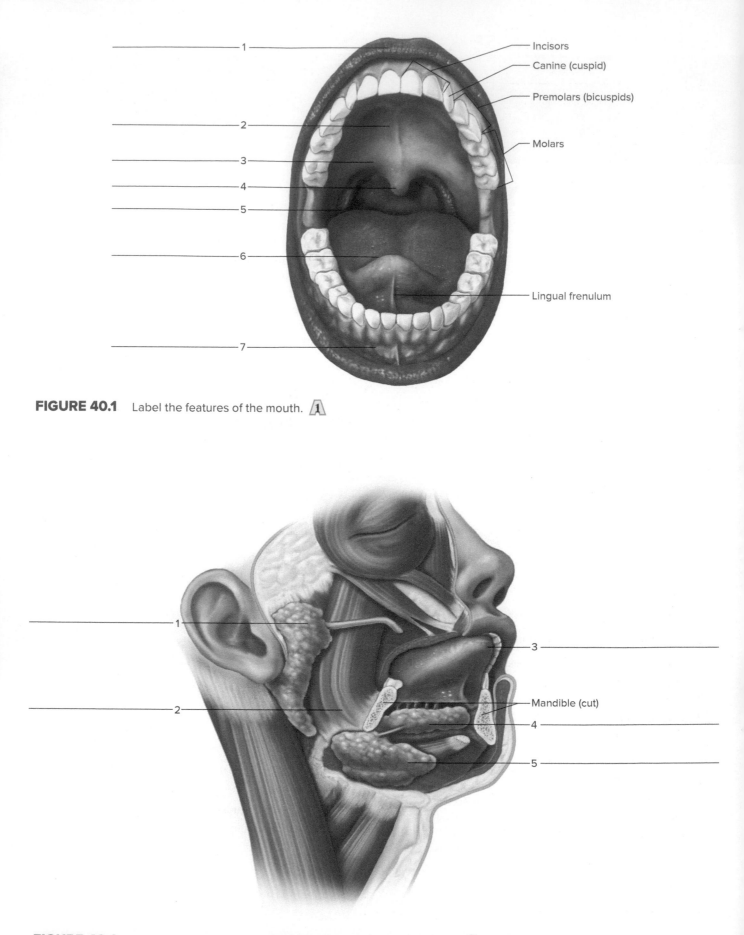

1 —————————————
Incisors
Canine (cuspid)
2 —————————————
Premolars (bicuspids)
3 —————————————
4 —————————————
Molars
5 —————————————
6 —————————————
Lingual frenulum
7 —————————————

FIGURE 40.1   Label the features of the mouth.

1 —————————————
3 —————————————
2 —————————————
Mandible (cut)
4 —————————————
5 —————————————

FIGURE 40.2   Label the features associated with the major salivary glands.   APR

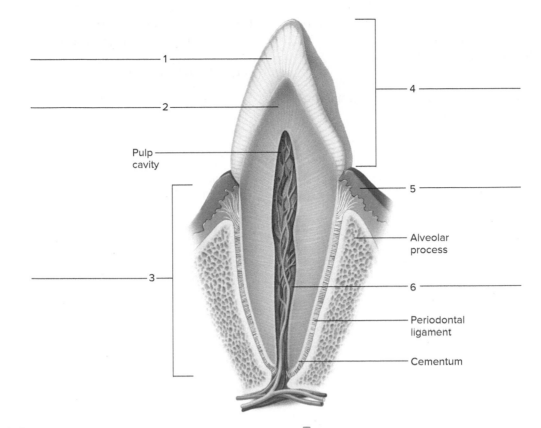

Pulp
cavity

4

5

Alveolar
process

6

Periodontal
ligament

Cementum

1

2

3

FIGURE 40.3   Label the features of this canine (cuspid) tooth. ⚠

4. Examine a sectioned tooth and a tooth model. Locate the following features:

crown
    enamel
    dentin

neck

root
    pulp cavity
    cementum
    root canal

5. Observe the head of the human torso model and locate the following:

parotid salivary gland
submandibular salivary gland
sublingual salivary gland

6. Examine a microscopic section of a salivary gland using low- and high-power magnification. Note that the mucous cells that produce mucus and the serous cells that produce enzymes are clustered around small ducts. Also note a larger secretory duct surrounded by lightly stained cuboidal epithelial cells (fig. 40.4).

7. Complete Part A of Laboratory Report 40.

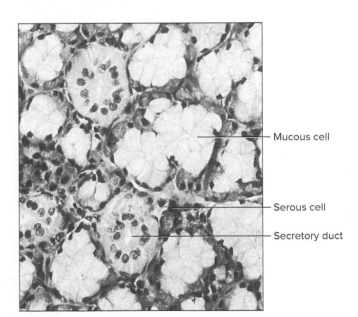

Mucous cell

Serous cell

Secretory duct

FIGURE 40.4   Micrograph of a salivary gland (300×).
Biophoto Associates/Science Source

PROCEDURE B—Pharynx and Esophagus

1. Review section 15.5 entitled "Pharynx and Esophagus" in chapter 15 of the textbook.
2. As a review activity, label figure 40.5.
3. Observe the human torso model and locate the following features:

   pharynx

   > nasopharynx
   > oropharynx
   > laryngopharynx

   esophagus
   lower esophageal sphincter (cardiac sphincter)

4. Have your laboratory partner take a drink from a cup of water. Carefully watch the movements in the anterior region of the neck. What steps in the swallowing process did you observe?

   _____

   _____

   _____

   _____

5. Examine a microscopic section of the esophagus wall using low-power magnification (fig. 40.6). The inner lining is composed of stratified squamous epithelium, and there are layers of muscle tissue in the wall. Locate some mucous glands in the submucosa. They appear as clusters of lightly stained cells.
6. Complete Part B of the laboratory report.

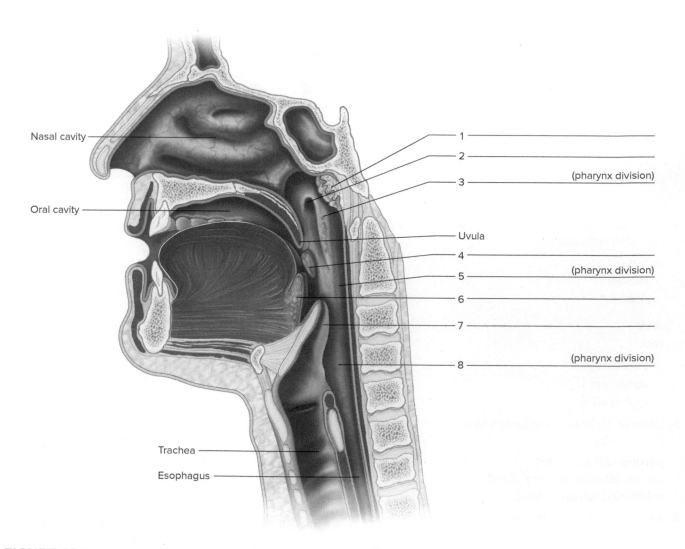

FIGURE 40.5 Label the features associated with the pharynx.

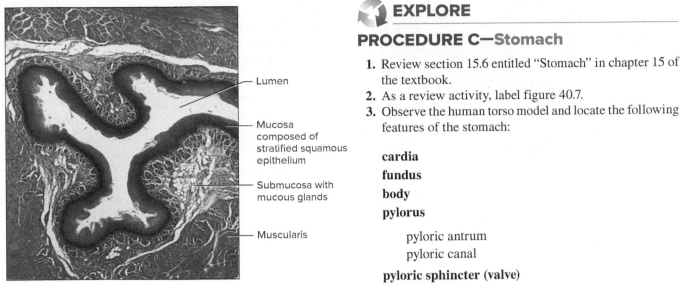

FIGURE 40.6 Micrograph of a cross section of the esophagus (10×). **APR** ©Ed Reschke

Lumen

Mucosa composed of stratified squamous epithelium

Submucosa with mucous glands

Muscularis

EXPLORE

PROCEDURE C—Stomach

1. Review section 15.6 entitled "Stomach" in chapter 15 of the textbook.
2. As a review activity, label figure 40.7.
3. Observe the human torso model and locate the following features of the stomach:

cardia

fundus

body

pylorus

    pyloric antrum

    pyloric canal

pyloric sphincter (valve)

gastric folds (rugae)

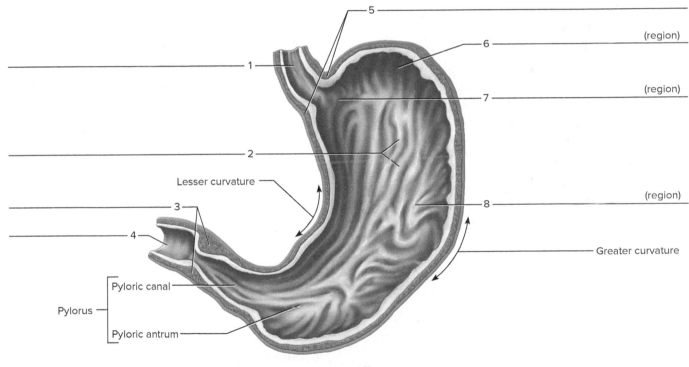

5

6 (region)

1

7 (region)

2

Lesser curvature

3

4

8 (region)

Greater curvature

Pyloric canal

Pylorus

Pyloric antrum

FIGURE 40.7 Label the stomach and associated structures. **APR**

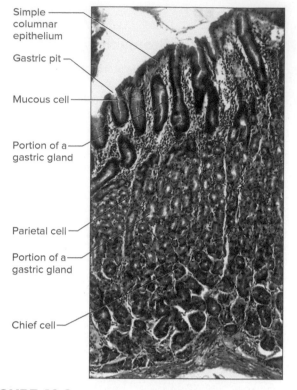

Simple
columnar
epithelium

Gastric pit

Mucous cell

Portion of a
gastric gland

Parietal cell

Portion of a
gastric gland

Chief cell

FIGURE 40.8 Micrograph of the mucosa of the stomach wall (60×). **APR** ©Ed Reschke

4. Examine a microscopic section of the stomach wall using low-power magnification (fig. 40.8). Note how the inner lining of simple columnar epithelium dips inward to form gastric pits. The gastric glands are tubular structures that open into the gastric pits. Near the deep ends of these glands, you should be able to locate some intensely stained (bluish) chief cells and some lightly stained (pinkish) parietal cells. What are the functions of these cells? _____

_____

_____

_____

_____

5. Complete Part C of the laboratory report.

EXPLORE

PROCEDURE D—Pancreas and Liver

1. Review sections 15.7 and 15.8 entitled "Pancreas" and "Liver and Gall Bladder" in chapter 15 of the textbook.
2. As a review activity, label figure 40.9.

3. Observe the human torso model and locate the following structures:
   pancreas
   pancreatic duct
   liver
   gallbladder
   hepatic ducts
   cystic duct
   bile duct (common bile duct)
   hepatopancreatic sphincter

EXPLORE

PROCEDURE E—Small and Large Intestines

1. Review sections 15.9 and 15.10 entitled "Small Intestine" and "Large Intestine" in chapter 15 of the textbook.
2. As a review activity, label figure 40.10.
3. Observe the human torso model and locate each of the following features:
   small intestine
   > duodenum
   > jejunum
   > ileum

   mesentery
   ileocecal sphincter (valve)
   large intestine
   > cecum
   > appendix (vermiform appendix)
   > ascending colon
   > transverse colon
   > descending colon
   > sigmoid colon
   > rectum
   > anal canal
   > anal columns

   anal sphincter muscles
   > external anal sphincter
   > internal anal sphincter

   anus

4. Using low-power magnification, examine a microscopic section of the small intestine wall. Identify the mucosa, submucosa, muscularis, and serosa. Note the villi that extend into the lumen of the tube. Study a single villus using high-power magnification. Note the core of connective tissue and the covering of simple columnar

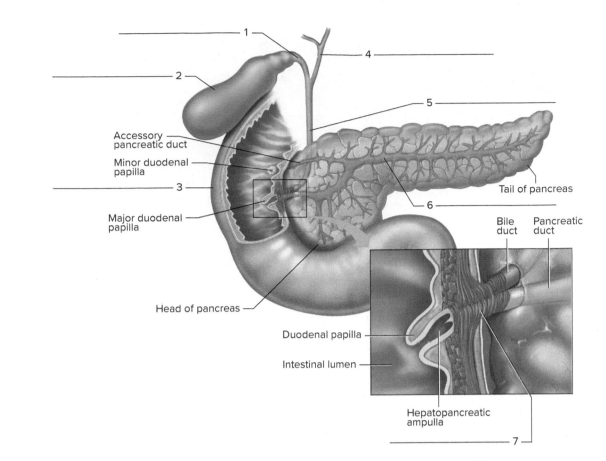

Accessory
pancreatic duct

Minor duodenal
papilla

Major duodenal
papilla

Tail of pancreas

Head of pancreas

Bile
duct

Pancreatic
duct

Duodenal papilla

Intestinal lumen

Hepatopancreatic
ampulla

FIGURE 40.9 Label the features associated with the liver, gallbladder, and pancreas. 🄰 APR

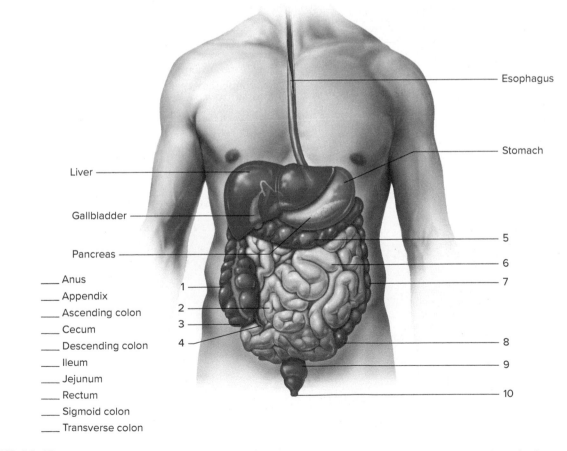

Esophagus

Stomach

Liver

Gallbladder

Pancreas

___ Anus

___ Appendix

___ Ascending colon

___ Cecum

___ Descending colon

___ Ileum

___ Jejunum

___ Rectum

___ Sigmoid colon

___ Transverse colon

FIGURE 40.10 Label this diagram of the small and large intestines by placing the correct numbers in the spaces provided. 🄰 APR

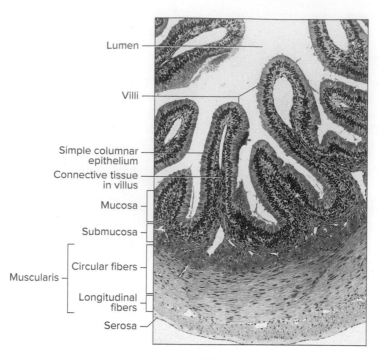

FIGURE 40.11   Micrograph of the small intestine wall (40×).   **APR**   ©Dennis Strete

epithelium that contains some lightly stained goblet cells (fig. 40.11). What is the function of these villi?

_____

_____

_____

_____

5. Prepare a labeled sketch of the wall of the small intestine in Part D of the laboratory report.

6. Examine a microscopic section of the large intestine wall. Note the lack of villi. Locate the four layers of the wall. Also note the tubular mucous glands that open on the surface of the inner lining and the numerous lightly stained goblet cells (fig. 40.12). What is the function of the mucus secreted by these glands?

_____

_____

7. Complete Part E of the laboratory report.

🧠 Critical Thinking Application

How is the structure of the small intestine better adapted for absorption than that of the large intestine?

_____

_____

_____

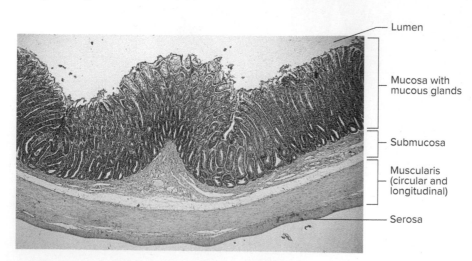

FIGURE 40.12   Micrograph of the large intestine wall (64×).   **APR**   ©Ed Reschke

The ⒜ corresponds to the indicated Learning Outcome(s) found at the beginning of the laboratory exercise.

Digestive Organs

◆ PART A ASSESSMENTS

Match the terms in column A with the descriptions in column B. Place the letter of your choice in the space provided. ⒜

Column A	Column B
a. Adenoids (pharyngeal tonsils)	_____ **1.** Bonelike substance beneath tooth enamel
b. Crown	_____ **2.** Smallest of major salivary glands
c. Dentin	_____ **3.** Tooth specialized for grinding
d. Incisor	_____ **4.** Chamber between tongue and palate
e. Lingual frenulum	_____ **5.** Projections on tongue surface
f. Molar	_____ **6.** Cone-shaped projection of soft palate
g. Oral cavity	_____ **7.** Attaches tooth to jaw
h. Papillae	_____ **8.** Chisel-shaped tooth
i. Periodontal ligament	_____ **9.** Space between the teeth, cheeks, and lips
j. Sublingual gland	_____ **10.** Anchors tongue to floor of mouth
k. Uvula	_____ **11.** Lymphatic tissue in posterior wall of pharynx near auditory tubes
l. Vestibule	_____ **12.** Portion of tooth projecting beyond gum

◆ PART B ASSESSMENTS

Complete the following:

1. The part of the pharynx superior to the soft palate is called the _____ . ⒜

2. The middle part of the pharynx is called the _____ . ⒜

3. The inferior portion of the pharynx is called the _____ . ⒜

4. _____ is the main secretion of the esophagus. ⒝

5. Summarize the functions of the esophagus. ⒝ _____

_____

_____

305

PART C ASSESSMENTS

Complete the following:

1. Name the four parts (regions) of the stomach. /1\ _____

2. Name the gastric cells that secrete digestive enzymes. /2\ _____

3. Name the gastric cells that secrete hydrochloric acid. /2\ _____

4. Name the most important digestive enzyme in gastric juice. /2\ _____

5. Summarize the functions of the stomach. /2\ _____

_____

_____

 PART D ASSESSMENTS

Prepare a sketch of a cross section of the small intestine. Label the four layers, villi, and lumen. Indicate the magnification used for the sketch. /3\

![icon] **PART E ASSESSMENTS**

Complete the following:

1. Name the three portions of the small intestine. ⚠️1 _____

2. Describe the function of the mesentery. ⚠️2 _____

3. Name five digestive enzymes secreted by the intestinal glands. ⚠️2 _____

_____

4. Name the valve located between the small and the large intestine. ⚠️1 _____

5. Summarize the functions of the small intestine. ⚠️2 _____

_____

6. Summarize the functions of the large intestine. ⚠️2 _____

_____

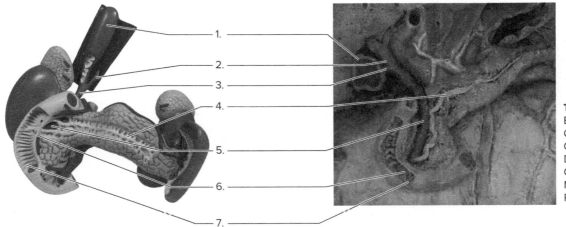

Terms:
Bile duct
Common hepatic duct
Cystic duct
Duodenum
Gallbladder
Major duodenal papilla
Pancreatic duct

FIGURE 40.13 Using the terms provided, label the indicated structures on the model and cadaver image associated with the liver, gallbladder, and pancreas. **(left):** ©2017 Denoyer-Geppert Science Company denoyer.com; **(right):** ©McGraw-Hill Education/APR

Notes

41

Action of a Digestive Enzyme

MATERIALS NEEDED

0.5% amylase solution*
Beakers (50 and 500 mL)
Distilled water
Funnel
Pipets (1 and 10 mL)
Pipet rubber bulbs
0.5% starch solution
Graduated cylinder (10 mL)
Test tubes
Test-tube clamps
Wax marker
Iodine-potassium-iodide
    solution

Medicine dropper
Ice
Water bath, 37°C
    (98.6°F)
Porcelain test plate
Benedict's solution
Hot plates
Test-tube rack
Thermometer

*For Alternative
    Procedure:*

Small, disposable cups
Distilled water

*The amylase should be free of sugar for best results; a low-maltose solution of amylase yields good results.

⚠ SAFETY

- Wear safety glasses when working with acids and when heating test tubes.
- Use test-tube clamps when handling hot test tubes.
- If an open flame is used for heating the test solutions, keep clothes and hair away from the flame.
- Use only a mechanical pipetting device (never your mouth). Use pipets with rubber bulbs or dropping pipets.
- Dispose of chemicals according to appropriate directions.
- Review the Laboratory Safety Guidelines in Appendix 1.
- Use appropriate disinfectant to wash laboratory tables before and after the procedures.
- If student saliva is used as a source of amylase, it is important that students wear disposable gloves and handle only their own materials.
- Wash your hands before leaving the laboratory.

PURPOSE OF THE EXERCISE

To investigate the action of amylase and the effect of heat on its enzymatic activity.

🔄 LEARNING OUTCOMES APR

After completing this exercise, you should be able to

1. Test a solution for the presence of starch or the presence of sugar.
2. Explain the action of amylase.
3. Test the effects of varying temperatures on the activity of amylase.

The digestive enzyme in salivary secretions is called *salivary amylase*. Pancreatic amylase is secreted among several other pancreatic enzymes. A bacterial extraction of amylase is available for laboratory experiments. This enzyme catalyzes the reaction of changing starch molecules into sugar (disaccharide) molecules, which is the first step in the digestion of complex carbohydrates.

As in the case of other enzymes, amylase is a protein catalyst. Its activity is affected by exposure to certain environmental factors, including various temperatures, pH, radiation, and electricity. As temperatures increase, faster chemical reactions occur as the collisions of molecules happen at a greater frequency. Eventually, temperatures increase to a point that the enzyme is denatured and the rate of the enzyme activity rapidly declines. As temperatures decrease, enzyme activity also decreases due to fewer collisions of the molecules; however, the colder temperatures do not denature the enzyme. Normal body temperature provides an environment for enzyme activity near the optimum for enzymatic reactions.

The pH of the environment in which enzymes are secreted also has a major influence on the activity of the enzyme reactions. The optimum pH for amylase activity is between 6.8 and 7.0, typical of salivary secretions. When salivary amylase arrives within the stomach, hydrochloric acid deactivates the enzyme, diminishing any further chemical digestion of remaining starch located in the stomach. Pancreatic amylase is secreted into the small intestine, where an optimum pH for amylase is once more provided. Other enzymes in the digestive system have different optimum pH ranges of activity compared to amylase. For example, pepsin from stomach secretions has an optimum activity around pH 2, whereas trypsin from pancreatic secretions operates best around pH 7–8.

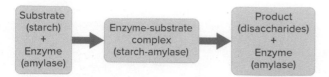

FIGURE 41.1 Lock-and-key model of enzyme action of amylase on starch digestion. **APR**

⟳ EXPLORE

PROCEDURE A—Amylase Activity

1. Study the lock-and-key model of amylase action on starch digestion (fig. 41.1).
2. Examine the digestive locations where starch is converted to glucose (fig. 41.2).
3. Mark three clean test tubes as *tubes 1, 2,* and *3,* and prepare the tubes as follows:

   Tube 1: **Add 6 mL of amylase solution.**
   Tube 2: **Add 6 mL of starch solution.**
   Tube 3: **Add 5 mL of starch solution and 1 mL of amylase solution.**

ALTERNATIVE PROCEDURE

Human saliva can be used as a source of amylase solutions instead of bacterial amylase preparations. Collect about 5 mL of saliva in a small, disposable cup. Add an equal amount of distilled water and mix together for the amylase solutions during the laboratory procedures. Be sure to follow all of the safety guidelines.

4. Shake the tubes well to mix the contents, and place them in a warm water bath, 37°C (98.6°F), for 10 minutes.
5. At the end of the 10 minutes, test the contents of each tube for the presence of starch. To do this, follow these steps:
   a. Place 1 mL of the solution to be tested in a depression of a porcelain test plate.
   b. Next, add 1 drop of iodine-potassium-iodide solution, and note the color of the mixture. If the solution becomes blue-black, starch is present.
   c. Record the results in Part A of Laboratory Report 41.
6. Test the contents of each tube for the presence of sugar (disaccharides, in this instance). To do this, follow these steps:
   a. Place 1 mL of the solution to be tested in a clean test tube.
   b. Add 1 mL of Benedict's solution.
   c. Place the test tube with a test-tube clamp in a beaker of boiling water for 2 minutes.
   d. Note the color of the liquid. If the solution becomes green, yellow, orange, or red, sugar is present. Blue indicates a negative test, whereas green indicates a positive test with the least amount of sugar, and red indicates the greatest amount of sugar present.
   e. Record the results in Part A of the laboratory report.
7. Complete Part A of the laboratory report.

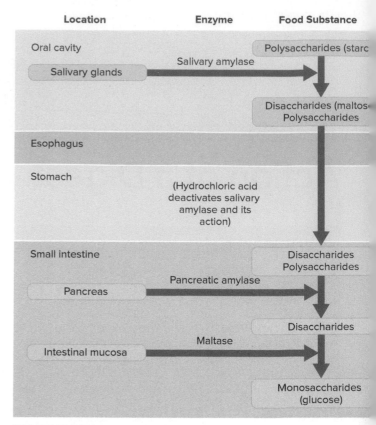

FIGURE 41.2 Flowchart of digestion of starch.

⟳ EXPLORE

PROCEDURE B—Effect of Heat

1. Mark three clean test tubes as *tubes 4, 5,* and *6.*
2. Add 1 mL of amylase solution to each of the tubes, and expose each solution to a different test temperature for 3 minutes as follows:

   Tube 4: **Place in beaker of ice water (about 0°C [32°F]).**
   Tube 5: **Place in warm water bath (about 37°C [98.6°F]).**
   Tube 6: **Place in beaker of boiling water (about 100°C [212°F]). Use a test-tube clamp.**

3. *It is important that the 5 mL of starch solution added to tube 4 be at ice-water temperature before it is added to the 1 mL of amylase solution.* Add 5 mL of starch solution to each tube, shake to mix the contents, and return the tubes to their respective test temperatures for 10 minutes.
4. At the end of the 10 minutes, test the contents of each tube for the presence of starch and the presence of sugar by following the same directions as in steps 5 and 6 of Procedure A.
5. Complete Part B of the laboratory report.

Name ______________________________

Date ______________________________

Section ______________________________

The [A] corresponds to the indicated Learning Outcome(s) found at the beginning of the laboratory exercise.

LABORATORY REPORT

41

Action of a Digestive Enzyme

PART A ASSESSMENTS

1. Test results: [1]

Tube		Starch	Sugar
1	Amylase solution		
2	Starch solution		
3	Starch-amylase solution		

2. Complete the following:

a. Explain the reason for including tube 1 in this experiment. [2]

____

____

____

b. What is the importance of tube 2? [2]

____

____

____

c. What do you conclude from the results of this experiment? [2]

____

____

____

PART B ASSESSMENTS

Complete the following:

1. Test results: 3

Tube		Starch	Sugar
4	0°C (32°F)		
5	37°C (98.6°F)		
6	100°C (212°F)		

2. Complete the following:

   a. What do you conclude from the results of this experiment?   3

   _____

   _____

   _____

   _____

   _____

   b. If digestion failed to occur in one of the tubes in this experiment, how can you tell whether the amylase was destroyed by the factor being tested or the amylase activity was simply inhibited by the test treatment?   3

   _____

   _____

   _____

   _____

   _____

Critical Thinking Application

What test result would occur if the amylase you used contained sugar? _____ Would your results be valid? Explain your answer.   2

_____

_____

_____

_____

42

Respiratory Organs

MATERIALS NEEDED

Textbook
Human skull (sagittal section)
Human torso model
Larynx model
Thoracic organs model
Compound light microscope
Prepared microscope slides of the following:
    Trachea (cross section)
    Lung, human (normal)

For Demonstrations:
Animal lung with trachea (fresh or preserved)
Bicycle pump
Prepared microscope slides of the following:
    Lung tissue (smoker)
    Lung tissue (emphysema)

⚠ SAFETY

- Wear disposable gloves when working on the fresh or preserved animal lung demonstration.
- Wash your hands before leaving the laboratory.

PURPOSE OF THE EXERCISE

To review the structure and function of the respiratory organs and to examine the tissues of some of these organs.

LEARNING OUTCOMES APR

After completing this exercise, you should be able to

1. Locate the major organs and structural features of the respiratory system.
2. Describe the functions of these organs.
3. Sketch and label features of tissue sections of the trachea and lung.

The organs of the respiratory system include the nose, nasal cavity, paranasal sinuses, pharynx, larynx, trachea, bronchial tree, and lungs. They mainly process incoming air and transport it to and from the atmosphere outside the body and the air sacs of the lungs.

In the alveoli (air sacs), gas exchanges take place between the air and the blood of nearby capillaries. The blood, in turn, transports gases to and from the alveoli and the body cells. This entire process of transporting and exchanging gases between the atmosphere and the body cells is called *respiration.*

EXPLORE

PROCEDURE A—Respiratory Organs

1. Review section 16.2 entitled "Organs and Structures of the Respiratory System" in chapter 16 of the textbook.
2. As a review activity, label figures 42.1 and 42.2.
3. Examine the sagittal section of the human skull and locate the following features:

nasal cavity
    nostril (naris)
    nasal septum
    nasal conchae

paranasal sinuses
    maxillary sinus
    frontal sinus
    ethmoidal sinus (ethmoidal air cells)
    sphenoidal sinus

4. Observe the larynx model, the thoracic organs model, and the human torso model. Locate the features listed in step 3. Also locate the following features:

pharynx
    nasopharynx
    oropharynx
    laryngopharynx

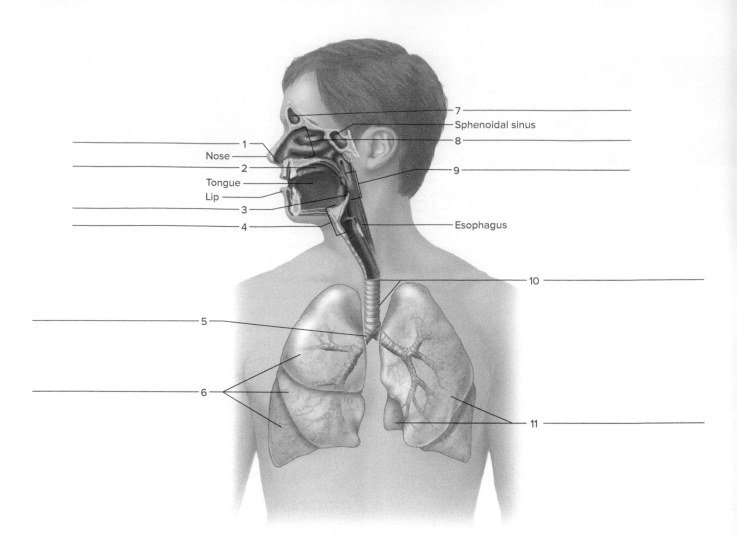

FIGURE 42.1 Label the major features of the respiratory system. **APR**

7 _____
Sphenoidal sinus
8 _____
1 _____
Nose
2 _____
9 _____
Tongue
Lip
3 _____
4 _____
Esophagus
10 _____
5 _____
6 _____
11 _____

larynx (palpate your larynx)
  vocal cords
    false vocal cords (vestibular folds)
    true vocal cords (vocal folds)
  thyroid cartilage (Adam's apple)
  cricoid cartilage
  epiglottic cartilage
  glottis
trachea (palpate your trachea)
bronchi
  right and left main (primary) bronchi
  lobar (secondary) bronchi
  segmental (tertiary) bronchi
bronchioles
lung
  lobes (right lung, three; left lung, two)
visceral pleura
parietal pleura
pleural cavity

5. Complete Part A of Laboratory Report 42.

DEMONSTRATION

Observe the animal lung and the attached trachea. Identify the larynx, major laryngeal cartilages, trachea, and incomplete cartilaginous rings of the trachea. Open the larynx and locate the vocal folds. Examine the visceral pleura on the surface of a lung. A bicycle pump can be used to demonstrate lung inflation. Section the lung and locate some bronchioles and alveoli. Squeeze a portion of a lung between your fingers. How do you describe the texture of the lung?

_____

_____

_____

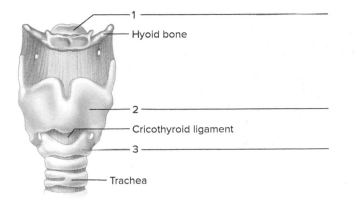

Hyoid bone
1
2
Cricothyroid ligament
3
Trachea

(a)

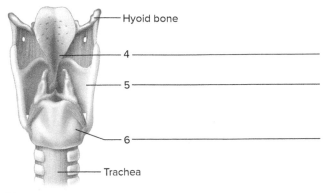

Hyoid bone
4
5
6
Trachea

(b)

FIGURE 42.2 Label the major features of the larynx: (*a*) anterior view; (*b*) posterior view. **2 APR**

PROCEDURE B—Respiratory Tissues

1. Obtain a prepared microscope slide of a trachea, and use low-power magnification to examine it. Notice the inner lining of ciliated pseudostratified columnar epithelium and the deep layer of hyaline cartilage, which represents a portion of an incomplete (C-shaped) tracheal ring (fig. 42.3).
2. Use high-power magnification to observe the cilia on the free surface of the epithelial lining. Locate the wineglass-shaped goblet cells, which secrete a protective mucus, in the epithelium (fig. 42.4).
3. Prepare a labeled sketch of a representative portion of the tracheal wall in Part B of the laboratory report.
4. Obtain a prepared microscope slide of human lung tissue. Examine it using low-power magnification, and note the many open spaces of the air sacs (alveoli). Look for a bronchiole—a tube with a relatively thick wall and a wavy inner lining. Locate the smooth muscle tissue in the wall of this tube (fig. 42.5). You also may see a section of cartilage as part of the bronchiole wall.
5. Use high-power magnification to examine the alveoli. Their walls are composed of simple squamous epithelium (fig. 42.6). The thin walls of the alveoli and the associated capillaries allow efficient gas exchange of oxygen and carbon dioxide. You also may see sections of blood vessels containing blood cells.
6. Prepare a labeled sketch of a representative portion of the lung in Part B of the laboratory report.
7. Complete Part C of the laboratory report.

Ciliated epithelium   Lumen of trachea

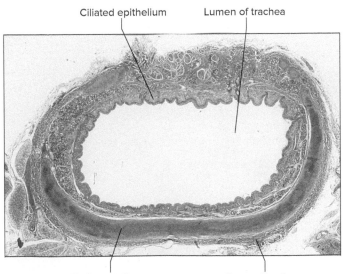

Hyaline cartilage   Connective tissue

FIGURE 42.3 Micrograph of a section of the tracheal wall (100×). **APR** Biophoto Associates/Science Source

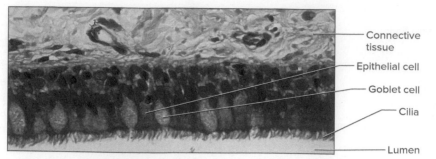

FIGURE 42.4 Micrograph of ciliated pseudostratified epithelium in the respiratory tract (275×). **APR** Biophoto Associates/
Science Source

Connective tissue
Epithelial cell
Goblet cell
Cilia
Lumen

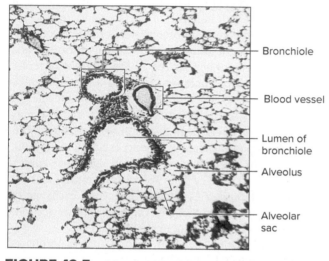

Bronchiole
Blood vessel
Lumen of bronchiole
Alveolus
Alveolar sac

FIGURE 42.5 Micrograph of human lung tissue (35×).
APR ©Victor B. Eichler

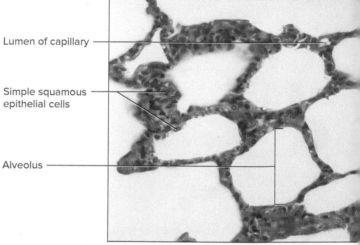

Lumen of capillary
Simple squamous epithelial cells
Alveolus

FIGURE 42.6 Micrograph of human lung tissue (250×).
APR ©Dwight R. Kuhn

DEMONSTRATION

Examine the prepared microscope slides of the lung tissue of a smoker and a person with emphysema using low-power magnification. How does the smoker's lung tissue compare with that of the normal lung tissue that you examined previously?

_____

_____

_____

How does the emphysema patient's lung tissue compare with the normal lung tissue?

_____

_____

Name _____

Date _____

Section _____

The Ⓐ corresponds to the indicated Learning Outcome(s) found at the beginning of the laboratory exercise.

LABORATORY REPORT

Respiratory Organs

PART A ASSESSMENTS

Match the terms in column A with the descriptions in column B. Place the letter of your choice in the space provided. ⚠ ⚠

Column A	Column B
a. Alveolus	_____ 1. Potential space between visceral and parietal pleurae
b. Cricoid cartilage	_____ 2. Most inferior portion of larynx
c. Epiglottis	_____ 3. Air-filled space in skull bone that opens into nasal cavity
d. Glottis	_____ 4. Microscopic air sac for gas exchange
e. Lung	_____ 5. Consists of large lobes
f. Nasal concha	
g. Pharynx	_____ 6. Opening between vocal cords
h. Pleural cavity	_____ 7. Fold of mucous membrane containing elastic fibers responsible for sounds
i. Sinus (paranasal sinus)	
j. Vocal cord (true)	_____ 8. Increases surface area of nasal mucous membrane
	_____ 9. Passageway for air and food
	_____ 10. Partially covers opening of larynx during swallowing

PART B ASSESSMENTS

Sketch and label a portion of the tracheal wall and a portion of lung tissue. ⚠

Tracheal wall (_____×)	Lung tissue (_____×)

PART C ASSESSMENTS

Complete the following:

1. What is the function of the mucus secreted by the goblet cells? 🔼 _____

   _____

   _____

2. Describe the function of the cilia in the respiratory tubes. 🔼 _____

   _____

   _____

3. How is breathing affected if the smooth muscle of the bronchial tree relaxes? 🔼 _____

   _____

   _____

   _____

   _____

4. How is the breathing affected if the smooth muscle of bronchial tree contracts? 🔼 _____

   _____

   _____

Critical Thinking Application

What is the functional advantage of the alveolar walls being so thin? 🔼 _____

_____

_____

_____

_____

43

Breathing and Respiratory Volumes

MATERIALS NEEDED

Textbook
Spirometer, handheld (dry portable)
Alcohol swabs
Disposable mouthpieces
Nose clips
Meterstick

For Demonstration:
Lung function model

⚠ SAFETY

- Clean the spirometer with an alcohol swab before each use.
- Place a new disposable mouthpiece on the stem of the spirometer before each use.
- Dispose of the alcohol swabs and mouthpieces according to directions from your laboratory instructor.

PURPOSE OF THE EXERCISE

To review the mechanisms of breathing and to measure or calculate certain respiratory air volumes and respiratory capacities.

LEARNING OUTCOMES APR

After completing this exercise, you should be able to

1. Differentiate between the mechanisms responsible for inspiration and expiration.
2. Match the respiratory air volumes and respiratory capacities with their definitions.
3. Measure respiratory air volumes using a spirometer.
4. Calculate respiratory capacities using the data obtained from respiratory air volumes.

Breathing involves the movement of air from outside the body through the bronchial tree and into the alveoli and the reversal of this air movement to allow gas (oxygen and carbon dioxide) exchange between air and blood. These movements are caused by changes in the size of the thoracic cavity that result from skeletal muscle contractions and from the elastic recoil of stretched tissues.

The volumes of air that move into and out of the lungs during various phases of breathing are called *respiratory air volumes* and *capacities*. A *volume* is a single measurement, whereas a *capacity* represents two or more combined volumes. Respiratory volumes can be measured by using an instrument called a *spirometer*. The values obtained vary with a person's age, sex, height, weight, stress, and physical fitness. Various respiratory capacities can be calculated by combining two or more of the respiratory volumes.

EXPLORE

PROCEDURE A—Breathing Mechanisms APR

1. Review the concept headings entitled "Inspiration" and "Expiration" in section 16.3 of chapter 16 of the textbook.
2. Complete Part A of Laboratory Report 43.

DEMONSTRATION

Observe the mechanical lung function model (fig. 43.1). It consists of a heavy plastic bell jar with rubber sheeting clamped over its wide open end. Its narrow upper opening is plugged with a rubber stopper through which a Y tube is passed. Small rubber balloons are fastened to the arms of the Y.

What happens to the balloons when the rubber sheeting is pulled downward? _____
_____
_____
_____

What happens when the sheeting is pushed upward?
_____
_____
_____

How do you explain these changes? _____
_____
_____
_____
_____
_____

What part of the respiratory system is represented by the rubber sheeting? _____
By the bell jar? _____
By the Y tube? _____
By the balloons? _____

Air moves into and out of the lungs in much the same way, as can be demonstrated using a syringe (fig. 43.2).

WARNING *If you begin to feel dizzy or light-headed while performing Procedure B, stop the exercise and breathe normally.*

EXPLORE

PROCEDURE B—Respiratory Air Volumes and Capacities

1. Review the concept heading entitled "Respiratory Air Volumes and Capacities" in section 16.3 of chapter 16 of the textbook.
2. Complete Part B of the laboratory report.
3. Get a handheld spirometer (fig. 43.3). Point the needle to zero by rotating the adjustable dial. Before using the instrument, clean it with an alcohol swab and place a new disposable mouthpiece over its stem. The instrument should be held with the dial upward, and air should be blown only into the disposable mouthpiece (fig. 43.4). If air tends to exit the nostrils during exhalation, use a nose clip or pinch your nose as a prevention when exhaling into the spirometer. Movement of the needle indicates the air volume that leaves the lungs. The exhalations should be slowly and forcefully performed. Too rapid an exhalation can result in erroneous data or damage to the spirometer.
4. *Tidal volume (TV)* (about 500 mL) is the volume of air that enters (or leaves) the lungs during a *respiratory cycle* (one inspiration plus the following expiration). *Resting tidal volume* is the volume of air that enters (or leaves) the lungs during normal, quiet breathing (fig. 43.5). To measure this volume, follow these steps:
   a. Sit quietly for a few moments.
   b. Position the spirometer dial so that the needle points to zero.
   c. Place the mouthpiece between your lips and exhale three ordinary expirations into it after inhaling through the nose each time. *Do not force air out of your lungs; exhale normally.*

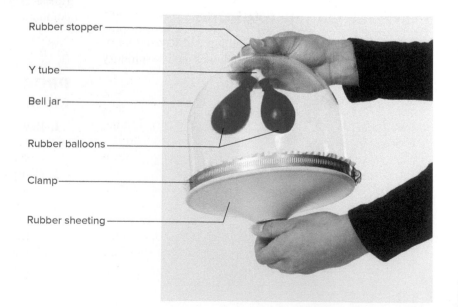

Rubber stopper
Y tube
Bell jar
Rubber balloons
Clamp
Rubber sheeting

FIGURE 43.1 A lung function model. ©J and J Photography

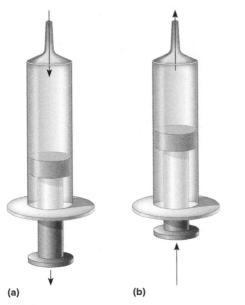

FIGURE 43.2 Moving the plunger of a syringe as in (*a*) results in air movements into the barrel of the syringe; moving the plunger of the syringe as in (*b*) results in air movements out of the barrel of the syringe. Air movements in and out of the lung function model and the lungs during breathing occur for similar reasons.

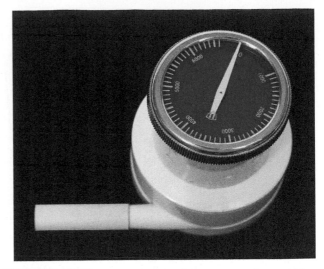

FIGURE 43.3 A handheld spirometer can be used to measure respiratory air volumes. ©J and J Photography

d. Divide the total value indicated by the needle by 3, and record this amount as your resting tidal volume in the table in Part C of the laboratory report.

5. *Expiratory reserve volume (ERV)* (about 1,100 mL) is the volume of air in addition to the tidal volume that leaves the lungs during forced expiration (fig. 43.5). To measure this volume, follow these steps:

a. Breathe normally for a few moments. Set the needle to zero.

b. At the end of an ordinary expiration, place the mouthpiece between your lips and exhale all of the air you can force from your lungs through the spirometer. Use a nose clip or pinch your nose to prevent air from exiting your nostrils.

c. Record the results as your expiratory reserve volume in Part C of the laboratory report.

6. *Vital capacity (VC)* (about 4,600 mL) is the maximum volume of air that can be exhaled after taking the deepest breath possible (fig 43.5). To measure this volume, follow these steps:

a. Breathe normally for a few moments. Set the needle at zero.

b. Breathe in and out deeply a couple of times; then take the deepest breath possible.

c. Place the mouthpiece between your lips and exhale all the air out of your lungs slowly and forcefully. Use a nose clip or pinch your nose to prevent air from exiting your nostrils.

FIGURE 43.4 Demonstration of a handheld spirometer. Air should be blown slowly and forcefully only into a disposable mouthpiece. *Use a nose clip or pinch your nose when measuring expiratory reserve volume and vital capacity volume if air exits the nostrils.* ©J and J Photography

d. Record the value as your vital capacity in Part C of the laboratory report. Compare your result with that expected for a person of your sex, age, and height, as listed in tables 43.1 and 43.2. Use the meterstick to determine your height in centimeters or multiply your height in inches times 2.54 to calculate your height in centimeters. Considerable individual variations from the expected will be noted due to parameters other than sex, age, and height, which can include fitness level, health, medications, and others.

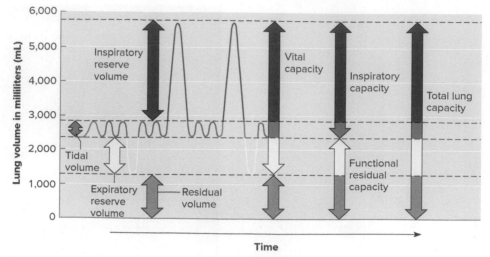

FIGURE 43.5 Graphic representation of respiratory volumes and capacities.

Critical Thinking Application

It can be noted from the data in tables 43.1 and 43.2 that vital capacities gradually decrease with age. Propose an explanation for this normal correlation.

_____

_____

_____

_____

_____

7. *Inspiratory reserve volume (IRV)* (about 3,000 mL) is the volume of air in addition to the tidal volume that enters the lungs during forced inspiration (fig. 43.5). Calculate your inspiratory reserve volume by subtracting your tidal volume (TV) and your expiratory reserve volume (ERV) from your vital capacity (VC):

$$IRV = VC - (TV + ERV)$$

8. *Inspiratory capacity (IC)* (about 3,500 mL) is the maximum volume of air a person can inhale following exhalation of the tidal volume (fig. 43.5). Calculate your inspiratory capacity by adding your tidal volume (TV) and your inspiratory reserve volume (IRV):

$$IC = TV + IRV$$

9. *Functional residual capacity (FRC)* (about 2,300 mL) is the volume of air that remains in the lungs following exhalation of the tidal volume (fig. 43.5). Calculate your functional residual capacity (FRC) by adding your expiratory reserve volume (ERV) and your residual volume (RV), which you can assume is 1,200 mL:

$$FRC = ERV + 1,200$$

10. *Residual volume (RV)* (about 1,200 mL) is the volume of air that always remains in the lungs after the most forceful expiration. Although it is part of *total lung capacity* (about 5,800 mL), it cannot be measured with a spirometer. The residual air allows gas exchange and the alveoli to remain open during the respiratory cycle (fig. 43.5).

11. Complete Part C of the laboratory report.

LEARNING EXTENSION

Determine your *minute respiratory volume*. To do this, follow these steps:

1. Sit quietly for a while, and then to establish your breathing rate, count the number of times you breathe in 1 minute. This might be inaccurate because conscious awareness of breathing rate can alter the results. You might ask a laboratory partner to record your breathing rate sometime when you are not expecting it to be recorded. A normal breathing rate is about 12–15 breaths per minute.

2. Calculate your minute respiratory volume by multiplying your breathing rate by your tidal volume:

_____ X _____ = _____
(breathing   (tidal   (minute respiratory
rate)   volume)   volume)

3. This value indicates the total volume of air that moves into your respiratory passages during each minute of ordinary breathing.

TABLE 43.1   Predicted Vital Capacities (in Milliliters) for Females

Height in Centimeters

Age	146	148	150	152	154	156	158	160	162	164	166	168	170	172	174	176	178	180	182	184	186	188	190	192	194
16	2950	2990	3030	3070	3110	3150	3190	3230	3270	3310	3350	3390	3430	3470	3510	3550	3590	3630	3670	3715	3755	3800	3840	3880	3920
18	2920	2960	3000	3040	3080	3120	3160	3200	3240	3280	3320	3360	3400	3440	3480	3520	3560	3600	3640	3680	3720	3760	3800	3840	3880
20	2890	2930	2970	3010	3050	3090	3130	3170	3210	3250	3290	3330	3370	3410	3450	3490	3525	3565	3605	3645	3695	3720	3760	3800	3840
22	2860	2900	2940	2980	3020	3060	3095	3135	3175	3215	3255	3290	3330	3370	3410	3450	3490	3530	3570	3610	3650	3685	3725	3765	3800
24	2830	2870	2910	2950	2985	3025	3065	3100	3140	3180	3220	3260	3300	3335	3375	3415	3455	3490	3530	3570	3610	3650	3685	3725	3765
26	2800	2840	2880	2920	2960	3000	3035	3070	3110	3150	3190	3230	3265	3300	3340	3380	3420	3455	3495	3530	3570	3610	3650	3685	3725
28	2775	2810	2850	2890	2930	2965	3000	3040	3070	3115	3155	3190	3230	3270	3305	3345	3380	3420	3460	3495	3535	3570	3610	3650	3685
30	2745	2780	2820	2860	2895	2935	2970	3010	3045	3085	3120	3160	3195	3235	3270	3310	3345	3385	3420	3460	3495	3535	3570	3610	3645
32	2715	2750	2790	2825	2865	2900	2940	2975	3015	3050	3090	3125	3160	3200	3235	3270	3310	3350	3385	3425	3460	3495	3535	3570	3610
34	2685	2725	2760	2795	2835	2870	2910	2945	2980	3020	3055	3090	3130	3165	3200	3240	3275	3310	3350	3385	3425	3460	3495	3535	3570
36	2655	2695	2730	2765	2805	2840	2875	2915	2950	2985	3020	3060	3095	3130	3165	3205	3240	3275	3310	3350	3385	3420	3460	3495	3530
38	2630	2665	2700	2735	2770	2810	2845	2880	2915	2950	2990	3025	3060	3095	3130	3170	3205	3240	3275	3310	3350	3385	3420	3455	3490
40	2600	2635	2670	2705	2740	2775	2810	2850	2885	2920	2955	2990	3025	3060	3095	3135	3170	3205	3240	3275	3310	3345	3380	3420	3455
42	2570	2605	2640	2675	2710	2745	2780	2815	2850	2885	2920	2955	2990	3025	3060	3100	3135	3170	3205	3240	3275	3310	3345	3380	3415
44	2540	2575	2610	2645	2680	2715	2750	2785	2820	2855	2890	2925	2960	2995	3030	3060	3095	3130	3165	3200	3235	3270	3305	3340	3375
46	2510	2545	2580	2615	2650	2685	2715	2750	2785	2820	2855	2890	2925	2960	2995	3030	3060	3095	3130	3165	3200	3235	3270	3305	3340
48	2480	2515	2550	2585	2620	2650	2685	2715	2750	2785	2820	2855	2890	2925	2960	2995	3030	3060	3095	3130	3160	3195	3230	3265	3300
50	2455	2485	2520	2555	2590	2625	2655	2690	2720	2755	2785	2820	2855	2890	2925	2955	2990	3025	3060	3090	3125	3155	3190	3225	3260
52	2425	2455	2490	2525	2555	2590	2625	2655	2690	2720	2755	2790	2820	2855	2890	2925	2955	2990	3020	3055	3090	3125	3155	3190	3220
54	2395	2425	2460	2495	2530	2560	2590	2625	2655	2690	2720	2755	2790	2820	2855	2885	2920	2950	2985	3020	3050	3085	3125	3150	3180
56	2365	2400	2430	2465	2495	2525	2560	2590	2625	2655	2690	2720	2755	2790	2820	2855	2885	2920	2950	2980	3015	3045	3080	3110	3145
58	2335	2370	2400	2430	2460	2495	2525	2560	2590	2625	2655	2690	2720	2750	2785	2815	2850	2880	2920	2945	2975	3010	3040	3075	3105
60	2305	2340	2370	2400	2430	2460	2495	2525	2560	2590	2625	2655	2690	2720	2750	2780	2810	2845	2875	2915	2940	2970	3000	3035	3065
62	2280	2310	2340	2370	2405	2435	2465	2495	2525	2560	2590	2620	2655	2685	2715	2745	2775	2810	2840	2870	2900	2935	2965	2995	3025
64	2250	2280	2310	2340	2370	2400	2430	2465	2495	2525	2555	2585	2620	2650	2680	2710	2740	2770	2805	2835	2865	2895	2925	2955	2990
66	2220	2250	2280	2310	2340	2370	2400	2430	2460	2495	2525	2555	2585	2615	2645	2675	2705	2735	2765	2800	2825	2860	2890	2920	2950
68	2190	2220	2250	2280	2310	2340	2370	2400	2430	2460	2490	2520	2550	2580	2610	2640	2670	2700	2730	2760	2795	2820	2850	2880	2910
70	2160	2190	2220	2250	2280	2310	2340	2370	2400	2425	2455	2485	2515	2545	2575	2605	2635	2665	2695	2725	2755	2780	2810	2840	2870
72	2130	2160	2190	2220	2250	2280	2310	2335	2365	2395	2425	2455	2480	2510	2540	2570	2600	2630	2660	2685	2715	2745	2775	2805	2830
74	2100	2130	2160	2190	2220	2245	2275	2305	2335	2360	2390	2420	2450	2475	2505	2535	2565	2590	2620	2650	2680	2710	2740	2765	2795

Baldwin, E. D. A. Cournand, and D. W. Richards Jr., "Pulmonary Insufficiency; Physiological Classification, Clinical Methods of Analysis. Standard Values in Normal Subjects," Medicine, 1948 Sept.; 27(3):243–278.

TABLE 43.2 Predicted Vital Capacities (in Milliliters) for Males

Height in Centimeters

Age	146	148	150	152	154	156	158	160	162	164	166	168	170	172	174	176	178	180	182	184	186	188	190	192	194
16	3765	3820	3870	3920	3975	4025	4075	4130	4180	4230	4285	4335	4385	4440	4490	4540	4590	4645	4695	4745	4800	4850	4900	4955	5005
18	3740	3790	3840	3890	3940	3995	4045	4095	4145	4200	4250	4300	4350	4405	4455	4505	4555	4610	4660	4710	4760	4815	4865	4915	4965
20	3710	3760	3810	3860	3910	3960	4015	4065	4115	4165	4215	4265	4320	4370	4420	4470	4520	4570	4625	4675	4725	4775	4825	4875	4930
22	3680	3730	3780	3830	3880	3930	3980	4030	4080	4135	4185	4235	4285	4335	4385	4435	4485	4535	4585	4635	4685	4735	4790	4840	4890
24	3635	3685	3735	3785	3835	3885	3935	3985	4035	4085	4135	4185	4235	4285	4330	4380	4430	4480	4530	4580	4630	4680	4730	4780	4830
26	3605	3655	3705	3755	3805	3855	3905	3955	4000	4050	4100	4150	4200	4250	4300	4350	4395	4445	4495	4545	4595	4645	4695	4740	4790
28	3575	3625	3675	3725	3775	3820	3870	3920	3970	4020	4070	4115	4165	4215	4265	4310	4360	4410	4460	4510	4555	4605	4655	4705	4755
30	3550	3595	3645	3695	3740	3790	3840	3890	3935	3985	4035	4080	4130	4180	4230	4275	4325	4375	4425	4470	4520	4570	4615	4665	4715
32	3520	3565	3615	3665	3710	3760	3810	3855	3905	3950	4000	4050	4095	4145	4195	4240	4290	4340	4385	4435	4485	4530	4580	4625	4675
34	3475	3525	3570	3620	3665	3715	3760	3810	3855	3905	3950	4000	4045	4095	4140	4190	4225	4285	4330	4380	4425	4475	4520	4570	4615
36	3445	3495	3540	3585	3635	3680	3730	3775	3825	3870	3920	3965	4010	4060	4105	4155	4200	4250	4295	4340	4390	4435	4485	4530	4580
38	3415	3465	3510	3555	3605	3650	3695	3745	3790	3840	3885	3930	3980	4025	4070	4120	4165	4210	4260	4305	4350	4400	4445	4495	4540
40	3385	3435	3480	3525	3575	3620	3665	3710	3760	3805	3850	3900	3945	3990	4035	4085	4130	4175	4220	4270	4315	4360	4410	4455	4500
42	3360	3405	3450	3495	3540	3590	3635	3680	3725	3770	3820	3865	3910	3955	4000	4050	4095	4140	4185	4230	4280	4325	4370	4415	4460
44	3315	3360	3405	3450	3495	3540	3585	3630	3675	3725	3770	3815	3860	3905	3950	3995	4040	4085	4130	4175	4220	4270	4315	4360	4405
46	3285	3330	3375	3420	3465	3510	3555	3600	3645	3690	3735	3780	3825	3870	3915	3960	4005	4050	4095	4140	4185	4230	4275	4320	4365
48	3255	3300	3345	3390	3435	3480	3525	3570	3615	3655	3700	3745	3790	3835	3880	3925	3970	4015	4060	4105	4150	4190	4235	4280	4325
50	3210	3255	3300	3345	3390	3430	3475	3520	3565	3610	3650	3695	3740	3785	3830	3870	3915	3960	4005	4050	4090	4135	4180	4225	4270
52	3185	3225	3270	3315	3355	3400	3445	3490	3530	3575	3620	3660	3705	3750	3795	3835	3880	3925	3970	4010	4055	4100	4140	4185	4230
54	3155	3195	3240	3285	3325	3370	3415	3455	3500	3540	3585	3630	3670	3715	3760	3800	3845	3890	3930	3975	4020	4060	4105	4145	4190
56	3125	3165	3210	3255	3295	3340	3380	3425	3465	3510	3550	3595	3640	3680	3725	3765	3810	3850	3895	3940	3980	4025	4065	4110	4150
58	3080	3125	3165	3210	3250	3290	3335	3375	3420	3460	3500	3545	3585	3630	3670	3715	3755	3800	3840	3880	3925	3965	4010	4050	4095
60	3050	3095	3135	3175	3220	3260	3300	3345	3385	3430	3470	3500	3555	3595	3635	3680	3720	3760	3805	3845	3885	3930	3970	4015	4055
62	3020	3060	3110	3150	3190	3230	3270	3310	3350	3390	3440	3480	3520	3560	3600	3640	3680	3730	3770	3810	3850	3890	3930	3970	4020
64	2990	3030	3080	3120	3160	3200	3240	3280	3320	3360	3400	3440	3490	3530	3570	3610	3650	3690	3730	3770	3810	3850	3900	3940	3980
66	2950	2990	3030	3070	3110	3150	3190	3230	3270	3310	3350	3390	3430	3470	3510	3550	3600	3640	3680	3720	3760	3800	3840	3880	3920
68	2920	2960	3000	3040	3080	3120	3160	3200	3240	3280	3320	3360	3400	3440	3480	3520	3560	3600	3640	3680	3720	3760	3800	3840	3880
70	2890	2930	2970	3010	3050	3090	3130	3170	3210	3250	3290	3330	3370	3410	3450	3480	3520	3560	3600	3640	3680	3720	3760	3800	3840
72	2860	2900	2940	2980	3020	3060	3100	3140	3180	3210	3250	3290	3330	3370	3410	3450	3490	3530	3570	3610	3650	3680	3720	3760	3800
74	2820	2860	2900	2930	2970	3010	3050	3090	3130	3170	3200	3240	3280	3320	3360	3400	3440	3470	3510	3550	3590	3630	3670	3710	3740

Baldwin, E. D., A. Cournand, and D. W. Richards Jr., "Pulmonary Insufficiency; Physiological Classification, Clinical Methods of Analysis, Standard Values in Normal Subjects", Medicine, 1948 Sept.; 27(3):243–278.

The ⒶΑ corresponds to the indicated Learning Outcome(s) found at the beginning of the laboratory exercise.

Breathing and Respiratory Volumes

PART A ASSESSMENTS

Complete the following statements:

1. Breathing can also be called _____.

2. The weight of air causes a force called _____ pressure.

3. The weight of air at sea level is sufficient to support a column of mercury within a tube _____ mm high.

4. If the pressure inside the lungs decreases, outside air is forced into the airways by _____. Ⓐ1

5. Impulses are conducted to the diaphragm by the _____ nerves. Ⓐ1

6. When the diaphragm contracts, the size of the thoracic cavity _____. Ⓐ1

7. The ribs are raised by the contraction of the _____ muscles, which increases the size of the thoracic cavity. Ⓐ1

8. Only a thin film of lubricating serous fluid separates the parietal pleura from the _____ attached to the surface of a lung.

9. A mixture of lipoproteins, called _____, acts to reduce the tendency of alveoli to collapse.

10. The force responsible for normal resting expiration comes from _____ of tissues and from surface tension. Ⓐ1

11. Muscles that help to force out more than the normal volume of air by pulling the ribs downward and inward include the _____. Ⓐ1

12. The diaphragm can be forced to move higher than normal by the contraction of the _____ wall muscles. Ⓐ1

PART B ASSESSMENTS

Match the air volumes in column A with the definitions in column B. Place the letter of your choice in the space provided. Ⓐ2

Column A	Column B
a. Expiratory reserve volume	_____ 1. Volume of air in addition to tidal volume that leaves the lungs during forced expiration
b. Functional residual capacity	
c. Inspiratory capacity	_____ 2. Vital capacity plus residual volume
d. Inspiratory reserve volume	_____ 3. Volume of air that remains in lungs after the most forceful expiration
e. Residual volume	_____ 4. Volume of air that enters or leaves lungs during a respiratory cycle
f. Tidal volume	
g. Total lung capacity	_____ 5. Volume of air in addition to tidal volume that enters lungs during forced inspiration
h. Vital capacity	
	_____ 6. Maximum volume of air a person can exhale after taking the deepest possible breath
	_____ 7. Maximum volume of air a person can inhale following exhalation of the tidal volume
	_____ 8. Volume of air remaining in the lungs following exhalation of the tidal volume

PART C ASSESSMENTS

1. Test results for respiratory air volumes and capacities: 🔺 🔺

Respiratory Volume or Capacity	Expected Value* (approximate)	Test Result	Percent of Expected Value (test result/expected value × 100)
Tidal volume (resting) (TV)	500 mL		
Expiratory reserve volume (ERV)	1,100 mL		
Vital capacity (VC)	4,600 mL (enter yours from table 43.1 or 43.2)		
Inspiratory reserve volume (IRV)	3,000 mL		
Inspiratory capacity (IC)	3,500 mL		
Functional residual capacity (FRC)	2,300 mL		

*The values listed are most characteristic for a healthy, tall, young adult male. In general, adult females have smaller bodies and therefore smaller lung volumes and capacities. If your expected value for vital capacity is considerably different than 4,600 mL, your other values will vary accordingly.

2. Complete the following:

   a. How do your test results compare with the expected values?

   _____

   _____

   _____

   _____

   b. How does your vital capacity compare with the average value for a person of your sex, age, and height?

   _____

   _____

   c. What measurement in addition to vital capacity is needed before you can calculate your total lung capacity? 🔺

   _____

3. If your experimental results are considerably different from the predicted vital capacities, propose reasons for the differences. As you write this paragraph, consider factors such as smoking, physical fitness, respiratory disorders, stress, and medications. (Your instructor might have you make some class correlations from class data.)

   _____

   _____

   _____

   _____

   _____

44

Urinary Organs

MATERIALS NEEDED

Textbook
Human torso model
Kidney model
Preserved pig (or sheep) kidney
Dissecting tray
Dissecting instruments
Long knife
Compound light microscope
Prepared microscope slides of the following:
  Kidney section
  Ureter
  Urinary bladder

⚠ SAFETY

- Wear disposable gloves when working on the kidney dissection.
- Dispose of the kidney and gloves as directed by your laboratory instructor.
- Wash the dissecting tray and instruments as instructed.
- Wash your laboratory table.
- Wash your hands before leaving the laboratory.

PURPOSE OF THE EXERCISE

To review the structure of urinary organs, to dissect a kidney, and to observe the major structures of a nephron.

🔄 LEARNING OUTCOMES APR

After completing this exercise, you should be able to

1. Locate and identify the major organs of the urinary system.
2. Locate and identify the major structures of a kidney.
3. Identify and sketch the structures of a nephron.

4. Trace the path of filtrate through a renal nephron.
5. Identify and sketch the structures of a ureter and the urinary bladder.

The two kidneys are the primary organs of the urinary system. They are located in the abdominal cavity, against the posterior wall and behind the parietal peritoneum (retroperitoneal). Masses of adipose tissue associated with the kidneys hold them in place at a vertebral level between T12 and L3. The right kidney is slightly more inferior due to the large mass of the liver near its superior border. The movements of the muscular walls of the ureters force urine by means of peristaltic waves into the urinary bladder, which temporarily stores urine. The urethra conveys urine to the outside of the body.

Each kidney contains about 1 million nephrons, which serve as the basic structural and functional units of the kidney. A glomerular capsule, proximal convoluted tubule, nephron loop, and distal convoluted tubule compose the microscopic, multicellular structure of a relatively long nephron tubule. Several nephrons merge and drain into a common collecting duct. Approximately 80% of the nephrons are cortical nephrons with short nephron loops, whereas the remaining represent juxtamedullary nephrons, with long nephron loops extending deeper into the renal medulla. An elaborate network of blood vessels surrounds the entire nephron. Glomerular filtration, tubular reabsorption, and tubular secretion represent three processes resulting in urine as the final product.

A variety of functions occur in the kidneys. They remove metabolic wastes from the blood; help regulate blood volume, blood pressure, and pH of blood; control water and electrolyte concentrations; and secrete renin and erythropoietin.

🔄 EXPLORE

PROCEDURE A—Kidney

1. Review 17.2 section entitled "Kidneys" in chapter 17 of the textbook.
2. As a review activity, label figures 44.1 and 44.2.
3. Complete Part A of Laboratory Report 44.

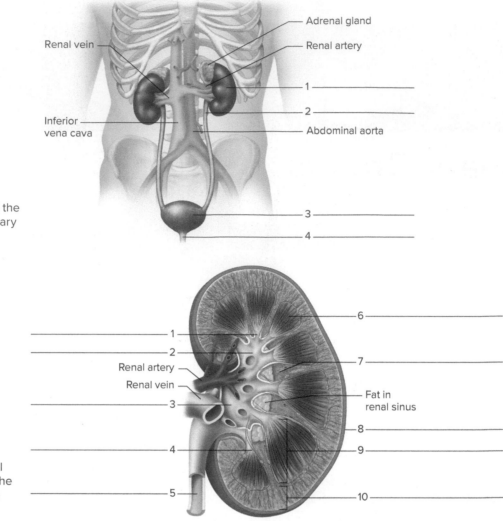

FIGURE 44.1 Label the major organs of the urinary system. ⓘ APR

FIGURE 44.2 Label the major structures in the longitudinal section of a kidney. ② APR

4. Observe the human torso model and the kidney model. Locate the following:

kidneys
ureters
urinary bladder
urethra
renal sinus
renal pelvis
    major calyces
    minor calyces
renal medulla
    renal pyramids
    renal papillae
renal cortex
renal columns
nephrons

5. Obtain a pig or sheep kidney along with a dissecting tray, dissecting instruments, and disposable gloves, and follow these steps:
   a. Rinse the kidney with water to remove as much of the preserving fluid as possible and place it in a dissecting tray.

   b. Carefully remove any adipose tissue from the surface of the specimen.
   c. Locate the following external anatomical features:

   renal (fibrous) capsule
   hilum of kidney
   renal artery
   renal vein
   ureter

   d. Use a long knife to cut the kidney in half longitudinally along the frontal plane, beginning on the convex border.
   e. Rinse the interior of the kidney with water and, using figure 44.3 as a reference, locate the following:

   renal pelvis
       major calyces
       minor calyces
   renal cortex
   renal columns (extensions of renal cortical tissue between renal pyramids)
   renal medulla
       renal pyramids
       renal papillae

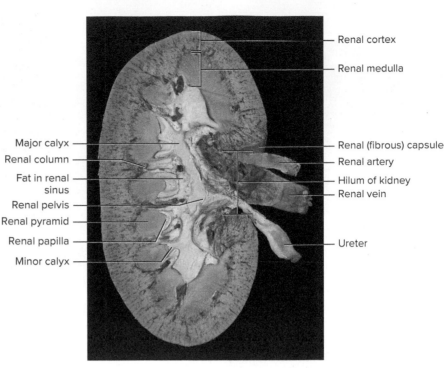

FIGURE 44.3

Longitudinal section of a pig kidney that has a triple injection of latex (*red* in the renal artery, *blue* in the renal vein, and *yellow* in the ureter and renal pelvis). **APR**

©J and J Photography

Renal cortex
Renal medulla
Renal (fibrous) capsule
Renal artery
Hilum of kidney
Renal vein
Ureter

Major calyx
Renal column
Fat in renal sinus
Renal pelvis
Renal pyramid
Renal papilla
Minor calyx

EXPLORE

PROCEDURE B—Renal Blood Vessels and Nephrons

1. Review the concept headings entitled "Renal Blood Supply" and "Nephrons" in section 17.2 of chapter 17 of the textbook.
2. As a review activity, label figure 44.4.
3. Complete Part B of the laboratory report.
4. Obtain a microscope slide of a kidney section, and examine it using low-power magnification. Locate the *renal capsule*, the *renal cortex* (which appears somewhat granular and may be more darkly stained than the other renal tissues), and the *renal medulla* fig. 44.5).

5. Examine the renal cortex using high-power magnification. Locate a *renal corpuscle*. These structures appear as isolated circular areas. Identify the *glomerulus*, the capillary cluster inside the corpuscle, and the *glomerular capsule*, which appears as a clear area surrounding the glomerulus. A glomerulus and a glomerular capsule compose a renal corpuscle. Also note the numerous sections of renal tubules that occupy the spaces between renal corpuscles (fig. 44.5a). The renal cortex contains the proximal and distal convoluted tubules.
6. Prepare a labeled sketch of a representative section of the renal cortex in Part C of the laboratory report.
7. Examine the renal medulla using high-power magnification. Identify longitudinal sections of various collecting

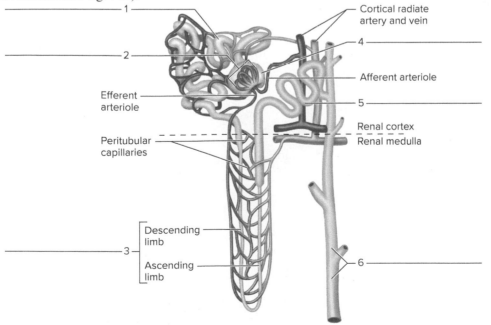

Cortical radiate artery and vein
Afferent arteriole
Renal cortex
Renal medulla

Efferent arteriole
Peritubular capillaries
Descending limb
Ascending limb

FIGURE 44.4 Label the major structures of the nephron and associated structures. ⚠

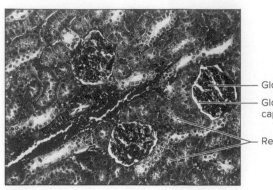

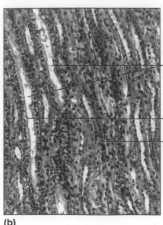
Collecting ducts

Epithelial cell

Blood vessel

(a)

(b)

FIGURE 44.5 (*a*) Micrograph of a section of the renal cortex (220×). (*b*) Micrograph of a section of the renal medulla (80×). (*Note*: With about a million nephrons in a kidney, there are many different orientations visible in micrograph sections of renal cortex and renal medulla.) **APR** **(a):** Manfred Kage/Science Source; **(b):** Al Telser/McGraw-Hill Education

ducts. These ducts are lined with simple epithelial cells, which vary in shape from squamous to cuboidal (fig. 44.5*b*).

8. Prepare a labeled sketch of a representative section of the renal medulla in Part C of the laboratory report.

EXPLORE

PROCEDURE C—Ureter and Urinary Bladder

1. Review section 17.4 entitled "Urine Elimination" in chapter 17 of the textbook.

2. Obtain a microscope slide of a cross section of a ureter, and examine it using low-power magnification. Locate the *mucous coat* layer next to the lumen. Examine the middle *muscular coat* composed of longitudinal and circular smooth muscle cells, responsible for the peristaltic waves that propel urine from the kidneys to the urinary bladder. The outer *fibrous coat,* composed of connective tissue, secures the ureter in the retroperitoneal position (fig. 44.6).

3. Examine the mucous coat using high-power magnification. The specialized tissue is transitional epithelium,

which allows changes in its thickness when unstretched and stretched.

4. Prepare a labeled sketch of a ureter in Part D of the laboratory report.

5. Obtain a microscope slide of a segment of the wall of a urinary bladder, and examine it using low-power magnification. Examine the *mucous coat* next to the lumen and the *submucous coat* composed of connective tissue just beneath the mucous coat. Examine the *muscular coat* composed of bundles of smooth muscle cells interlaced in many directions. This thick, muscular layer is called the *detrusor muscle* and functions for elimination of urine (fig. 44.7). An outer *serous coat* is composed of connective tissue. The serous coat on the upper surface of the urinary bladder consists of the parietal peritoneum.

6. Examine the mucous coat using high-power magnification. The tissue is transitional epithelium, which allows changes in its thickness from unstretched when the bladder is empty to stretched when the bladder distends with urine.

7. Prepare a labeled sketch of a segment of the urinary bladder wall in Part D of the laboratory report.

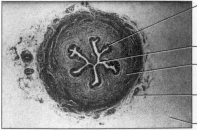

Mucous coat
(contains transitional
epithelium next to lumen)

Lumen

Muscular coat
(of smooth muscle)

Fibrous coat
(of connective tissue)

Adipose tissue

FIGURE 44.6 Micrograph of a cross section of a ureter (75×). **APR**
Biophoto Associates/Science Source

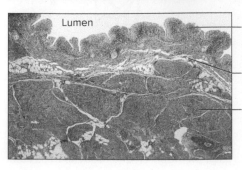

Lumen

Mucous coat
(contains transitional
epithelium next to lumen)

Submucous coat
(of connective tissue)

Muscular coat
(of smooth muscle)
(detrusor muscle)

FIGURE 44.7 Micrograph of a segment of the human urinary bladder wall (6×). **APR** Alvin Telser/ McGraw-Hill Education

The ⚠ corresponds to the indicated Learning Outcome(s) found at the beginning of the laboratory exercise.

Urinary Organs

🔁 PART A ASSESSMENTS

Match the terms in column A with the descriptions in column B. Place the letter of your choice in the space provided. ⚠

Column A	Column B
a. Calyces	_____ **1.** Superficial region around the renal medulla
b. Hilum of kidney	_____ **2.** Extensions of renal pelvis to renal papillae containing urine
c. Nephron	_____ **3.** Conical mass of tissue within renal medulla
d. Renal column	_____ **4.** Projection with tiny openings into minor calyx
e. Renal cortex	_____ **5.** Medial depression for blood vessels and ureter to enter kidney chamber
f. Renal papilla	
g. Renal pelvis	_____ **6.** Hollow chamber within kidney
h. Renal pyramid	_____ **7.** Microscopic functional subunit of kidney
i. Renal sinus	_____ **8.** Cortical tissue extensions between renal pyramids
	_____ **9.** Superior, funnel-shaped sac at end of ureter inside renal sinus

🔁 PART B ASSESSMENTS

1. Distinguish between a renal corpuscle and a renal tubule. ⚠ _____

_____

_____

2. Number the following structures to indicate their respective positions in relation to the nephron. Assign number 1 to the structure nearest the glomerulus. ⚠

_____ Ascending limb of the nephron loop

_____ Collecting duct

_____ Descending limb of the nephron loop

_____ Distal convoluted tubule

_____ Glomerular capsule

_____ Proximal convoluted tubule

_____ Renal papilla

PART C ASSESSMENTS

Sketch a representative section of the renal cortex and the renal medulla. Label the glomerulus, the glomerular capsule, and sections of renal tubules in the renal cortex and a longitudinal section and a cross section of a collecting duct in the renal medulla. ⚠3

Renal cortex (_____ ×)

Renal medulla (_____ ×)

PART D ASSESSMENTS

Sketch a cross section of a ureter and label the three layers and the lumen. Sketch a segment of a urinary bladder and label the four layers and the lumen. ⚠5

Ureter (_____ ×)

Urinary bladder (_____ ×)

Label the following images.

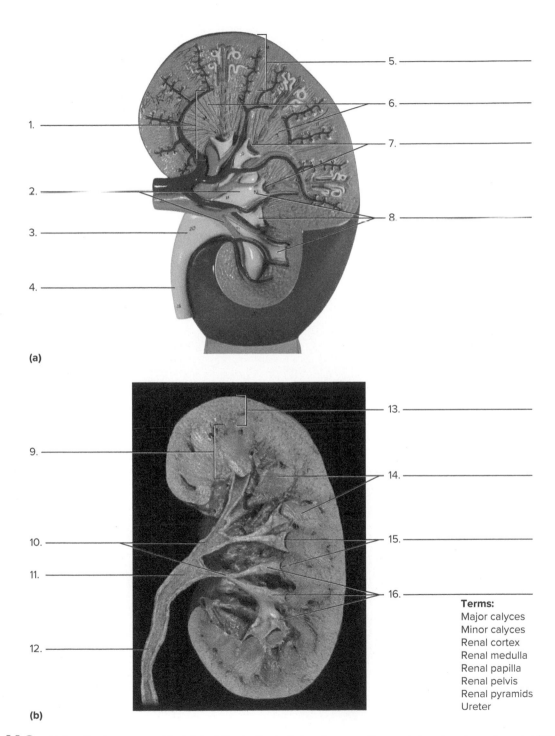

(a)

(b)

Terms:
Major calyces
Minor calyces
Renal cortex
Renal medulla
Renal papilla
Renal pelvis
Renal pyramids
Ureter

FIGURE 44.8   Using the terms provided, label the indicated structures on the model and cadaver image of the kidney.

(a): ©2017 Denoyer-Geppert Science Company denoyer.com; (b): ©McGraw-Hill Education/APR

Notes

45

Urinalysis

MATERIALS NEEDED

Normal and abnormal simulated urine specimens can be substituted for collected urine.
Disposable urine-collecting container
Paper towel
Urinometer cylinder (jar)
Urinometer hydrometer
Laboratory thermometer
pH test paper
Reagent strips (individual or combination test strips such as Chemstrip or Multistix) to test for the presence of the following:
Glucose
Protein
Ketones
Bilirubin
Hemoglobin/occult blood
Compound microscope
Microscope slide
Coverslip
Centrifuge
Centrifuge tube
Graduated cylinder, 10 mL
Medicine dropper
Sedi-stain

⚠ SAFETY

- Consider using normal and abnormal simulated urine samples available from various laboratory supply houses.
- Use safety glasses, laboratory coats, and disposable gloves when working with body fluids.
- Work only with your own urine sample.
- Use an appropriate disinfectant to wash the laboratory table before and after the procedures.
- Place glassware in a disinfectant when finished.
- Dispose of contaminated items as directed by your laboratory instructor.
- Wash your hands before leaving the laboratory.

PURPOSE OF THE EXERCISE

To perform the observations and tests commonly used to analyze the characteristics and composition of urine.

LEARNING OUTCOMES

After completing this exercise, you should be able to

1. Evaluate the color, transparency, and specific gravity of a urine sample.
2. Measure the pH of a urine sample.
3. Test a urine sample for the presence of glucose, protein, ketones, bilirubin, and hemoglobin.
4. Perform a microscopic study of urine sediment.
5. Summarize the results of these observations and tests.

U rine is the product of three processes of the nephrons within the kidneys: glomerular filtration, tubular reabsorption, and tubular secretion. As a result of these processes, various waste substances are removed from the blood, and body fluid and electrolyte balance are maintained. Consequently, the composition of urine varies considerably because of differences in dietary intake and physical activity from day to day and person to person. The normal urinary output ranges from 1.0–1.8 liters per day. Typically, urine consists of 95% water and 5% solutes. The volume of urine produced by the kidneys varies with such factors as fluid intake, environmental temperature, relative humidity, respiratory rate, and body temperature.

An analysis of urine composition and volume often is used to evaluate the functions of the kidneys and other organs. This procedure, called *urinalysis,* is a clinical assessment and a diagnostic tool for certain pathological conditions and general overall health. A urinalysis and a complete blood analysis complement each other for an evaluation of certain diseases, such as diabetes mellitus, and general health.

A urinalysis involves three aspects: *physical characteristics, chemical analysis,* and a *microscopic examination.* Physical characteristics of urine that are noted include volume, color, transparency, and odor. The chemical analysis of solutes in urine addresses urea and other nitrogenous wastes; electrolytes; pigments; and possible glucose, protein, ketones, bilirubin, and hemoglobin. The specific gravity and pH of urine are greatly influenced by the components and amounts of solutes. An examination of microscopic solids, including cells, casts, and crystals, assists in the diagnosis of injury, various diseases, and urinary infections.

> **WARNING** *While performing the following tests, you should wear disposable latex gloves so that skin contact with the urine is avoided. Observe all safety procedures listed for this laboratory exercise. (Normal and abnormal simulated urine specimens can be used instead of real urine.)*

EXPLORE

PROCEDURE A—Physical and Chemical Urinalysis

1. Proceed to the restroom with a clean, disposable container if collected urine is to be used for this laboratory exercise. The first small volume of urine should not be collected because it contains abnormally high levels of microorganisms from the urethra. Collect a midstream sample of about 50 mL of urine. The best collections are the first specimen in the morning or one taken 3 hours after a meal. Refrigerate samples if they are not used immediately.

2. Place a sample of urine in a clean, transparent container. Describe the *color* of the urine. Normal urine varies from light yellow to amber, depending on the presence of urochromes, end-product pigments produced during the decomposition of hemoglobin. Dark urine indicates a high concentration of pigments.

   Abnormal urine colors include yellow-brown or green, due to elevated concentrations of bile pigments, and red to dark brown, due to the presence of blood. Certain foods, such as beets or carrots, and various drug substances also can cause color changes in urine, but in such cases, the colors have no clinical significance. Enter the results of this and the following tests in Part A of Laboratory Report 45.

3. Evaluate the *transparency* of the urine sample (judge whether the urine is clear, slightly cloudy, or very cloudy). Normal urine is clear enough to see through. You can read newsprint through slightly cloudy urine; you can no longer read newsprint through cloudy urine. Cloudy urine indicates the presence of various substances, including mucus, bacteria, epithelial cells, fat droplets, and inorganic salts.

4. Gently wave your hand over the urine sample toward your nose to detect the *odor*. Normal urine should have a slight ammonia-like odor due to the nitrogenous wastes. Some vegetables (such as asparagus), drugs (such as certain vitamins), and diseases (such as diabetes mellitus) will influence the odor of urine.

5. Determine the *specific gravity* of the urine sample, which indicates the solute concentration. Specific gravity is the ratio of the weight of something to the weight of an equal volume of pure water. For example, mercury (at 15°C) weighs 13.6 times as much as an equal volume of water; thus, it has a specific gravity of 13.6. Although urine is mostly water, it has substances dissolved in it and is slightly heavier than an equal volume of water. The specific gravity of pure (distilled) water is 1.000, whereas the specific gravity of urine is higher, varying from 1.003 to 1.035 under normal circumstances. If the specific gravity is too low, the urine contains few solutes and represents dilute urine, a likely result of excessive fluid intake or the use of diuretics. A specific gravity above the normal range represents a higher concentration of solutes, likely from a limited fluid intake. Concentrated urine over an extended time increases the risk of the formation of kidney stones (renal calculi).

   To determine the specific gravity of a urine sample, follow these steps:
   a. Pour enough urine into a clean urinometer cylinder to fill it about three-fourths full. Any foam that appears should be removed with a paper towel.
   b. Use a laboratory thermometer to measure the temperature of the urine.
   c. Gently place the urinometer hydrometer into the urine, and *make sure that the float is not touching the sides or the bottom of the cylinder* (fig. 45.1).

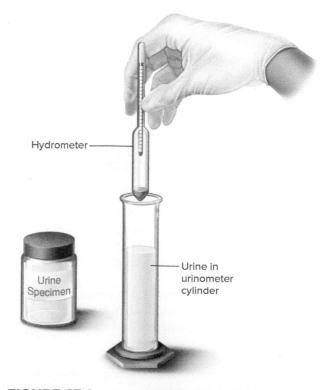

Hydrometer

Urine Specimen

Urine in urinometer cylinder

FIGURE 45.1 Float the hydrometer in the urine, making sure that it does not touch the sides or the bottom of the cylinder.

d. Position your eye at the level of the urine surface. Determine which line on the stem of the hydrometer intersects the lowest level of the concave surface (meniscus) of the urine.

e. Liquids tend to contract and become denser as they are cooled or to expand and become less dense as they are heated, so it may be necessary to make a temperature correction to obtain an accurate specific gravity measurement. To do this, add 0.001 to the hydrometer reading for each 3 degrees of urine temperature above 25°C or subtract 0.001 for each 3 degrees below 25°C. Enter this calculated value in the table in Part A of the laboratory report as the test result.

6. Individual or combination reagent strips can be used to perform a variety of urine tests (fig. 45.2). In each case, directions for using the strips are found on the strip container. *Be sure to read them.*

To perform each test, follow these steps:

a. Obtain a urine sample and the proper reagent test strip.

b. Read the directions on the strip container.

c. Dip the strip in the urine sample, as directed on the container.

d. Remove the strip at an angle and let it touch the inside rim of the urine container to remove any excess liquid.

e. Wait for the length of time indicated by the directions on the container before you compare the color of the test strip with the standard color scale provided with the container. The value or amount represented by the matching color should be used as the test result for the following tests performed and recorded in Part A of the laboratory report.

7. Perform the *pH test*. The pH of normal urine varies from 4.6 to 8.0, but most commonly, it is near 6.0 (slightly acidic). The pH of urine may decrease as a result of a diet high in protein, or it may increase with a vegetarian diet. Significant daily variations within the broad normal range are results of concentrations of excesses from variable diets.

FIGURE 45.2 An example of a reagent test strip dipped into urine to determine a variety of urine components.
© McGraw-Hill Education

8. Perform the *glucose test*. Normally, there is no glucose in urine. However, glucose may appear in the urine temporarily following a meal high in carbohydrates. Glucose also may appear in the urine as a result of uncontrolled diabetes mellitus.

9. Perform the *protein test*. Normally, proteins of large molecular size are not present in urine. However, those of small molecular sizes, such as albumins, may appear in trace amounts, particularly following strenuous exercise. Increased amounts of proteins also may appear as a result of kidney diseases in which the glomeruli are damaged or as a result of high blood pressure.

10. Perform the *ketone test*. Ketones are products of fat metabolism. Usually, they are not present in urine. However, they may appear in the urine if the diet fails to provide adequate carbohydrates, as in the case of prolonged fasting or starvation, or as a result of insulin deficiency (diabetes mellitus).

11. Perform the *bilirubin test*. Bilirubin, which results from hemoglobin decomposition in the liver, normally is absent in urine. It may appear, however, as a result of liver disorders that cause obstructions of the biliary tract. Urochrome, a normal yellow component of urine, is a result of additional breakdown of bilirubin.

12. Perform the *hemoglobin/occult blood test*. Hemoglobin occurs in the red blood cells, and because such cells normally do not pass into the renal tubules, hemoglobin is not found in normal urine. Its presence in urine usually indicates a disease process, a transfusion reaction, an injury to the urinary organs, or menstrual blood.

13. Perform the *leukocytes test*. White blood cells in the urine are indicative of an infection of the urinary tract, urinary bladder, or kidney; normally, they are not present.

14. Complete Part A of the laboratory report.

EXPLORE

PROCEDURE B—Microscopic Sediment Analysis

1. A urinalysis usually includes an analysis of urine sediment—the microscopic solids present in a urine sample. This sediment normally includes mucus; certain crystals; and a variety of cells, such as the epithelial cells that line the urinary tubes and an occasional white blood cell. Other types of solids, such as casts or red blood cells, may indicate a disease or injury if they are present in excess. (Casts are cylindrical masses of cells or other substances that form in the renal tubules and are flushed out by the flow of urine.)

To observe urine sediment, follow these steps:

a. Thoroughly stir or shake your urine sample to suspend the sediment, which tends to settle to the bottom of the container.

b. Pour 10 mL of urine into a clean centrifuge tube and centrifuge it for 5 minutes at a slow speed (1,500 rpm). Be sure to balance the centrifuge with an even number of tubes filled to the same levels.

c. Carefully decant 9 mL (leave 1 mL) of the liquid from the sediment in the bottom of the centrifuge tube, as directed by your laboratory instructor. Resuspend the 1 mL of sediment.

d. Use a medicine dropper to remove some of the sediment and place it on a clean microscope slide.

e. Add a drop of Sedi-stain to the sample, and add a coverslip.

f. Examine the sediment with low-power (reduce the light when using low power) and high-power magnifications.

g. With the aid of figure 45.3, identify the types of solids present.

2. In Part B of the laboratory report, make a sketch of each type of sediment that you observed.

3. Complete Part B of the laboratory report.

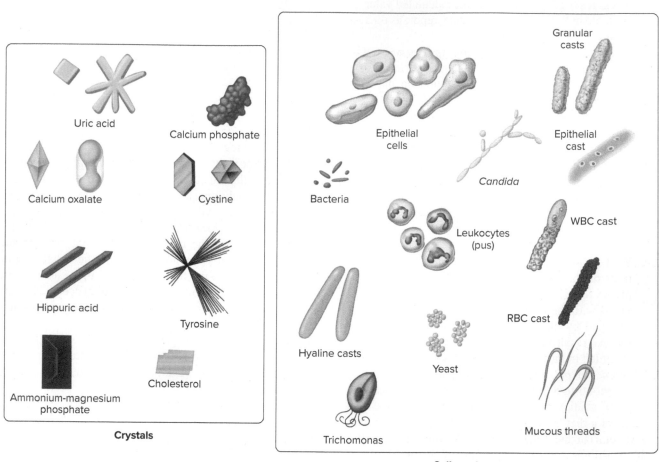

Crystals

Cells and casts

FIGURE 45.3 Types of urine sediment. Healthy individuals lack many of these sediments and possess only occasional to trace amounts of others. (*Note:* Shades of white to purple sediments are most characteristic when using Sedi-stain.)

Name _____

Date _____

Section _____

The Ⓐ corresponds to the indicated Learning Outcome(s) found at the beginning of the laboratory exercise.

Urinalysis

♻ PART A ASSESSMENTS

1. Enter your observations, test results, and evaluations in the following table. Ⓐ1 Ⓐ2 Ⓐ3

Urine Characteristics	Observations and Test Results	Normal Values	Evaluations
Color		Light yellow to amber	
Transparency		Clear	
Odor		Slight ammonia-like	
Specific gravity (corrected for temperature)		1.003–1.035	
pH		4.6–8.0	
Glucose		Negative (0)	
Protein		Negative to trace	
Ketones		Negative (0)	
Bilirubin		Negative (0)	
Hemoglobin/occult blood		Negative (0)	
Leukocytes		Negative to trace	
(Other)			

2. Summarize the results of the physical and chemical analyses of urine. Ⓐ5 _____

_____

_____

_____

_____

_____

_____

Critical Thinking Application

Why do you think it is important to refrigerate a urine sample if an analysis cannot be performed immediately after collecting it?

_____

_____

_____

_____

PART B ASSESSMENTS

1. Make a sketch for each type of sediment you observed. Label any from those shown in figure 45.3. ⃟4⃟

2. Summarize the results of the microscopic sediment analysis of urine. ⃟5⃟ _____

_____

_____

_____

_____

_____

_____

46

Male Reproductive System

MATERIALS NEEDED

Textbook
Human torso model
Model of the male reproductive system
Anatomical chart of the male reproductive system
Compound light microscope
Prepared microscope slides of the following:
    Testis section
    Epididymis, cross section
    Penis, cross section

PURPOSE OF THE EXERCISE APR

To review the structures and functions of the male reproductive organs and to examine some of their features.

LEARNING OUTCOMES APR

After completing this exercise, you should be able to

(1) Locate and identify the organs of the male reproductive system.

(2) Describe the functions of these organs.

(3) Sketch and label the major features from microscopic sections of the testis, epididymis, and penis.

The organs of the male reproductive system are specialized to produce and maintain the male sex cells, to transport these cells together with supporting fluids to the female reproductive tract, and to produce and secrete male sex hormones.

These organs include the testes, in which sperm and male sex hormones are produced, and sets of internal and external accessory structures. The internal structures include various tubes and glands, whereas the external structures are the scrotum and the penis.

EXPLORE

PROCEDURE A—Male Reproductive Organs

1. Review the concept headings entitled "Testes," "Male Internal Accessory Reproductive Organs," and "Male External Accessory Reproductive Organs" in section 19.2 of chapter 19 of the textbook.

2. As a review activity, label figures 46.1 and 46.2.

3. Observe the human torso model, the model of the male reproductive system, and the anatomical chart of the male reproductive system. Locate the following features:

testes
    seminiferous tubules
    interstitial cells (cells of Leydig)
epididymis
ductus deferens (vas deferens)
ejaculatory duct
seminal vesicles
prostate gland
bulbourethral glands
scrotum
penis
    corpora cavernosa
    corpus spongiosum
    glans penis
    external urethral orifice
    prepuce (foreskin)

4. Complete Part A of Laboratory Report 46.

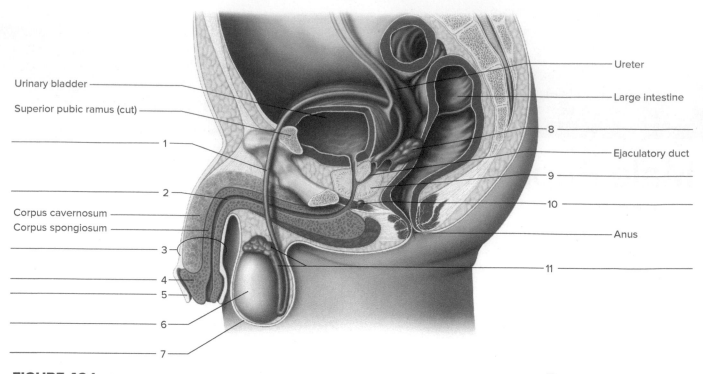

Urinary bladder

Superior pubic ramus (cut)

1

2

Corpus cavernosum

Corpus spongiosum

3

4

5

6

7

Ureter

Large intestine

8

Ejaculatory duct

9

10

Anus

11

FIGURE 46.1 Label the major structures of the male reproductive system in this sagittal view. 🄰 **APR**

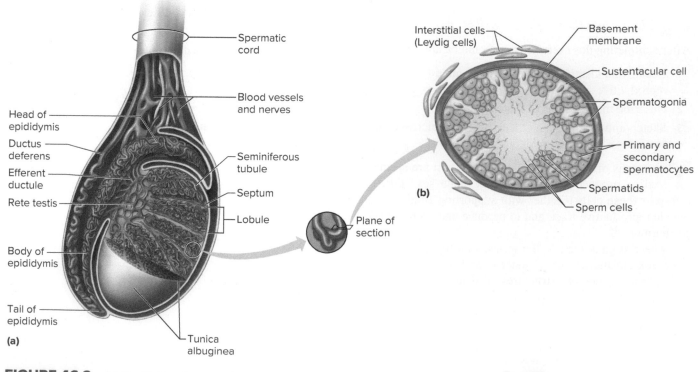

Spermatic cord

Blood vessels and nerves

Head of epididymis

Ductus deferens

Efferent ductule

Rete testis

Seminiferous tubule

Septum

Lobule

Body of epididymis

Tail of epididymis

Plane of section

Tunica albuginea

(a)

Interstitial cells (Leydig cells)

Basement membrane

Sustentacular cell

Spermatogonia

Primary and secondary spermatocytes

Spermatids

Sperm cells

(b)

FIGURE 46.2 (a) Sagittal section of a testis. (b) Cross section of a seminiferous tubule. 🄰 **APR**

PROCEDURE B—Microscopic Anatomy

1. Obtain a microscope slide of a human testis section and examine it using low-power magnification (fig. 46.3). Locate the thick *fibrous capsule* on the surface and the numerous sections of *seminiferous tubules* inside.

2. Focus on some of the seminiferous tubules using high-power magnification (fig. 46.4). Locate the *basement membrane* and the layer of *spermatogonia* (undifferentiated *spermatogenic cells*) just beneath the basement membrane. Identify some *supporting cells* (sustentacular cells; Sertoli cells), which have pale, oval-shaped nuclei. Spermatogonia give rise to spermatogenic cells that are in various stages of spermatogenesis as they are forced toward the lumen. Near the lumen of the tube, find some darkly stained, elongated heads of developing sperm cells. In the spaces between adjacent seminiferous tubules, locate some isolated *interstitial cells* (cells of Leydig) of the endocrine system. Interstitial cells produce the hormone testosterone, transported by the blood.

3. Prepare a labeled sketch of a representative section of the testis in Part B of the laboratory report.

4. Obtain a microscope slide of a cross section of an *epididymis* (fig. 46.5). Examine its wall using high-power magnification. Note the elongated, *pseudostratified columnar epithelial cells* that compose most of the inner lining. These cells have nonmotile steriocilia (elongated microvilli) on their free surfaces that absorb excess fluid secreted by the testes. Also note the thin layer of smooth muscle and connective tissue surrounding the tube.

5. Prepare a labeled sketch of the epididymis wall in Part B of the laboratory report.

6. Obtain a microscope slide of a *penis* cross section, and examine it with low-power magnification (fig. 46.6). Identify the following features:

   corpora cavernosa
   corpus spongiosum
   tunica albuginea
   urethra
   skin

7. Prepare a labeled sketch of a penis cross section in Part B of the laboratory report.

8. Complete Part B of the laboratory report.

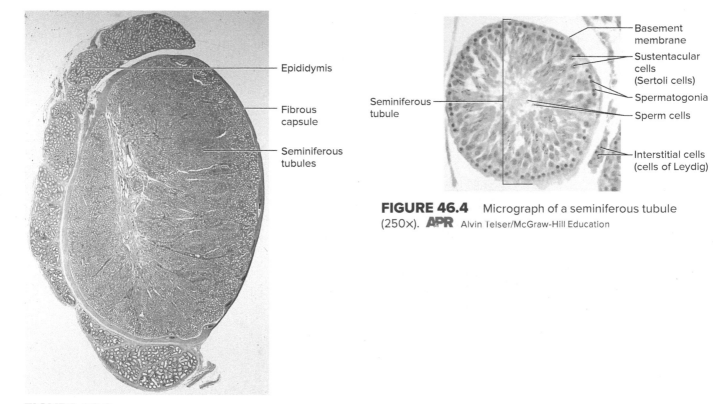

FIGURE 46.3   Micrograph of a human testis (1.7×).
Biophoto Associates/Science Source

Epididymis

Fibrous capsule

Seminiferous tubules

Seminiferous tubule

Basement membrane

Sustentacular cells (Sertoli cells)

Spermatogonia

Sperm cells

Interstitial cells (cells of Leydig)

FIGURE 46.4   Micrograph of a seminiferous tubule (250×). **APR** Alvin Telser/McGraw-Hill Education

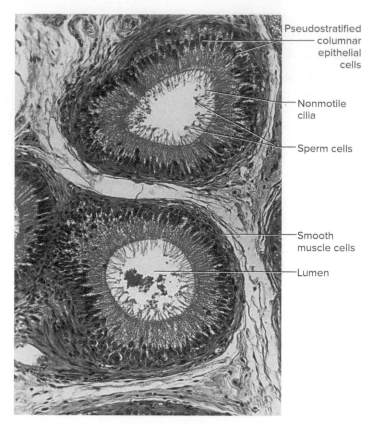

FIGURE 46.5 Micrograph of a cross section of a human epididymis (200×).
Ed Reschke

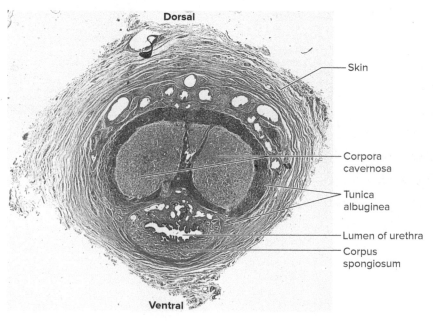

FIGURE 46.6 Micrograph of a cross section of the body of the penis (5×).
Michael Peres

Name _____

Date _____

Section _____

The  corresponds to the indicated Learning Outcome(s) found at the beginning of the laboratory exercise.

LABORATORY
REPORT

46

Male Reproductive System

⟳ PART A ASSESSMENTS

Complete the following statements:

1. Both testes are contained within the cavity of the saclike _____ .

2. Connective tissue subdivides a testis into many lobules, which contain _____ . ⓐ

3. The _____ is a highly coiled tube on the surface of the testis. ⓐ

4. The _____ cells of the epithelium that line the seminiferous tubule give rise to sperm cells. ⓐ

5. Specialized cells called _____ lic in the spaces between the seminiferous tubules and produce and secrete male sex hormones.

6. Undifferentiated sperm cells in the male embryo are called _____ . ⓐ

7. _____ is the process by which sperm cells are formed. ⓐ

8. The number of chromosomes normally present in a human sperm cell is _____ . ⓐ

9. The anterior end of a sperm head, called the _____ , contains enzymes that aid in the penetration of an egg cell at the time of fertilization. ⓐ

10. Sperm cells undergo maturation while they are stored in the _____ . ⓐ

11. The secretion of the seminal vesicles is rich in the monosaccharide called _____ . ⓐ

12. The single _____ gland surrounds the urethra near the urinary bladder. ⓐ

13. The secretion of the _____ glands lubricates the end of the penis in preparation for sexual intercourse. ⓐ

14. The shaft of the penis has three columns of erectile tissue—a pair of dorsally located _____ and a single, ventral _____ .

15. The sensitive, cone-shaped end of the penis is called the _____ . ⓐ

1. In the space that follows, sketch a representative area of the organ indicated. Label any of the structures that were observed, and indicate the magnification used for each sketch. ⟨3⟩

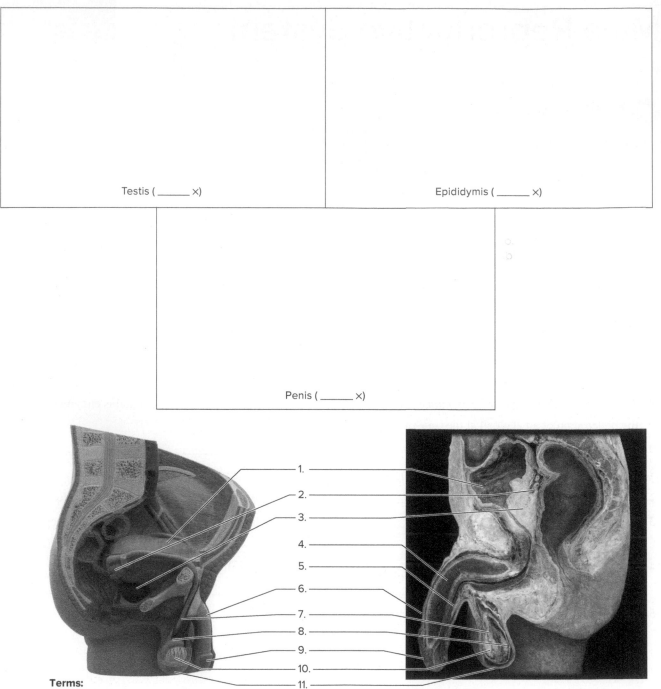

Testis (_____ ×)

Epididymis (_____ ×)

Penis (_____ ×)

1. _____
2. _____
3. _____
4. _____
5. _____
6. _____
7. _____
8. _____
9. _____
10. _____
11. _____

Terms:
Bladder
Corpus cavernosum
Corpus spongiosum
Ductus deferens
Epididymis
Glans penis
Penis
Prostate gland
Scrotum
Seminal vesicle
Testis

FIGURE 46.7   Using the terms provided, label the indicated structures on the model and cadaver image of the male reproductive organs. **(a)** 2017 Denoyer-Geppert Science Company denoyer.com, **(b)** McGraw-Hill Education/APR

47

Female Reproductive System

MATERIALS NEEDED

Textbook
Human torso model
Model of the female reproductive system
Anatomical chart of the female reproductive system
Compound light microscope
Prepared microscope slides of the following:
    Ovary section with maturing follicles
    Uterine tube, cross section
    Uterine wall section

PURPOSE OF THE EXERCISE **APR**

To review the structures and functions of the female reproductive organs and to examine some of their features.

LEARNING OUTCOMES **APR**

After completing this exercise, you should be able to

1. Locate and identify the organs of the female reproductive system.
2. Describe the functions of these organs.
3. Sketch and label the major features from microscopic sections of the ovary, uterine tube, and uterine wall.

The organs of the female reproductive system are specialized to produce and maintain the female sex cells, to transport these cells to the site of fertilization, to provide a favorable environment for a developing offspring, to move the offspring to the outside, and to produce female sex hormones.

These organs include the ovaries, which produce the oocytes (egg cells) and female sex hormones, and sets of internal and external accessory structures. The internal accessory structures include the uterine tubes, uterus, and vagina. The external accessory structures are the labia majora, labia minora, clitoris, and vestibular glands.

EXPLORE

PROCEDURE A—Female Reproductive Organs

1. Review the concept headings entitled "Ovaries," "Female Internal Accessory Reproductive Organs," "Female External Accessory Reproductive Organs," and "Mammary Glands" in section 19.5 of chapter 19 of the textbook.
2. As a review activity, label figures 47.1 and 47.2.
3. Observe the human torso model, the model of the female reproductive system, and the anatomical chart of the female reproductive system. Locate the following features:

ovaries
    medulla
    cortex
uterine tubes (oviducts; Fallopian tubes)
    infundibulum
    fimbriae
uterus
    body
    cervix
    uterine wall
        endometrium
        myometrium
        perimetrium
vagina
    vaginal orifice
    hymen
    mucosal layer
vulva (external accessory structures)
    mons pubis
    labia majora
    labia minora
    vestibule
    vestibular glands
    clitoris

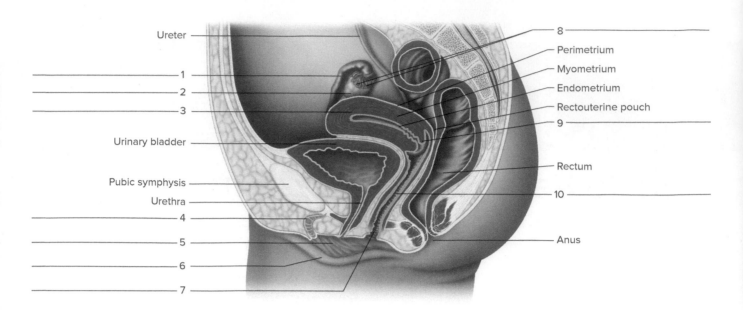

Ureter

1

2

3

Urinary bladder

Pubic symphysis

Urethra

4

5

6

7

8

Perimetrium

Myometrium

Endometrium

Rectouterine pouch

9

Rectum

10

Anus

FIGURE 47.1 Label the structures of the female reproductive system in this sagittal view. 🅰 **APR**

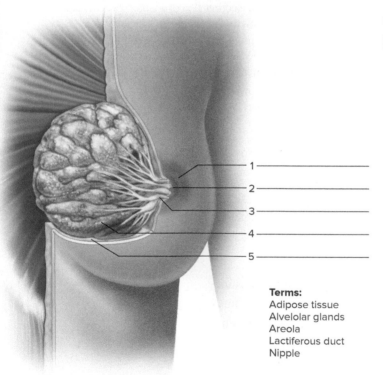

1

2

3

4

5

Terms:
Adipose tissue
Alvelolar glands
Areola
Lactiferous duct
Nipple

FIGURE 47.2 Using the terms provided, label the structures of the breast (anterior view). 🅰 **APR**

breasts

nipple

areola

alveolar glands (compose milk-producing parts of mammary glands)

lactiferous ducts

adipose tissue

4. Complete Part A of Laboratory Report 47.

🔁 **EXPLORE**

PROCEDURE B—Microscopic Anatomy

1. Obtain a microscope slide of an ovary section with maturing follicles, and examine it with low-power magnification (fig. 47.3). Locate the outer layer, or *cortex,* of the ovary, composed of densely packed cells with developing follicles, and the inner layer, or *medulla,* which largely consists of loose connective tissue.

2. Focus on the cortex of the ovary using high-power magnification (fig. 47.4). Note the thin layer of small cuboidal cells on the free surface. These cells constitute the *germinal epithelium.* Also locate some *primordial follicles* just beneath the germinal epithelium. Each follicle consists of a single, relatively large *primary oocyte* with a prominent nucleus and a covering of *follicular cells.*

3. Use low-power magnification to search the ovarian cortex for maturing follicles in various stages of development (fig. 47.3). Locate and compare primordial follicles and *primary follicles.* A certain primary follicle can develop into a *pre-antral follicle* and then into a *mature antral follicle* just before ovulation (figs. 47.3 and 47.4).

4. Prepare two labeled sketches in Part B of the laboratory report to illustrate the changes that occur in a follicle as it matures.

5. Obtain a microscope slide of a cross section of a uterine tube. Examine it using low-power magnification. The shape of the lumen is very irregular because of folds of the mucosa layer.

6. Focus on the inner lining of the uterine tube using high-power magnification (fig. 47.5). The lining is composed of *simple columnar epithelium,* and some of the epithelial cells are ciliated on their free surfaces.

7. Prepare a labeled sketch of a representative region of the wall of the uterine tube in Part B of the laboratory report.

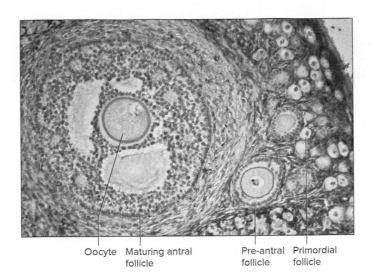

FIGURE 47.3   Micrograph of the ovary showing various stages of follicular development in the cortex (250×). **APR**
Biophoto Associates/Science Source

Oocyte   Maturing antral follicle   Pre-antral follicle   Primordial follicle

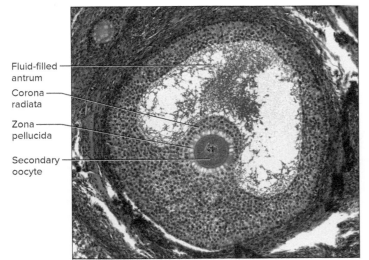

Fluid-filled antrum
Corona radiata
Zona pellucida
Secondary oocyte

FIGURE 47.4   Micrograph of a mature antral follicle (250×). **APR**   Al Telser/McGraw-Hill Education

8. Obtain a microscope slide of the uterine wall section (fig. 47.6). Examine it using low-power magnification, and locate the following:

     endometrium (inner mucosal layer)

     myometrium (middle, thick muscular layer)

     perimetrium (outer serosal layer)

9. Prepare a labeled sketch of a representative section of the uterine wall in Part B of the laboratory report.
10. Complete Part B of the laboratory report.

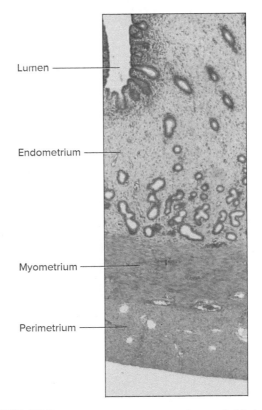

Lumen
Endometrium
Myometrium
Perimetrium

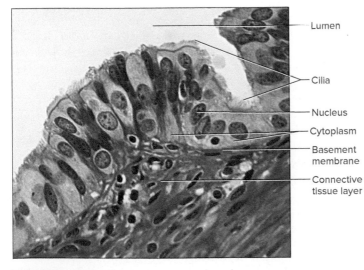

Lumen
Cilia
Nucleus
Cytoplasm
Basement membrane
Connective tissue layer

FIGURE 47.5   Micrograph of a cross section of the uterine tube (800×). **APR**   Ed Reschke/Getty Images

FIGURE 47.6   Micrograph of the uterine wall (10×). **APR**   Carol D. Jacobson/McGraw-Hill Education

Notes

Name _____

Date _____

Section _____

The ▲ corresponds to the indicated Learning Outcome(s) found at the beginning of the laboratory exercise.

LABORATORY
REPORT

47

Female Reproductive System

🔁 PART A ASSESSMENTS

Complete the following statements:

1. The ovaries are located in the lateral wall of the _____ cavity. ▲1

2. The ovarian cortex appears granular because of the presence of _____. ▲1

3. A primary oocyte is closely surrounded by epithelial cells called _____ cells. ▲1

4. When a primary oocyte divides, a secondary oocyte and a(n) _____ are produced. ▲2

5. Primordial follicles are stimulated to develop into primary follicles by the hormone called _____. ▲2

6. _____ is the process by which a secondary oocyte is released from the ovary. ▲2

7. Uterine tubes are also called oviducts or _____. ▲1

8. The _____ is the funnel-shaped expansion at the end of a uterine tube. ▲1

9. A portion of the uterus called the _____ extends downward into the upper portion of the vagina. ▲1

10. The inner mucosal layer of the uterus is called the _____. ▲1

11. The myometrium is largely composed of _____ tissue. ▲1

12. The vaginal orifice is partially closed by a thin membrane called the _____. ▲1

13. The group of external accessory organs that surround the openings of the urethra and vagina comprise the _____. ▲1

14. The rounded mass of fatty tissue overlying the pubic symphysis of the female is called the _____. ▲1

15. The female organ that corresponds to the male penis is the _____. ▲1

16. The _____ of the female correspond to the bulbourethral glands of the male. ▲1

🔁 PART B ASSESSMENTS

1. Complete each of the following:

   a. Describe the fate of a mature antral follicle. ▲2 _____

   _____

   _____

   b. Describe the function of the cilia in the lining of the uterine tube. ▲2 _____

   _____

   _____

c. Briefly describe the changes that occur in the uterine lining during the reproductive cycle. 2 _____

_____

_____

2. In the space that follows, sketch a representative area of the organ indicated. Label any of the structures observed, and indicate the magnification used for each sketch. 3

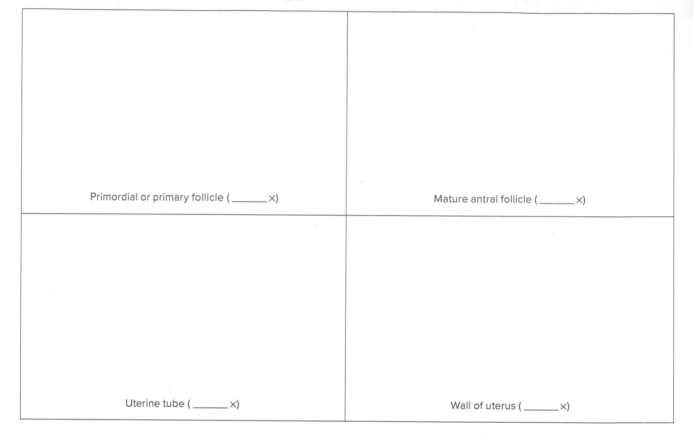

Primordial or primary follicle (_____ ×)	Mature antral follicle (_____ ×)
Uterine tube (_____ ×)	Wall of uterus (_____ ×)

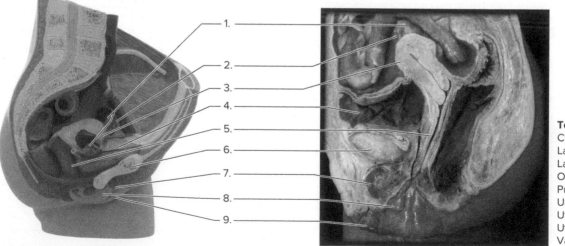

Terms:
Clitoris
Labia majora
Labia minora
Ovaries
Pubic symphysis
Urinary bladder
Uterine tube
Uterus
Vagina

FIGURE 47.7 Using the terms provided, label the indicated structures on the model and cadaver image of the organs associated with the female reproductive system. **(a)** 2017 Denoyer-Geppert Science Company denoyer.com, **(b)** McGraw-Hill Education/APR

48

Genetics

MATERIALS NEEDED

Textbook
Pennies (or other coins)
Dice
PTC paper
Astigmatism chart
Ichikawa's or Ishihara's color plates for colorblindness test

PURPOSE OF THE EXERCISE

To observe some selected human traits, to use pennies and dice to demonstrate laws of probability, and to solve some genetic problems using a Punnett square.

LEARNING OUTCOMES

After completing this exercise, you should be able to

1. Examine and record twelve genotypes and phenotypes of selected human traits.
2. Demonstrate the laws of probability using tossed pennies and dice and interpret the results.
3. Predict genotypes and phenotypes of complete dominance, codominance, and sex-linked problems using Punnett squares.

Many examples of human genetics in this laboratory exercise are basic, external features to observe. Some traits are based upon simple Mendelian genetics. As our knowledge of genetics continues to develop, traits such as tongue roller and free earlobe may no longer be considered examples of simple Mendelian genetics. We may need to continue to abandon some simple Mendelian models. New evidence may involve polygenic inheritance, effects of other genes, or environmental factors as more appropriate explanations. Therefore, the analysis of family genetics is not always feasible or meaningful.

Genetics is the study of the inheritance of characteristics. The genes that transmit this information are coded in segments of DNA in chromosomes. Homologous chromosomes possess the same gene at the same *locus*. These genes may exist in variant forms, called *alleles*.

If a person possesses two identical alleles, the person is said to be *homozygous* for that particular trait. If a person possesses two different alleles, the person is said to be *heterozygous* for that particular trait. The particular combination of these gene variants (alleles) represents the person's *genotype;* the appearance or physical manifestation of the individual that develops as a result of the way the genes are expressed represents the person's *phenotype*.

If one allele determines the phenotype by masking the expression of the other allele in a heterozygous individual, the allele is termed *dominant*. The allele whose expression is masked is termed *recessive*. If the heterozygous condition determines an intermediate phenotype, the inheritance represents *incomplete dominance*. However, different alleles are *codominant* if both are expressed in the heterozygous condition. Some characteristics inherited on the sex chromosomes result in phenotype frequencies that might be more prevalent in males or females. Such characteristics are called *sex-linked (X-linked or Y-linked) characteristics*.

As a result of meiosis during the formation of eggs and sperm, a mother and father each transmit an equal number of chromosomes (the haploid number 23) to form the zygote (diploid number 46). An offspring will receive one allele from each parent. These gametes combine randomly in the formation of each offspring. Hence, the *laws of probability* can be used to predict possible genotypes and phenotypes of offspring. A genetic tool called a *Punnett square* simulates all possible combinations (probabilities) in offspring genotypes and resulting phenotypes.

EXPLORE

PROCEDURE A—Human Genotypes and Phenotypes

A complete set of genetic instructions in one human cell constitutes one's *genome*. The human genome contains about 2.9 billion base pairs representing approximately 20,500 protein-encoding genes. These instructions represent our genotypes and are expressed as phenotypes sometimes clearly observable on our bodies. Some of these traits are listed in table 48.1 and are discernable in figure 48.1.

TABLE 48.1 Examples of Some Common Human Phenotypes

Dominant Traits and Genotypes	Recessive Traits and Genotypes
Tongue roller (*R___*)	Nonroller (*rr*)
Freckles (*F___*)	No freckles (*ff*)
Widow's peak (*W___*)	Straight hairline (*ww*)
Dimples (*D___*)	No dimples (*dd*)
Free earlobe (*E___*)	Attached earlobe (*ee*)
Normal skin coloration (*M___*)	Albinism (*mm*)
Astigmatism (*A___*)	Normal vision (*aa*)
Curly hair (*C___*)	Straight hair (*cc*)
PTC taster (*T___*)	Nontaster (*tt*)
Blood type A (I^A ___), B(I^B ___), or AB ($I^A I^B$)	Blood type O (*ii*)
Normal color vision ($X^C X^C$), ($X^C X^c$), or ($X^C Y$)	Red-green colorblindness ($X^c X^c$) or ($X^c Y$)

A dominant trait might be homozygous or heterozygous, so only one capital letter is used along with a blank for the possible second dominant or recessive allele. For a recessive trait, two lowercase letters represent the homozygous recessive genotype for that characteristic. Dominant does not always correlate with the predominance of the allele in the gene pool; dominant means one allele will determine the appearance of the phenotype.

1. **Tongue roller/nonroller:** The dominant allele (*R*) determines the person's ability to roll the tongue into a U-shaped trough. The homozygous recessive condition (*rr*) prevents this tongue rolling (fig. 48.1). Record your results in the table in Part A of Laboratory Report 48.
2. **Freckles/no freckles:** The dominant allele (*F*) determines the appearance of freckles. The homozygous recessive condition (*ff*) does not produce freckles (fig. 48.1). Record your results in the table in Part A of the laboratory report.
3. **Widow's peak/straight hairline:** The dominant allele (*W*) determines the appearance of a hairline above the forehead that has a distinct downward point in the center, called a widow's peak. The homozygous recessive condition (*ww*) produces a straight hairline (fig. 48.1). A receding hairline would prevent this phenotype determination. Record your results in the table in Part A of the laboratory report.
4. **Dimples/no dimples:** The dominant allele (*D*) determines the appearance of a distinct dimple in one or both cheeks upon smiling. The homozygous recessive condition (*dd*) results in the absence of dimples (fig. 48.1). Record your results in the table in Part A of the laboratory report.
5. **Free earlobe/attached earlobe:** The dominant allele (*E*) codes for the appearance of an inferior earlobe that

hangs freely below the attachment to the head. The homozygous recessive condition (*ee*) determines the earlobe attaching directly to the head at its inferior border (fig. 48.1). Record your results in the table in Part A of the laboratory report.
6. **Normal skin coloration/albinism:** The dominant allele (*M*) determines the production of some melanin, producing normal skin coloration. The homozygous recessive condition (*mm*) determines albinism due to the inability to produce or use the enzyme tyrosinase in pigment cells. An albino does not produce melanin in the skin, hair, or the middle tunic (choroid coat, ciliary body, and iris) of the eye. The absence of melanin in the middle tunic allows the pupil to appear slightly red to nearly black. Remember that the pupil is an opening in the iris filled with transparent aqueous humor. An albino human has pale white skin, flax-white hair, and a pale blue iris. Record your results in the table in Part A of the laboratory report.
7. **Astigmatism/normal vision:** The dominant allele (*A*) results in an abnormal curvature to the cornea or the lens. As a consequence, some portions of the image projected on the retina are sharply focused, and other portions are blurred. The homozygous recessive condition (*aa*) generates normal cornea and lens shapes and normal vision. Use the astigmatism chart and directions to assess this possible defect described in Laboratory Exercise 32. Other eye defects, such as nearsightedness (myopia) and farsightedness (hyperopia), are different genetic traits due to genes at other locations. Record your results in the table in Part A of the laboratory report.
8. **Curly hair/straight hair:** The dominant allele (*C*) determines the appearance of curly hair. Curly hair is somewhat flattened in cross section, as the hair follicle of a similar shape served as a mold for the root of the

Dominant Traits

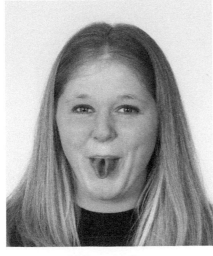

(a) Tongue roller

Recessive Traits

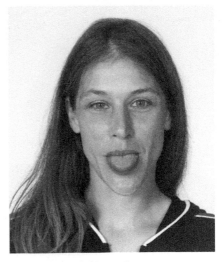

(b) Nonroller

(c) Freckles

(d) No freckles

(e) Widow's peak

(f) Straight hairline

FIGURE 48.1 Representative genetic traits comparing dominant and recessive phenotypes: (*a*) tongue roller; (*b*) nonroller; (*c*) freckles; (*d*) no freckles; (*e*) widow's peak; (*f*) straight hairline; (*g*) dimples; (*h*) no dimples; (*i*) free earlobe; (*j*) attached earlobe. ©J and J Photography

(g) Dimples

(h) No dimples

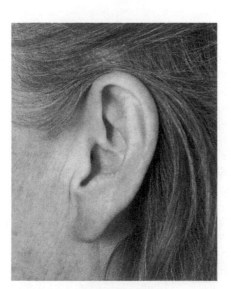

(i) Free earlobe

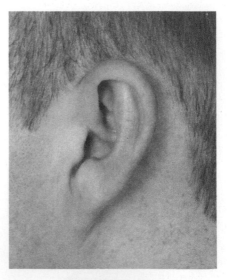

(j) Attached earlobe

FIGURE 48.1 *Continued*

hair during its formation. The homozygous recessive condition (*cc*) produces straight hair. Straight hair is nearly round in cross section from being molded into this shape in the hair follicle. In some populations (Caucasians), the heterozygous condition (*Cc*) expresses the intermediate wavy hair phenotype (incomplete dominance). This trait determination assumes no permanents or hair straightening procedures have been performed. Such hair alterations do not change the hair follicle shape, and future hair growth results in original genetic hair conditions. Record your unaltered hair appearance in the table in Part A of the laboratory report.

9. **PTC taster/nontaster:** The dominant allele (*T*) determines the ability to experience a bitter sensation when PTC paper is placed on the tongue. About 70% of people possess this dominant gene. The homozygous recessive condition (*tt*) makes a person unable to notice the substance. Place a piece of PTC (phenylthiocarbamide) paper on the upper tongue surface and chew it slightly to see if you notice a bitter sensation from this harmless chemical. The nontaster of the PTC paper does not detect any taste at all from this substance. Record your results in the table in Part A of the laboratory report.

10. **Blood type A, B, or AB/blood type O:** There are three alleles (I^A, I^B, and i) in the human population affecting RBC membrane structure. These alleles are located on a single pair of homologous chromosomes, so a person possesses either two of the three alleles or two of the same allele. All of the possible combinations of these alleles of genotypes and the resulting phenotypes are depicted in table 48.2. The expression of the blood type AB is a result of both codominant alleles located in the same individual. Possibly you have already determined your blood type in Laboratory Exercise 34 or have it recorded on a blood donor card. (If simulated blood-typing kits were used for Laboratory Exercise 34, those results would not be valid for your genetic factors.) Record your results in the table in Part A of the laboratory report.

11. **Sex determination:** A person with sex chromosomes XX displays a female phenotype. A person with sex chromosomes XY displays a male phenotype. Record your results in the table in Part A of the laboratory report.

12. **Normal color vision/red-green colorblindness:** This condition is a sex-linked (X-linked) characteristic. The alleles for color vision are also on the X chromosome, but they are absent on the Y chromosome. As a result, a female might possess both alleles (C and c), one on each of the X chromosomes. The dominant allele (C) determines normal color vision; the homozygous recessive condition (cc) results in red-green colorblindness. However, a male would possess only one of the two alleles for color vision because there is only a single X chromosome in a male. Hence, a male with even a single recessive gene for colorblindness possesses the defect. Note all the possible genotypes and phenotypes for this condition (table 48.1). Review the color vision test in Laboratory Exercise 32 using the color plate in figure 32.4 and color plates in Ichikawa's or Ishihara's book. Record your results in the table in Part A of the laboratory report.

13. Complete Part A of the laboratory report.

EXPLORE

PROCEDURE B—Laws of Probability

The laws of probability provide a mathematical way to determine the likelihood of events occurring by chance. This prediction is often expressed as a ratio of the number of results from experimental events to the number of results considered possible. For example, when tossing a coin, there is an equal chance of the results displaying heads or tails. Hence the probability is one-half of obtaining either a heads or a tails (there are two possibilities for each toss). When all of the probabilities of all possible outcomes are considered for the result, they will always add up to 1. To predict the probability of two or more events occurring in succession,

TABLE 48.2 Genotypes and Phenotypes (Blood Types)

Genotypes	Phenotypes (Blood Types)
$I^A I^A$ or $I^A i$	A
$I^B I^B$ or $I^B i$	B
$I^A I^B$ (codominant)	AB
ii	O

multiply the probabilities of each individual event. For example, the probability of tossing a die and displaying a 4 twice in a row is $1/6 \times 1/6 = 1/36$ (there are six possibilities for each toss). Each toss in a sequence is an *independent event* (chance has no memory). The same laws apply when parents have multiple children (each fertilization is an independent event). Perform the following experiments to demonstrate the laws of probability:

1. Use a single penny (or other coin) and toss it 20 times. Predict the number of heads and tails that would occur from the 20 tosses. Record your prediction and the actual results observed in Part B of the laboratory report.

2. Use a single die (*pl.* dice) and toss it 24 times. Predict the number of times a number below 3 (numbers 1 and 2) would occur from the 24 tosses. Record your prediction and the actual results in Part B of the laboratory report.

3. Use two pennies and toss them simultaneously 32 times. Predict the number of times two heads, a heads and a tails, and two tails would occur. Record your prediction and the actual results in Part B of the laboratory report.

4. Use a pair of dice and toss them simultaneously 32 times. Predict the number of times for both dice coming up with odd numbers, one die an odd and the other an even number, and both dice coming up with even numbers. Record your prediction and the actual results in Part B of the laboratory report.

5. Obtain class totals for all of the coins and dice tossed by adding your individual results to a class tally location, as on the blackboard.

6. A Punnett square can be used to visually demonstrate the probable results for two pennies tossed simultaneously. For the purpose of a genetic comparison, an h (heads) will represent one "allele" on the coin; a t (tails) will represent a different "allele" on the coin.

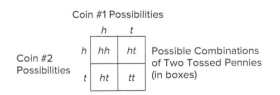

7. Complete Part B of the laboratory report.

PROCEDURE C—Genetic Problems

1. A Punnett square can be constructed to demonstrate a visual display of the predicted offspring from parents with known genotypes. Recall that in complete dominance, a dominant allele is expressed in the phenotype, as it can mask the other recessive allele on the homologous chromosome pair. During meiosis, the homologous chromosomes with their alleles separate (Mendel's Law of Segregation) into different gametes. An example of such a cross might be a homozygous dominant mother for dimples (*DD*) who has offspring with a father homozygous recessive (*dd*) for the same trait. The results of such a cross, according to the laws of probability, would be represented by the following Punnett square:

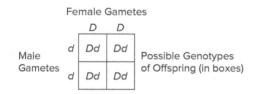

Results: Genotypes: 100% *Dd* (all heterozygous)

Phenotypes: 100% dimples

In another example, assume that both parents are heterozygous (*Dd*) for dimples. The results of such a

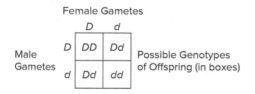

cross, according to the laws of probability, would be represented by the following Punnett square:

Results: Genotypes: 25% *DD* (homozygous dominant); 50% *Dd* (heterozygous); 25% *dd* (homozygous recessive) (1:2:1 genotypic ratio)

Phenotypes: 75% dimples; 25% no dimples (3:1 phenotypic ratio)

2. Work genetic problems 1 and 2 in Part C of the laboratory report.
3. The ABO blood type inheritance represents an example of codominance. Review table 48.2 for the genotypes and phenotypes for the expression of this trait. A Punnett square can be constructed to predict the offspring

of parents of known genotypes. In this example, assume the genotype of the mother is $I^A I^B$ and that of the father is *ii*. The results of such a cross would be represented by the following Punnett square:

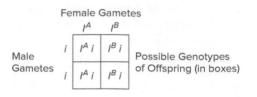

Results: Genotypes: 50% $I^A i$ (heterozygous for A); 50% $I^B i$ (heterozygous for B) (1:1 genotypic ratio)

Phenotypes: 50% blood type A; 50% blood type B (1:1 phenotypic ratio)

Note: In this cross, all of the children would have blood types unlike either parent.

4. Work genetic problems 3 and 4 in Part C of the laboratory report.
5. Review the inheritance of red-green colorblindness, an X-linked characteristic, in table 48.1. A Punnett square can be constructed to predict the offspring of parents of known genotypes. In this example, assume the genotype of the mother is heterozygous $X^C X^c$ (normal color vision but a carrier for the colorblindness defect) and that of the father is $X^C Y$ (normal color vision; no allele on the Y chromosome). The results of such a cross would be represented by the following Punnett square:

Results: Genotypes: 25% $X^C X^C$; 25% $X^C X^c$; 25% $X^C Y$; 25% $X^c Y$ (1:1:1:1 genotypic ratio)

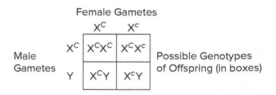

Phenotypes for sex determination: 50% females; 50% males (1:1 phenotypic ratio)

Phenotypes for color vision: Females 100% normal color vision (However, 50% are heterozygous carriers for colorblindness.)

Phenotypes for color vision: Males 50% normal; 50% with red-green colorblindness (1:1 phenotypic ratio)

Note: In X-linked inheritance, the colorblind males received the recessive gene from their mothers.

6. Complete Part C of the laboratory report.

Name _____

Date _____

Section _____

The  corresponds to the indicated Learning Outcome(s) found at the beginning of the laboratory exercise.

LABORATORY
REPORT

48

Genetics

PART A ASSESSMENTS

1. Enter your test results for genotypes and phenotypes in the table. Circle your phenotype and genotype for each of the twelve traits. **1**

Trait	Dominant Phenotype	Genotype	Recessive Phenotype	Genotype
Tongue movement	Roller	$R\_\_$	Nonroller	rr
Freckles	Freckles	$F\_\_$	No freckles	ff
Hairline	Widow's peak	$W\_\_$	Straight	ww
Dimples	Dimples	$D\_\_$	No dimples	dd
Earlobe	Free	$E\_\_$	Attached	ee
Skin coloration	Normal (some melanin)	$M\_\_$	Albinism	mm
Vision	Astigmatism	$A\_\_$	Normal	aa
Hair shape*	Curly	$C\_\_$	Straight	cc
Taste	PTC taster	$T\_\_$	Nontaster for PTC	tt
Blood type	A, B, or AB	$I^A\_\_$; $I^B\_\_$; or $I^A I^B$	O	ii
Sex		XX or XY		
Color vision	Normal	$X^C X\_\_$ or $X^C Y$	Red-green colorblindness	$X^c X^c$ or $X^c Y$

*In some populations (Caucasians), the heterozygous condition (Cc) results in the appearance of wavy hair, which actually represents an example of incomplete dominance for this trait.

2. Choose at least three dominant phenotypes that you circled in question 1, and analyze the genotypes for those traits. If it's feasible to observe your biological parents and siblings for any of these traits, are you able to determine if any of your dominant genotypes are homozygous dominant or heterozygous? _____ If so, which ones? _____ Explain the rationale for your response.

_____

_____

_____

_____

_____

_____

PART B ASSESSMENTS

1. Single penny tossed 20 times and counting heads and tails: $\boxed{2}$
   Probability (prediction): ____/20 heads ____/20 tails
   (*Note: Traditionally, probabilities are converted to the lowest fractional representation.*)
   Actual results: ____ heads ____ tails
   Class totals: ____ heads ____ tails

2. Single die tossed 24 times and counting the number of times a number below 3 occurs: $\boxed{2}$
   Probability: ____/24 number below 3 (numbers 1 and 2)
   Actual results: ____ number below 3
   Class totals: ____ number below 3 ____ total tosses by class members

3. Two pennies tossed simultaneously 32 times and counting the number of two heads, a heads and a tails, and two tails: $\boxed{2}$
   Probability: ____/32 of two heads ____/32 of a heads and a tails ____/32 of two tails
   Actual results: ____ two heads ____ heads and tails ____ two tails
   Class totals: ____ two heads ____ heads and tails ____ two tails

4. Two dice tossed simultaneously 32 times and counting the number of two odd numbers, an odd and an even number, and two even numbers: $\boxed{2}$
   Probability: ____/32 of two odd numbers ____/32 of an odd and an even number ____/32 of two even numbers
   Actual results: ____ two odd numbers ____ an odd and an even number ____ two even numbers
   Class totals: ____ two odd numbers ____ an odd and an even number ____ two even numbers

5. Use the example of the two dice tossed 32 times, and construct a Punnett square to represent the possible combinations that could be used to determine the probability (prediction) of odd and even numbers for the resulting tosses. Your construction should be similar to the Punnett square for the two coins tossed that is depicted in Procedure B of the laboratory exercise. $\boxed{2}$

6. Complete the following:
   a. Are the class totals closer to the predicted probabilities than your results? _____ Explain your response.

   _____
   _____
   _____
   _____

   b. Does the first toss of the penny or the first toss of the die have any influence on the next toss? _____ Explain your response.

   _____
   _____
   _____
   _____

c. Assume a family has two boys or two girls. They wish to have one more child but hope for the child to be of the opposite sex from the two they already have. What is the probability that the third child will be of the opposite sex? _____ Explain your response.

_____

_____

_____

d. What is the probability (prediction) that a couple without children will eventually have four children, all girls? _____ Explain your response.

_____

_____

_____

PART C ASSESSMENTS

For each of the genetic problems, (a) determine the parents' genotypes, (b) determine the possible gametes for each parent, (c) construct a Punnett square, and (d) record the resulting genotypes and phenotypes as ratios from the cross. Problems 1 and 2 involve examples of complete dominance; problems 3 and 4 are examples of codominance; problem 5 is an example of sex-linked (X-linked) inheritance.

1. Determine the results from a cross of a mother who is heterozygous (*Rr*) for tongue rolling with a father who is homozygous recessive (*rr*). ◿3◿

2. Determine the results from a cross of a mother and a father who are both heterozygous for freckles. ◿3◿

3. Determine the results from a mother who is heterozygous for blood type B and a father who is homozygous dominant for blood type A. ◿3◿

4. Determine the results from a mother who is heterozygous for blood type A and a father who is heterozygous for blood type B.

5. Colorblindness is an example of X-linked inheritance. Hemophilia is another example of X-linked inheritance, also from a recessive allele (*h*). The dominant allele (*H*) determines whether the person possesses normal blood clotting. A person with hemophilia has a permanent tendency for hemorrhaging due to a deficiency of one of the clotting factors (VIII—antihemophilic factor). Determine the offspring from a cross of a mother who is a carrier (heterozygous) for the disease and a father with normal blood coagulation.

Critical Thinking Application

Assume that the genes for hairline and earlobes are on different pairs of homologous chromosomes. Determine the genotypes and phenotypes of the offspring from a cross if both parents are heterozygous for both traits. (1) First determine the genotypes for each parent. (2) Determine the gametes, but remember each gamete has one allele for each trait (gametes are haploid). (3) Construct a Punnett square, with 16 boxes, that has four different gametes from each parent along the top and the left edges. (This is an application to demonstrate Mendel's Law of Independent Assortment.) (4) List the results of genotypes and phenotypes as ratios.

Appendix 1

Laboratory Safety Guidelines

Carefully review the following safety guidelines before starting a laboratory exercise. Safety guideline reminders are also included in the appropriate laboratory exercise sections. Avoid tardiness to laboratory sessions because specific directions and precautions are often given at the beginning of the period. Your instructor should update you about safety procedures, as your school might have safety regulations or modifications in addition to those listed here.

1. Become familiar with all room exits and the location and operating procedures of all safety equipment (first aid kit, Material Safety Data Sheets (MSDS), eyewash station, safety shower, fire extinguisher, and fire blanket).
2. Read all laboratory exercises prior to starting the procedures. Do not work alone in the laboratory.
3. Do not smoke, chew, eat, drink, apply cosmetics or lip balm, handle contact lenses (regular eyeglasses should be worn), or work with open wounds in the laboratory. Consider all materials you use as potentially hazardous.
4. Use safety glasses and laboratory coats when handling dangerous chemicals or heating dangerous materials. Volatile materials should never be heated using an open flame. When heating materials in a test tube, never point the test tube in the direction of a person or leave heated materials unattended.
5. Use a commercially prepared disinfectant or a 10% bleach solution to clean laboratory work surfaces *before* and *after* any procedures when using animals, body fluids, and dissection specimens.
6. Wear enclosed shoes and disposable gloves when working with dangerous chemicals, body fluids, and dissection specimens. (Replace latex gloves with vinyl or nitrile gloves if allergies to latex develop.)

7. If body fluids are being studied, *work only with your own.* Special precautions to prevent contact with body fluids may include wearing gloves, gowns, aprons, or masks, as directed by your instructor.
8. Special precautions during dissections may include wearing laboratory coats and goggles. Always use cutting blades in a direction away from yourself and laboratory partners.
9. Restrain all loose clothing, long hair, and dangling jewelry during laboratory procedures. Keep all unnecessary materials away from the work area to reduce clutter and the possibility of an accident.
10. If you have special needs, are taking medications, experience allergic reactions, are pregnant, or are uncomfortable with the procedures, inform your instructor.
11. Use only a mechanical pipetting device (*never your mouth*).
12. Immediately report any accidents, spills, or damaged equipment to your instructor.
13. Use only disposable lancets and needles, and never attempt to bend, cut, or recap them when finished. Place sharp items in a *puncture-resistant container* that is marked "Biohazard Sharps Container."
14. Dispose of chemicals, waste material, body fluids, and dissection specimens according to appropriate directions. Any reusable glassware or utensils that have been contaminated with body fluids should be placed in a disinfectant (10% bleach solution) and later autoclaved.
15. Thoroughly wash your hands with soap and warm water immediately after removing gloves and before leaving the laboratory.
16. Practice any modified or additional safety guidelines required by your instructor.

Periodic Table of Elements

1 1A																	18 8A
1 **H** Hydrogen 1.008	2 2A											13 3A	14 4A	15 5A	16 6A	17 7A	**2** **He** Helium 4.003
3 **Li** Lithium 6.941	**4** **Be** Beryllium 9.012											**5** **B** Boron 10.81	**6** **C** Carbon 12.01	**7** **N** Nitrogen 14.01	**8** **O** Oxygen 16.00	**9** **F** Fluorine 19.00	**10** **Ne** Neon 20.18
11 **Na** Sodium 22.99	**12** **Mg** Magnesium 24.31	3 3B	4 4B	5 5B	6 6B	7 7B	8	9 8B	10	11 1B	12 2B	**13** **Al** Aluminum 26.98	**14** **Si** Silicon 28.09	**15** **P** Phosphorus 30.97	**16** **S** Sulfur 32.07	**17** **Cl** Chlorine 35.45	**18** **Ar** Argon 39.95
19 **K** Potassium 39.10	**20** **Ca** Calcium 40.08	**21** **Sc** Scandium 44.96	**22** **Ti** Titanium 47.88	**23** **V** Vanadium 50.94	**24** **Cr** Chromium 52.00	**25** **Mn** Manganese 54.94	**26** **Fe** Iron 55.85	**27** **Co** Cobalt 58.93	**28** **Ni** Nickel 58.69	**29** **Cu** Copper 63.55	**30** **Zn** Zinc 65.39	**31** **Ga** Gallium 69.72	**32** **Ge** Germanium 72.59	**33** **As** Arsenic 74.92	**34** **Se** Selenium 78.96	**35** **Br** Bromine 79.90	**36** **Kr** Krypton 83.80
37 **Rb** Rubidium 85.47	**38** **Sr** Strontium 87.62	**39** **Y** Yttrium 88.91	**40** **Zr** Zirconium 91.22	**41** **Nb** Niobium 92.91	**42** **Mo** Molybdenum 95.94	**43** **Tc** Technetium (98)	**44** **Ru** Ruthenium 101.1	**45** **Rh** Rhodium 102.9	**46** **Pd** Palladium 106.4	**47** **Ag** Silver 107.9	**48** **Cd** Cadmium 112.4	**49** **In** Indium 114.8	**50** **Sn** Tin 118.7	**51** **Sb** Antimony 121.8	**52** **Te** Tellurium 127.6	**53** **I** Iodine 126.9	**54** **Xe** Xenon 131.3
55 **Cs** Cesium 132.9	**56** **Ba** Barium 137.3	**57** **La** Lanthanum 138.9	**72** **Hf** Hafnium 178.5	**73** **Ta** Tantalum 180.9	**74** **W** Tungsten 183.9	**75** **Re** Rhenium 186.2	**76** **Os** Osmium 190.2	**77** **Ir** Iridium 192.2	**78** **Pt** Platinum 195.1	**79** **Au** Gold 197.0	**80** **Hg** Mercury 200.6	**81** **Tl** Thallium 204.4	**82** **Pb** Lead 207.2	**83** **Bi** Bismuth 209.0	**84** **Po** Polonium (210)	**85** **At** Astatine (210)	**86** **Rn** Radon (222)
87 **Fr** Francium (223)	**88** **Ra** Radium (226)	**89** **Ac** Actinium (227)	**104** **Rf** Rutherfordium (257)	**105** **Db** Dubnium (260)	**106** **Sg** Seaborgium (263)	**107** **Bh** Bohrium (262)	**108** **Hs** Hassium (265)	**109** **Mt** Meitnerium (266)	**110** **Ds** Darmstadtium (269)	**111** **Rg** Roentgenium (272)	**112** **Cn** Copernicium	(113)	**114** **Fl** Flerovium	(115)	**116** **Lv** Livermorium	(117)	(118)

Atomic number — **9 F** Fluorine — Atomic mass **19.00**

Metals
Metalloids
Nonmetals

58 **Ce** Cerium 140.1	**59** **Pr** Praseodymium 140.9	**60** **Nd** Neodymium 144.2	**61** **Pm** Promethium (147)	**62** **Sm** Samarium 150.4	**63** **Eu** Europium 152.0	**64** **Gd** Gadolinium 157.3	**65** **Tb** Terbium 158.9	**66** **Dy** Dysprosium 162.5	**67** **Ho** Holmium 164.9	**68** **Er** Erbium 167.3	**69** **Tm** Thulium 168.9	**70** **Yb** Ytterbium 173.0	**71** **Lu** Lutetium 175.0
90 **Th** Thorium 232.0	**91** **Pa** Protactinium (231)	**92** **U** Uranium 238.0	**93** **Np** Neptunium (237)	**94** **Pu** Plutonium (242)	**95** **Am** Americium (243)	**96** **Cm** Curium (247)	**97** **Bk** Berkelium (247)	**98** **Cf** Californium (249)	**99** **Es** Einsteinium (254)	**100** **Fm** Fermium (253)	**101** **Md** Mendelevium (256)	**102** **No** Nobelium (254)	**103** **Lr** Lawrencium (257)

The 1–18 group designation has been recommended by the International Union of Pure and Applied Chemistry (IUPAC) but is not yet in wide use.

Elements up to 122 have been reported. Of these, some of the higher-numbered elements have been observed only under laboratory conditions. Four such elements, 113, 115, 117, and 118, were officially confirmed in December 2015 and names have been proposed. If accepted, they will be called nihonium, moscovium, tennessine, and organesson, respectively.

Appendix  3

Preparation of Solutions

Amylase solution, 0.5%

Place 0.5 g of bacterial amylase in a graduated cylinder or volumetric flask. Add distilled water to the 100-mL level. Stir until dissolved. The amylase should be free of sugar for best results; a low-maltose solution of amylase yields good results. (Store amylase powder in a freezer until mixing this solution.)

Benedict's solution

Prepared solution is available from various suppliers.

Glucose solutions

1. *1.0% solution.* Place 1 g of glucose in a graduated cylinder or volumetric flask. Add distilled water to the 100-mL level. Stir until dissolved.
2. *10% solution.* Place 10 g of glucose in a graduated cylinder or volumetric flask. Add distilled water to the 100-mL level. Stir until dissolved.

Iodine-potassium-iodide (IKI solution)

Add 20 g of potassium iodide to 1 L of distilled water, and stir until dissolved. Then add 4.0 g of iodine, and stir again until dissolved. Solution should be stored in a dark, stoppered bottle.

Methylene blue

Dissolve 0.3 g of methylene blue powder in 30 mL of 95% ethyl alcohol. In a separate container, dissolve 0.01 g of potassium hydroxide in 100 mL of distilled water. Mix the two solutions. (Prepared solution is available from various suppliers.)

Physiological saline solution

Place 0.9 g of sodium chloride in a graduated cylinder or volumetric flask. Add distilled water to the 100-mL level. Stir until dissolved.

Ringer's solution (frog)

Dissolve the following salts in 1 L of distilled water:
    6.50 g sodium chloride
    0.20 g sodium bicarbonate
    0.14 g potassium chloride
    0.12 g calcium chloride

Sodium chloride solutions

1. *0.9% solution.* Place 0.9 g of sodium chloride in a graduated cylinder or volumetric flask. Add distilled water to the 100-mL level. Stir until dissolved.
2. *3.0% solution.* Place 3.0 g of sodium chloride in a graduated cylinder or volumetric flask. Add distilled water to the 100-mL level. Stir until dissolved.

Starch solutions

1. *0.5% solution.* Add 5 g of cornstarch to 1 L of distilled water. Heat until the mixture boils. Cool the liquid, and pour it through a filter. Store the filtrate in a refrigerator.
2. *1.0% solution.* Add 10 g of cornstarch to 1 L of distilled water. Heat until the mixture boils. Cool the liquid, and pour it through a filter. Store the filtrate in a refrigerator.
3. *10% solution.* Add 100 g of cornstarch to 1 L of distilled water. Heat until the mixture boils. Cool the liquid, and pour it through a filter. Store the filtrate in a refrigerator.

Assessments of Laboratory Reports

Many assessment models can be used for laboratory reports. A rubric, which can be used for performance assessments, contains a description of the elements (requirements or criteria) of success to various degrees. The term *rubric* originated from *rubrica terra,* which is Latin for "the application of red earth" to indicate anything of importance. A rubric used for assessment contains elements for judging student performance, with points awarded for varying degrees of success in meeting the learning outcomes. The content and the quality level necessary to attain certain points are indicated in the rubric. It is effective if the assessment tool is shared with the students before the laboratory exercise is performed.

Following are two sample rubrics that can easily be modified to meet the needs of a specific course. Some of the elements for these sample rubrics may not be necessary for every laboratory exercise. The generalized rubric needs to contain the possible assessment points that correspond to learning outcomes for a specific course. The point value for each element may vary. The specific rubric example contains performance levels for laboratory reports. The elements and the point values can easily be altered to meet the value placed on laboratory reports for a specific course.

Assessment: Generalized Laboratory Report Rubric

Element	Assessment Points Possible	Assessment Points Earned
1. Figures are completely and accurately labeled.		
2. Sketches are accurate, contain proper labels, and are of sufficient detail.		
3. Colored pencils were used extensively to differentiate structures on illustrations.		
4. Matching and fill-in-the-blank answers are completed and accurate.		
5. Short-answer/discussion questions contain complete, thorough, and accurate answers. Some elaboration is evident for some answers.		
6. Data collected are complete, are accurately displayed, and contain a valid explanation.		

Total Points: Possible _____   Earned _____

Assessment: Specific Laboratory Report Rubric

Element	Excellent Performance (4 points)	Proficient Performance (3 points)	Marginal Performance (2 points)	Novice Performance (1 point)	Points Earned
Figure labels	Labels completed with ≥ 90% accuracy.	Labels completed with 80%–89% accuracy.	Labels completed with 70%–79% accuracy.	Labels < 70% accurate.	
Sketches	Accurate use of scale, details illustrated, and all structures labeled accurately.	Minor errors in sketches. Missing or inaccurate labels on one or more structures.	Sketch not realistic. Missing or inaccurate labels on two or more structures.	Several missing or inaccurate labels.	
Matching and fill-in-the-blanks	All completed and accurate.	One or two errors or omissions.	Three or four errors or omissions.	Five or more errors or omissions.	
Short-answer and discussion questions	Answers complete, valid, and contain some elaboration. No misinterpretations noted.	Answers generally complete and valid. Only minor inaccuracies noted. Minimal elaboration.	Marginal answers to the questions and contain inaccurate information.	Many answers incorrect or fail to address the topic. May be misinterpretations.	
Data collection and analysis	Data complete and displayed with a valid interpretation.	Only minor data missing or a slight misinterpretation.	Some omissions. Not displayed or interpreted accurately.	Data incomplete or show serious misinterpretations.	

Total Points Earned _____

Index

iliac crest, 119, 173*f*–174*f*
iliacus muscle, 161
iliopsoas group, 161
iliotibial tract, 164*f*, 177*f*
ilium, 119
immunity, 291
immunofluorescence stains, 239*f*
incisors, 297, 298*f*
incomplete dominance, 353
incus, 211, 215*f*
independent assortment, Mendel's law, 362
independent events, 357
independent variables, 1
infections, 247, 291
inferior articular processes, 105
inferior mesenteric artery, 274, 275*f*
inferior mesenteric vein, 279
inferior nasal conchae, 94, 95*f*, 220*f*
inferior oblique muscle, 219
inferior rectus muscle, 219
inferior vena cava, 257, 258, 260*f*, 279, 328*f*
inferior vertebral notch, 105
infraspinatus muscle, 147, 149*f*
infraspinous fossa, 109, 111*f*
infundibulum, 207, 208*f*, 234, 347
inguinal ligament, 278*f*, 281*f*
inguinal region lymph nodes, 291
inner ear, 211
inner layer of eye, 221
innominate, 85
insertions, 130, 138
inspiratory capacity, 322
inspiratory reserve volume, 322
insula, 199
insulin, 233–234, 238
insulin-dependent diabetes mellitus, 234.
    See also diabetes mellitus
insulin resistance, 234
insulin shock, 237–238
integumentary system, 10, 73–74, 75*f*–76*f*
intercalated discs, 70*f*
intercostal spaces, 264
intercostal trunk, 292*f*
intermediate cuneiform, 121
internal anal sphincter, 302
internal carotid artery, 274, 276*f*
internal iliac artery, 274
internal iliac vein, 279
internal jugular veins, 279, 280*f*, 286*f*, 291
internal oblique muscle, 155, 156*f*
interneurons, 193
interphalangeal joints, 175*f*
interphase, 53, 54*f*
interstitial cells, 341, 343
intertubercular sulcus, 111, 112*f*
intervertebral discs, 66*f*, 101, 130
intervertebral foramina, 101
intestinal trunk, 292*f*
intestine. *See* large intestine; small intestine
intestines, 280*f*, 302–304
intraocular lens replacement, 223*f*
inversion, 131*f*, 132*t*
iodine test, 24
ionic bonds, 22*f*
ions, 22*f*
IPMAT acronym, 53
iris
    with albinism, 354
    examining of, 221, 223, 224*f*
    location of, 221
    reflexes involving, 230
iris diaphragm of microscope, 31, 32, 34
ischial spine, 119
ischial tuberosity, 119
ischiocavernosus muscle, 156
ischium, 119
islets of Langerhans, 237–238, 239*f*–240*f*
isotonic solutions, 47

J
jejunum, 302
joints
    structures and movements of, 129–130, 131*f*, 132*t*
    surface anatomy of, 173*f*–175*f*
jugular foramen, 100*f*
jugular notch, 105, 106*f*, 173*f*, 264*f*
jugular veins, 279, 280*f*, 286*f*, 291
juxtamedullary nephrons, 327

K
karyotypes, 56*f*
ketone test, 337
kidneys
    epithelial tissues from, 62*f*
    functions of, 327, 335
    location of, 12, 235*f*, 327
    structure of, 327–330, 330*f*
kilograms, 4*t*
kilometers, 4*t*
knee-jerk reflex, 193, 195*f*
knee joint, 129*f*
Korotkoff sounds, 288
kyphotic curve of spine, 174*f*

L
labia majora, 347
labia minora, 347
lacrimal apparatus, 219
lacrimal bones, 92*f*, 94, 99*f*
lacrimal gland, 219
lacrimal sac, 219
lactiferous ducts, 348
lacunae, 66*f*, 80, 81*f*
lambdoid suture, 91
lamellae, 80, 81*f*
laminae, 105
large intestine
    location of, 12, 303*f*
    structure of, 302, 304, 304*f*
    veins of, 280*f*
laryngeal prominence, 172*f*
laryngopharynx, 300, 313
larynx
    location of, 12
    structure of, 314, 315*f*
    surface anatomy of, 172*f*
lateral border of scapula, 109, 111*f*
lateral condyles, 121, 123*f*, 129*f*
lateral cuneiform, 121
lateral epicondyles, 111, 112*f*, 121
lateral funiculus, 189
lateral hemispheres, 199
lateral horn (spinal cord), 189
lateral malleolus, 121, 177*f*
lateral meniscus, 129*f*
lateral rectus muscle, 219, 221*f*
lateral rotation, 131*f*
lateral sulcus, 199
lateral ventricles, 199, 207
latissimus dorsi muscle, 147, 173*f*
laws of probability, 353, 357
left atrioventricular valve, 257, 259
left bundle branch, 263
left gastric artery, 275*f*
legs. *See* lower limbs
length, metric units, 4*t*
lens (corrective), 227
lens (eye)
    aging effects on, 227, 229
    defects of, 223, 228, 354
    dissection of, 223, 224*f*
    location of, 221
lens (microscope), 29–33
lesser curvature of stomach, 301*f*
lesser sciatic notch, 120*f*
lesser trochanter, 121
lesser tubercle of humerus, 111
letter *e* slides, 33
leukocytes, 246*f*
leukocytes test, 337
levator ani muscle, 156

levator palpebrae superioris, 219, 220*f*
levator scapulae muscle, 147
Leydig cells, 341, 343
LH, 234
ligaments
    of dura mater, 190
    inguinal, 278*f*, 281*f*
    of knee joint, 129*f*
    of lens, 223
    periodontal, 299*f*
    in skeletal system, 10
ligamentum arteriosum, 256*f*, 258*f*–259*f*
limb length, relation to height, 2–3
linea, 88
linea alba, 156*f*, 174*f*
linea aspera, 121, 122*f*
lingual frenulum, 297, 298*f*
lipids tests, 24
liters, 4*t*
lithium, 22*f*
liver
    blood vessels of, 275*f*, 280*f*
    location of, 10, 12*f*, 302, 303*f*
lobes
    of brain, 199, 205, 206*f*
    of lungs, 314
lobules of thymus, 293–294
loci, 353
lock-and-key model of amylase action, 310*f*
long bones, 80*f*
longitudinal fissure, 199, 205, 206*f*, 207, 208*f*
longitudinal sections, 14, 16*f*
lordotic curve, of spine, 174*f*
lower esophageal sphincter, 300
lower limbs
    arteries of, 274, 278*f*
    bones of, 85, 119, 121, 122*f*
    muscles of, 161, 162*f*–165*f*
    reflexes involving, 193–194, 195*f*, 196*f*
    veins of, 279, 281*f*
low-power objectives, 32, 33
lumbar artery, 275*f*
lumbar curvature, 101, 102*f*
lumbar trunk, 292*f*
lumbar vertebrae, 101, 104*f*, 105
lunate, 114
lung(s). *See also* respiratory system
    epithelial tissues from, 62*f*
    location of, 12*f*
    as major organs, 12
    respiratory volumes, 319–322
    structure of, 314–315, 316*f*
lung function models, 320
lymphatic nodules, 293
lymphatic sinuses, 293
lymphatic system, 10, 291–294
lymphatic trunks, 291, 292*f*
lymphatic vessels, 10, 291, 293*f*
lymph nodes, 10, 291–293
lymphocytes, 246*f*, 247, 293, 294
lysosomes, 40*f*

M
macrophages, 294
maculae, 213*f*
macula lutea, 221, 222*f*
magnification of microscopes, 32
major calyces, 328
male pelvic floor, 157*f*
male reproductive system, 12, 341–343, 344*f*
malleoli, 121, 177*f*
malleus, 211, 215*f*
mammary glands, 348
mammillary bodies, 199, 207, 208*f*
mandible, 94, 95*f*
mandibular fossa, 91, 93*f*, 99*f*
manubrium, 105, 264*f*
marrow, 79, 80*f*
mass, metric units, 4*t*
masseter muscle, 143
mastication, 143
mastoid fontanel, 96*f*

mastoid process, 91, 99*f*, 172*f*, 212
mature follicles, 348
maxillae, 91, 95*f*
maxillary sinuses, 91, 95*f*, 313
measurements, 2, 4*t*, 32–33
meatus, 88
mechanical stage, 30*f*, 32
medial adductor muscles, 161
medial border of scapula, 109, 111*f*
medial condyles, 121, 129*f*
medial cuneiform bone, 121, 127*f*
medial end of clavicle, 109
medial epicondyles, 111, 112*f*, 121, 122*f*
medial longitudinal arch, 176*f*
medial malleolus, 121, 177*f*
medial meniscus, 129*f*
medial rectus muscle, 219, 221*f*
medial rotation, 131*f*
median cubital vein, 279
median sacral crest, 105
mediastinum, 9
medulla
    adrenal, 234, 236, 237
    ovary, 347, 348
    renal, 327, 328, 329, 330*f*
    thymus, 293, 294*f*
medulla oblongata
    of sheep brain, 205, 206*f*, 207, 208*f*
    structure of, 199
medullary cavity, 79
meiosis, 53, 358
melanin, 74, 75*f*, 354
melanocytes, 74
membranous labyrinth, 211, 213*f*
Mendelian genetics, 353
Mendel's Law of Independent Assortment, 362
Mendel's Law of Segregation, 358
meninges, 189, 190, 205
menisci, 129*f*
meniscus level, 46, 47*f*
merocrine sweat glands, 76*f*
mesenteric arteries, 274, 275*f*
mesentery, 302
metacarpal bones, 85, 114
metacarpophalangeal joints, 175*f*
metaphase, 54*f*, 55
metatarsal bones, 85, 121
meters, 4*t*
metersticks, 2, 229
metric system, 2, 4*t*
microhematocrit reader, 364
micrometers
    description of, 4*t*
    microscope scales, 33, 34*f*
microscopes, 29–36
microscopic sediment analysis, 337–338
microtubules, 40*f*, 54*f*
microvilli, 40*f*, 62*f*
midbrain, 199, 207, 208*f*
middle cardiac vein, 256*f*
middle cerebral artery, 276*f*
middle deltoid muscle, 174*f*
middle ear, 85, 211, 215*f*
middle ear cavity, 9
middle layer of eye, 221
middle nasal concha, 91
middle phalanges, 114, 121
middle sacral artery, 275*f*
milligrams, 4*t*
milliliters, 4*t*
millimeters, 4*t*
milliseconds, 4*t*
minor calyces, 328
minute respiratory volume, 322
mitosis, 53–55
mitral valve, 257, 264
M line, 142*f*
mnemonics, 53, 114, 201
molars, 297, 298*f*
molecules, 21, 22*f*–23*f*
monocytes, 246*f*, 247
monosaccharides. *See* sugars
mons pubis, 347
motor neurons, 184*f*, 193, 194*f*